William Stimpson AND THE GOLDEN AGE
OF AMERICAN NATURAL HISTORY

William Stimpson

AND THE GOLDEN AGE OF

AMERICAN NATURAL HISTORY

RONALD SCOTT VASILE

NIU PRESS / DeKalb IL

Northern Illinois University Press, DeKalb 60115
© 2018 by Northern Illinois University Press
All rights reserved
Printed in the United States of America
27 26 25 24 23 22 21 20 19 18   1 2 3 4 5
978-0-87580-784-3 (paper)
978-1-60909-240-5 (e-book)
Book and cover design by Yuni Dorr

Library of Congress Cataloging-in-Publication Data
is available online at http://catalog.loc.gov

To Jennifer, FOR BELIEVING IN ME

TABLE OF CONTENTS

Acknowledgments

The researching and writing of this book has taken thirty years, so I have incurred more debts than I can possibly recount. It saddens me that several of the people listed below are not alive to see this book.

I owe the largest intellectual debt to Lester D. Stephens, Professor Emeritus at the University of Georgia. Since the inception of this endeavor he edited several drafts of the manuscript and provided many helpful suggestions and corrections. Even more importantly, he has always been there to provide encouragement and wise counsel. No budding historian has ever had a better mentor.

My graduate professors at the University of Illinois at Chicago, especially Perry Duis, Leo Schelbert, Robert Messer, and Gerald Danzer, were instrumental in helping to shape me as a historian.

Numerous historians have commented on portions of the manuscript and shared advice and suggestions, including Ronald L. Numbers, Helen Rozwadowski, David Hull, Keir Sterling, Robert H. Silliman, Toby A. Appel, Daniel Goldstein, Sally G. Kohlsedt, Janet Browne, Ralph Dexter, Mike Foster, Clifford M. Nelson, Judith A. Simonsen, Richard Rabinowitz, Eric L. Mills, and Pam Henson.

People who have served as sources of information on various aspects of natural history include Donald G. Mikulic, Joanne Klussendorf, and Kevin S. Cummings. Others who have shared their knowledge and sources were B. J. Earle, Jack White, Don Zochert, Michele Aldrich, Betsy Mendelsohn, Billy Stimpson, Lipke B. Holthuis, and Jack Fooden.

During my fourteen years at the Chicago Academy of Sciences several people were supportive of this work, including then Academy director Paul G. Heltne, who encouraged my work on Stimpson. Mark F. Spreyer, modern-day naturalist extraordinaire, has been a font of information and a true friend. M. Dodge Mumford was with me on a memorable pilgrimage to Stimpson's grave. Elizabeth Thompson (Lampert), Vicki Byre, and Mary Hennen also assisted in various ways.

I am grateful to the Smithsonian for two small travel grants in 1987 and 1992 to research in the Smithsonian Archives, arranged largely due to the efforts of Bill Deiss. He and other Archives staff members, including Bill Cox and Pamela Henson, were instrumental in locating and copying materials. Rafael

Lemaitre and the late Raymond G. Manning of the Smithsonian arranged for a travel grant in 2005 in order to publish Stimpson's Journal, and they also assisted with advice concerning crustaceans. Other Smithsonian scientists that have shared their expertise include Robert Hershler and Storrs Olson.

Steve Swanson at The Grove National Historic Landmark has allowed me free access to the voluminous and well-organized Kennicott archive and in the process has become a good friend.

Barbara Scudder Wilson provided the tintype of the Stimpson family and other information on Stimpson family genealogy, as did Sue Morin. Mary Gordon did likewise for the Gordon family.

Kenton Clymer, then interim director at NIU Press, showed interest in the manuscript and began the process of publication. His help has been invaluable. The current staff of NIU Press, especially Amy Farranto, Nathan Holmes, and Lori Propheter, have been unfailingly helpful in answering my questions and providing assistance.

I am grateful to many librarians, particularly the staff at the Downers Grove Public Library, for facilitating interlibrary loans.

Dawn Roberts, current Director of Collections at the Academy, provided access to the archives and aided me in getting images.

On a personal level, my wife Jennifer has been living with this book for almost as long as I have, and I am grateful for her many sacrifices that helped make this book a reality. My parents and my mother-in-law supported my education, and my three children, Samuel, Daniel, and Katherine, have been a source of joy and have listened to Stimpson stories for their entire lives.

I am also grateful to Joel Greenberg and two anonymous reviewers for critical feedback. Boyd Zenner of the University of Virginia Press also provided helpful feedback. Of course any errors in the book are my responsibility alone.

Abbreviations

AJC	*American Journal of Conchology*
AJS	*American Journal of Science and Arts*
ANB	*American National Biography*
Proc. ANSP	*Proceedings of the Academy of Natural Sciences of Philadelphia*
Proc. BSNH	*Proceedings of the Boston Society of Natural History*
CAS	Chicago Academy of Sciences
NA RG	National Archives, Record Group
SIA RU	Smithsonian Institution Archives, Record Unit
GNHL	Grove National Historic Landmark

Introduction

There are few nineteenth-century zoologists whose lives were as compelling, colorful, and eventful as that of William Stimpson. Outgoing, dedicated, and signally unlucky—on three occasions he suffered the loss of significant scientific collections and manuscripts by fire—Stimpson's story provides an extraordinary panorama of what it was like to collect and describe the world's unknown marine fauna during the foundational years of nineteenth-century natural science. His many and varied contributions to science shed light on a number of important trends in the American natural history community.

Stimpson was one of the leaders of that community from the 1850s until his death, and he was in many ways a transitional figure in the history of science between 1850 and 1870. He was the first to systematically dredge for marine invertebrates on America's Atlantic Coast, from Maine to Florida. Unlike some field naturalists, Stimpson also excelled at the "closet" work of taxonomy, accurately describing and classifying these organisms, many of them new to science. He worked mainly on marine invertebrates but also made important contributions to the study of the Great Lakes and freshwater mollusks as well.

Beginning in 1850 Stimpson began a two-year apprenticeship with the celebrated Swiss naturalist Louis Agassiz. Thanks to Agassiz's brilliant tutelage, Stimpson was one of the very first professionally trained zoologists in the United States, and he was also the first of Agassiz's American students to quarrel and break with him over issues relating to intellectual property. Later Agassiz students would look to Stimpson as the leader of the generation of naturalists born in the 1830s. Stimpson's lifelong relationship with Agassiz is one of the threads that runs throughout this book.

At the age of twenty Stimpson received the position of zoologist on the US North Pacific Exploring Expedition, an ambitious but little-known voyage around the world. One of America's most significant scientific explorations, it has received relatively little attention from historians, despite the loss of one of the ships on the expedition, including fifty of the crew. During a three-year cruise Stimpson became the first Western zoologist to collect, describe, and classify marine animals from Japan and other islands in the Pacific. Much of Stimpson's reputation as a naturalist stems from his work on this expedition.

On returning he spent nine years based in Washington at the Smithsonian Institution, working closely with that other giant of American natural history, Spencer Fullerton Baird. The story of American natural history during Stimpson's life must be viewed, at least in part, through the lens of the rivalry between Baird and Agassiz. Both men wanted to advance American science but they were radically different in temperament and in the way they interacted with their colleagues. Stimpson's position at the Smithsonian, although unofficial, provided unparalleled opportunities to examine specimens from all over the world, thanks to Baird's network of collectors. As one of the few to work closely with both Agassiz and Baird, Stimpson's career provides insights into these two very different men who reared a generation of naturalists.

More than anything else Stimpson loved being a field naturalist, encompassing marine invertebrates from the Atlantic and Pacific Oceans to the Great Lakes. His travels made him a pioneering explorer of the faunas of disparate regions, including Japan, California, Maine, and Florida. Stimpson's was the last generation of naturalists before intense specialization brought a narrowing of focus and naturalists gave way to zoologists. He was one of the first Americans who constituted a new breed of what Lynn Nyhart has called "scientific zoologists," uniting intensive fieldwork with comprehensive museum-based taxonomy. These scientists wove together many strands of natural history, including "'life-history studies' of animals, which undertook to understand all aspects of individual species, including their life-cycles, distribution, habits and behavior, and connections to the past."[1]

But Stimpson left his most lasting legacy through his labors in taxonomy. While it has been derided by some as mere bean counting, Robert Kohler has contributed to the realization that the naming, describing, and classifying of the world's flora and fauna, a key goal during Stimpson's life, created a necessary framework for theorizing, and that such work is extremely labor-intensive and every bit as creative as laboratory science.[2] Fortunate to live during what has been dubbed the Second Great Age of Discovery, Stimpson described and named over eight hundred currently recognized taxa across eleven phyla of marine invertebrates. Despite the breadth of his research Stimpson's life also illustrates the trend in American natural history towards specialization, especially in his embrace of the terms malacology (the study of mollusks) and carcinology (the study of crustaceans) as distinct subfields of zoology.

Stimpson's era was a critical period in the maturation of American natural history. The proliferation of academies of science and professional journals, coupled with the increasingly important role of the Smithsonian and the American Association for the Advancement of Science, led to increased career opportunities. It was also a time of intense nationalism and exploration, shaped by the rallying cry of manifest destiny. Nationalism helped form Stimpson's worldview, and he left some record of his views when for five pivotal years (1858–1863) he served as one of the zoological editors of America's leading scientific journal, the *American Journal of Science*. Although his name never appeared on the masthead, Stimpson's articles and reviews reveal him as a staunch patriot and defender of the efforts of American natural science in the face of perceived neglect and disrespect from some Europeans. During this period, natural history itself underwent revolutionary change in 1859 with Darwin's landmark *On the Origin of Species*, which gave natural science a theoretical foundation on which to explain life. While Stimpson did not play a major role in the debates over Darwin, he left enough evidence for us to conclude that he accepted Darwin's views and incorporated them into his own work.

Stimpson's peers held him in high esteem, electing him to the National Academy of Sciences at the age of thirty-six, which made him one of the youngest members at the time. He helped lead the movement to reform descriptive natural history by adopting higher standards for taxonomists, including the use of measurements and a more thorough knowledge of morphology and technical terminology. Stimpson harshly criticized those who did not meet these standards.

During his years in Washington Stimpson took on a seminal and leadership role in two scientific organizations, the informal and raucous Megatherium Club, whose members combined youthful bravado and scientific accomplishment, and its more formal counterpart, the Potomac Side Naturalists Club. While both were short-lived, they left an indelible mark on American natural history and provide us with stirring examples of the social side of science. Stimpson's exuberant personality and love of fieldwork served as an inspiration for literary figures as well, including a series of articles published in the *American Naturalist* and the popular nonfiction book *A Summer Cruise on the Coast of New England*.

Beginning in 1865 Stimpson largely set aside his own research to help establish the first permanent scientific presence in America's heartland. Through his leadership of the Chicago Academy of Sciences he helped make the museum into one of the most important repositories of natural history collections in the country. The rise of the Chicago Academy between 1865 and 1871 has hitherto been a missing chapter in the story of American natural history as well as the city of Chicago.

One final transition that occurred in Stimpson's lifetime was the beginning of the scientific study of deep-sea life. By the late 1860s naturalists had begun to collect animals from great depths, ushering in a new age of understanding of the ocean's fauna. Stimpson became the first to describe deep-sea crustaceans and participated in some of the earliest American deep-sea dredging, mostly in Florida.

Stimpson is virtually unknown in the three cities where he lived and worked: Boston/Cambridge, Washington, and Chicago. Most of what he accomplished for natural history has been forgotten and many of the specimens he collected have been lost, as were several significant unpublished manuscripts destroyed in three separate fires, culminating in the Great Chicago Fire of 1871. While nothing can restore these losses, this book will provide some measure of redemption for his character and contribution to American natural history.

CHAPTER 1

SEAFARING PRODIGY

The ocean knows no favorites. Her bounty is reserved for those who have the wit to learn her secrets, the courage to bear her buffets, and the will to persist, through good fortune and ill, in her rugged service.
—Samuel Eliot Morison[1]

For many a New England lad in the nineteenth century, the sea served as a source of recreation and, eventually, livelihood. Boston Harbor, crowded with a seemingly endless vista of "stately ships, rich in the association of distant lands," stood at the center of a vast foreign commerce.[2] In the mid-1840s one young man began taking a rowboat into the waters in and around the harbor. Over the next twenty years William Stimpson would come to know the ocean's inhabitants in a way few ever would.

The Stimpsons had come to America in the 1630s as part of the Great Migration from England, specifically the port city of Newcastle-upon-Tyne. For generations they lived in New England, mostly in Boston but some as far north as Maine, toiling as shoemakers, bookbinders, grocers, and merchants. Charles Stimpson, William's paternal grandfather, had risen to become secretary of an insurance company. In general, the men of the Stimpson family shared several characteristics: hard work, late marriage, and early death.[3]

Herbert Hathorne Stimpson, William's father, was Charles's sixth child. Born in Maine in 1802, Herbert grew up hearing stories about his grandfather Isaac Hall, who had fought at Lexington and Concord.[4] Herbert and his siblings were brought up in an atmosphere still containing a whiff of revolutionary fervor, and he and his younger brother Frederick marched in uniform as part of the "Boston Boys" after the outbreak of the War of 1812. Herbert was eventually apprenticed to a sheet-iron worker, and by 1829 he and Frederick ran a thriving business in stoves and ranges, selling to Ralph Waldo Emerson, among others. Herbert received the first of what would be several patents relating to cooking ranges in 1840 and gained renown as the developer of the "Stimpson range," famous in its day throughout New England. He also developed the first sheet-iron air-tight cooking stove.[5]

Little is known about William Stimpson's mother. Mary Ann Devereaux Brewer has been identified variously as having been born in New York, Virginia, and Rhode Island. At the age of twenty she married Herbert Stimpson, eight years her senior, on April 28, 1830. Their oldest child, William was born two years later in Roxbury, Massachusetts, on February 14, 1832. Three more children followed: Sarah in 1835, James in 1837 and Francis in 1839. William's mother died at the age of thirty-two.[6] Like so many others in the nineteenth century she succumbed to pulmonary tuberculosis. Losing his mother at such an early age must have been traumatic for young William. It is possible that he resembled her, as he did not look at all like his father, being taller, thinner, and with more delicate features.

Two years after becoming a widower Herbert remarried, this time to a woman nearly half his age. The former Mary Elizabeth Sawyer was only ten years older than her stepson William. The marriage took place in her hometown of Lancaster, Massachusetts, about forty miles west of Boston, and Herbert bought land in South Lancaster near the Nashua River. An avid sportsman, Herbert retreated here when he could, and young William often came along to hunt and fish with his father. He became a familiar sight to locals each summer, gun in one hand and fishing pole in the other, his pockets stuffed with bottles and boxes for transporting specimens. It was here that William discovered a love and appreciation for the natural world.[7]

In 1845 Herbert decided to move the family from Roxbury (later incorporated into Boston) to Cambridge, where he invested in railroads, joined several charitable organizations, and was active in the Episcopal Church. He possessed great energy, a love of social life, and a brilliant wit, all qualities later attributed to his son William.[8]

A small but growing village in the 1840s, Cambridge offered an abundance of natural areas, and William took long walks in the countryside. He soon became fascinated by the region's abundant land snails and began to collect and study them in earnest, amassing a large collection that he carefully organized and labeled. But it was the lure of the sea that truly captivated him, and whenever he had time away from school he haunted the beaches looking for shells. Stimpson loved fishing, not just for the sport of it but because he knew that it could also shed light on his study of shells. In some cases the only known example of a shell came from the stomachs of fishes. Gutting fish to examine the stomach contents was messy work, and several of Stimpson's early papers mention obtaining specimens through this method.

His more formal education came amid reformer Horace Mann's efforts to improve Massachusetts schools. Herbert Stimpson groomed his eldest son to follow in his footsteps, enrolling him in Boston's English High School, the oldest public high school in the country (1821) and one that focused on preparing students for the business world. As a second-class student in July 1847 Stimpson won first prize in mathematics and less than a year later he graduated at age sixteen, again sharing top honors in his class in mathematics and natural philosophy and winning the school's Lawrence Prize. He also garnered a coveted Franklin Medal, awarded in the form of a silver medallion to the top six students at each school.[9] The English school shared a building with the Boston Public Latin School, founded in 1635 to prepare young men for Harvard. In September 1848, William enrolled there, excelling in Latin, the international language of science.[10]

Stimpson's predilection for natural history was enhanced by proximity to Harvard. By the late 1840s Louis Agassiz was giving lectures at the Lawrence Scientific School there, along with the botanist Asa Gray and the anatomist Jeffries Wyman. Equally important to Cambridge's intellectual climate were men such as Ralph Waldo Emerson, Henry Wadsworth Longfellow, and Henry David Thoreau, all of whom were interested in nature. Immensely popular in his day, many of Longfellow's poems were paeans to Nature and Stimpson copied several of them in a notebook, moved by the beauty of the verses in "Woods in Winter" and "Sunrise on the Hills."[11]

Stimpson became tangibly aware of the larger scientific world through Augustus A. Gould's *Report on the Invertebrata of Massachusetts*. This landmark work, published in 1841, inspired a generation of young naturalists. The lavish illustrations and the comprehensive nature of Gould's work made an indelible impression, leading Stimpson to call on Gould at home to ask if he might receive a copy of the treasured tome. Gould's reaction was one of bemused inquisitiveness. "Struck by his [Stimpson's] diminutive figure and his youthful appearance, the Doctor could not at first comprehend the nature of his request. His curiosity was aroused, and he questioned his strange visitor. He found that he had already collected largely, and had mastered all of the works within his reach relating to his favorite pursuits. From that day forth until his death Dr. Gould became his counsellor and friend."[12]

Gould had a profound impact on Stimpson's early years as a naturalist, both by the example he set and through the guidance he offered. In August 1849 Gould sponsored him for membership in the Boston Society of Natural History. Founded in 1830, the Society had recently moved into a building just off Boston Common. Here Gould introduced him to other naturalists, and membership in the Society allowed Stimpson to spend many happy hours studying in the large library and examining the collections.[13]

One other occurrence was pivotal in assuring Stimpson's entry into the scientific world. Two weeks after he joined the Boston Society the second meeting of the American Association for the Advancement of Science was held in Cambridge, and Stimpson became a member at seventeen. As Sally Kohlstedt and others have noted, America was just beginning to develop a unified scientific community. That year's cholera epidemic, which took over six hundred lives in Boston alone, probably limited attendance, but meeting and mingling with men of science from various fields made a lasting impression on Stimpson.[14] One man in particular stood out. The brilliant Swiss naturalist Louis Agassiz had arrived in Boston three years earlier and had quickly captivated the public with his popular lectures and genuine love for nature.[15]

At the close of the sessions Stimpson and others took a cruise from Boston to Salem, where Agassiz demonstrated the use of the naturalists' dredge. A keen student of marine invertebrates, Agassiz attempted to cultivate men to undertake this grueling but rewarding work. American naturalists had been slow to use the dredge in a systematic way and most collected in the littoral zone close to shore. Stimpson had already done some dredging but Agassiz's example spurred him on to new efforts. Quick to see the potential of dredging, Stimpson soon set his sights on becoming the first to intensively explore America's East Coast with a dredge. Dredging soon became an all-encompassing passion.

Ironically for a man whose life would be identified with the sea, Stimpson's first published scientific contribution focused on a terrestrial species. On one of his many rambles around Boston he found a small land snail, and after some study he concluded that it represented a hitherto unknown species, a deficiency he quickly rectified with a short description written in Latin. The paper was read at the September 19, 1849, meeting of the BSNH.[16] He was a published author at seventeen and now decided to devote more time to natural history.

Stimpson soon left the study of land snails to his friend William G. Binney and turned his full attention to marine invertebrates. By 1849 he had filled the basement of his father's home with tanks of saltwater and populated them with different species of marine invertebrates whose life cycle he observed, making drawings of their anatomy. In fact, Stimpson was among the first to discover the principle of the self-regulating aquarium.[17] Through painstaking trial and error he managed to combine plants, animals, and seawater to achieve an equilibrium that allowed his miniature ecosystems to thrive for months without an infusion of new water.

He spent hours observing mollusks, carefully making drawings of them as they emerged from their shells. Stimpson did not mingle much with other boys his age, and his brother James later recalled that he never had to perform the brotherly task of searching for William at dinnertime; they always knew where to find him.[18] Stimpson's growing obsession with all things marine clearly rankled his father. Herbert Stimpson seems to have been disappointed that his oldest son had no interest in taking over the successful stove business. He simply did not understand his son's passion, and as a thoroughly practical man the elder Stimpson believed that it was time for his son to outgrow youthful pursuits and embark on a real career.

Taking matters into his own hands, he arranged for William to take a job at the engineering office of Fletcher and Parker. William did not particularly like the idea of becoming a civil engineer, but he stayed at the job for nearly eighteen months. He spent more time looking for shells than he did in surveying, however, a fact that his employers dutifully reported to his chagrined father.

Later that year Stimpson joined the crew of a fishing smack bound for the Newfoundland Banks. Whether he did so clandestinely or with his father's approval is unknown, but he clearly enjoyed the experience of being out on the sea, learning how to rig nets and trawls, skills that he would later put to good use in the name of science.

By June of 1850 Stimpson was dredging on a regular basis near Point Shirley, off Buzzard's Bay, near Bird Island, and in Salem Harbor. He found several new mollusks and published descriptions of them in the *Proceedings of the Boston Society of Natural History*. That society recognized his efforts by naming him curator of conchology (the study of mollusk shells), an unpaid position he would hold for the next three years. In his first yearly report Stimpson stressed the need to obtain specimens of the animals inhabiting New England shells, to show "our citizens the great variety of animal forms which exist in close proximity with, and yet concealed from them." Stimpson would contribute numerous examples through his own fieldwork.[19]

The meetings of the Boston Society were an education in and of themselves, giving Stimpson the chance to listen and learn from established naturalists such as Agassiz, Waldo Burnett, Charles Girard, and others. They in turn took Stimpson's measure, and Agassiz liked what he saw enough to begin the next phase of Stimpson's scientific apprenticeship. In October 1850, Stimpson, with his father's grudging approval, became a student of Agassiz's. His two years with that celebrated savant would be both edifying and tumultuous. Agassiz's popularity cannot be overestimated. With a seemingly inexhaustible supply of energy and bonhomie he traveled across the country, energizing scientific endeavors everywhere he went. Agassiz made science socially fashionable, and thousands attended his lectures at the Lowell Institute.[20]

At forty-three, Agassiz had recently married a prominent young Bostonian fifteen years his junior. Powerfully built, with broad shoulders and large hands, Agassiz's most prominent physical feature was an oversize head. One naturalist described him as "a dumpy Dutch-looking little sort of a man but with a splendid head to look at and a very pleasant face."[21] An unabashed extrovert, Agassiz never seemed to stop talking and clearly enjoyed being the center of attention. He was without question the most famous and influential naturalist in America.

Agassiz enthusiastically adopted the cause of advancing American science. Europeans had long denigrated American scientific efforts, and Agassiz meant to change that perception by training Americans using his methods. Tellingly, his praise for American efforts was leavened by a somewhat superior and condescending attitude. He saw America passing from childhood to maturity "with the faults of spoiled children, and yet with the nobility of character and the enthusiasm of youth."[22]

While an excellent naturalist, Agassiz's most lasting contributions to science came as an educator. His fame had attracted other Americans eager to learn from him, and Joseph LeConte, William Louis Jones, and Henry James Clark all arrived about the same time as Stimpson.[23] Regarding his role as a mentor, one historian has described Agassiz as "grandiose, narcissistic, exploitative, and manifestly unfair. He was also inspiring, passionate [and] often caring."[24] One student claimed that Agassiz's lectures were limited in scope and that they never changed from year to year, while some Harvard professors questioned Agassiz's instructional methods. Nathaniel Shaler perhaps said it best when he wrote, "Agassiz was the worst instructor I have ever known, but in diverse ways the greatest educator.[25]

Agassiz did not teach from books. For him the book of Nature was always open and one merely needed to observe it to understand its riches. The students began their education in a similar fashion. "Agassiz brought me a small fish, placing it before me with the rather stern requirement that I study it, but should on no account talk to any one concerning it, nor read anything relating to fishes, until I had his permission to do so. To my inquiry, 'What shall I do?' he said in effect: 'Find out what you can without damaging the specimen; when I think that you have done the work I will question you."[26] After an hour or perhaps a day the student was convinced that he had discovered all there was to know about his often slimy and smelly specimen. With rapt attention Agassiz would listen to the report, puffing on a cigar and invariably responding with the words, "That is not quite right," admonishing the disappointed student to look again. This indoctrination could last days or weeks, and helped weed out less-than-serious students.

The next step in Agassiz's training was to sort a large number of related but unlabeled specimens in order to distinguish similarities and differences.[27] The students that persevered credited it with honing their observational skills, and none of them ever forgot the experience. After two years with Agassiz one student "felt ready to go anywhere in the world . . . with nothing but his note book and study out anything quite alone."[28] Stimpson was one of many who would owe much of their success in science to the human dynamo that was Agassiz.

It must have been quite a heady experience for Stimpson to find himself in Agassiz's world. He seemed to know everyone, and the most distinguished literary and political figures of the day practically prostrated themselves in his presence. Unfortunately we have no firsthand accounts that would shed light as to the degree that Stimpson entered into Agassiz's social circle. But there was a darker side to Agassiz, and Stimpson was surely aware that his teacher was at the time embroiled in two very public and nasty disputes with men who had worked for him, Edward Desor and Charles Girard.[29]

Agassiz probably admired Stimpson's pluck in forging ahead with natural history despite his father's opposition, much as Agassiz himself had done. Stimpson's knowledge and abilities quickly became apparent, and Agassiz appreciated the fact that his pupil showed an interest in the lower marine invertebrates. Agassiz emphasized the importance of making careful drawings of anatomical details, and Stimpson showed a real flair for executing these technical drawings.[30]

In training his students Agassiz gave special importance to embryology, which he believed the "necessary basis for all classification."[31] One of Stimpson's first attempts came when he kept a gastropod alive for seven months so that he could record its three distinct stages of development. In another instance he kept marine worms in captivity until they laid eggs, and he then made drawings of the different stages of development.[32] He even tried an experiment, cutting a rare sea star in twenty pieces, each of which eventually became a perfect animal.[33] Above all, Agassiz taught comparative zoology. He trained the first generation of professional zoologists in America and Stimpson was among his earliest students.[34]

Stimpson spent most of the summer of 1851 dredging at Eastport and Grand Manan off the coast of Maine, providing him with the raw materials to publish twelve papers between January 1851 and June 1852. In them he showed his all-around abilities by collectively describing thirty-six new species encompassing crustaceans, mollusks, ascidians, annelids, and echinoderms. Among them were his first decapod crustacean, the rare *Axius serratus*, a burrowing mud shrimp. He also corrected a mistake that had entered the literature years before when he recognized that what had been described as two different species of mollusk were actually the juvenile and adult forms of the same species, proving the point by collecting a series of specimens showing the entire life cycle.[35]

He wrote most of his descriptions in Latin, as was the custom at that time with many naturalists, especially marine invertebrate specialists. Another long-standing tradition led him to name species in honor of men that had helped supply him with specimens, including Gould, Joseph P. Couthouy, and General Joseph G. Totten.[36]

By the time Stimpson began his career most of the largest mollusks had been described. He sought out what others had overlooked, animals too insignificant to be noticed except by the eye of a trained field naturalist. One of the best examples is the Beautiful Caecum, a mollusk that he found in Buzzard's Bay and described in 1851. This tiny and unusual gastropod inhabits a shell that is less than one-eighth of an inch long, about the size of an en-dash. In describing three new species Stimpson showed for the first time that this genus existed in the United States.[37]

He also dabbled in paleontology, having come across an accumulation of fossil shells at Point Shirley in Chelsea, Massachusetts. He had dredged living specimens of all but one of the fourteen fossil species within a mile of the locality, all of them deep-water species found most abundantly in northern waters. Stimpson thought that the fossils furnished evidence in favor of Charles Lyell's theory that the deposits had been formed by melting icebergs. Stimpson's contribution was important enough to be reprinted in the *Annual of Scientific Discovery* for 1852.[38]

In addition to fieldwork, Stimpson continued to attend the meetings of the BSNH. The collegial atmosphere that generally prevailed did not prevent differences of opinion from being aired. Throughout early 1851, he and others engaged in discussions on a group of echinoderms, the sea cucumbers. William O. Ayres had recently decided to focus on this group and stated that he did not see evidence for the hypotheses that certain types of echinoderms were found at greater depths than others. Stimpson disagreed, siding with Agassiz's view that the "highest" or most developed forms were generally found in shallow waters, while the more primitive types were found in deep waters.[39] Ayres also believed that the sea stars off the American coast were all different species from those found off the coast of Europe while Stimpson maintained (correctly) that many were in fact identical. It took conviction for the nineteen-year-old Stimpson to disagree publicly with a man fifteen years his senior, but he stood by his observations.[40]

After a huge storm in April 1851, Stimpson discovered literally hundreds of sea cucumbers thrown upon the beach, including one that he described as new. In a paper that year he also described twenty-five species of marine invertebrates not previously observed in Massachusetts Bay.[41] By the end of the year he had expanded the paper into a slim book entitled *A Revision of the Synonymy of the Testaceous Mollusks of New England*.[42] This represented an ambitious effort to update Gould's 1841 book. Much had changed in the decade since its publication, a shift signaled in Stimpson's embrace of the word malacology, as opposed to Gould's use of the older term conchology. Mollusks had largely been classified by their shells, while malacology emphasized the animals inhabiting the shells. For many mollusks the animal itself had never been described, but Stimpson included notes on the morphological characters of some of these animals.

Stimpson did much more than collect—he observed and recorded animal behavior. In one case he watched as a clam made surprising leaps and swam about for long periods, a feat performed by "suddenly drawing in the umbrella-shaped foot at the same time that water is expelled from the posterior opening by the closing of the valves."[43] Such observations were more sophisticated than the activities of conchologists who were content to merely classify dead shells.

Stimpson also utilized the microscope, learning from one of the masters in the field, Dr. Waldo I. Burnett. The twenty-three-year-old Burnett had already graduated from Harvard Medical School, studied in Europe, published numerous articles on a variety of scientific and medical subjects, *and* been elected to the prestigious American Academy of Arts and Sciences. Together he and Stimpson examined spermatozoa from clams, leading them to conclude that the sperm glands were arranged differently among similar species. As Leonard Warren has noted in his biography of Joseph Leidy, very few American naturalists were using a microscope for scientific work in 1850. In the coming decades it would serve as a sign of increasing professionalism in both science and medicine.[44]

Pointing out that most of the progress in studying marine invertebrates had been made by Europeans, Stimpson was especially influenced by the work of the British naturalist Edward Forbes, who had popularized the use of the dredge.[45] One of Forbes's major contributions was the idea that marine animals were distributed in zones of depth, and Stimpson adopted Forbes's terminology. Through his dredging Stimpson added a wealth of information concerning the geographical distribution of 344 species of mollusks, as well as the bathymetrical data, or horizontal and vertical range of species. He had ascertained for the first time localities and depths for many species and added eighty-four species to those listed by Gould. Stimpson had already acquired a network of collectors to aid his research and cited sixteen men who supplied him with specimens, including physicians, boat captains, wood carvers, and fellow naturalists.

The monograph received good reviews. Agassiz called it "an excellent revision ... particularly valuable for the extensive observations he [Stimpson] has collected upon their geographical distribution and the depths at which they occur." He urged the naturalist James Dwight Dana to give the book a favorable review in the *American Journal of Science*, as "it deserves it fully, for the great accuracy and care with which the facts there condensed have been gathered." Dana obliged, stating that Stimpson's book was "valuable in representing the existing state of the science, especially as regards New England species."[46]

In addition to praise, however, the book also generated controversy. Stimpson had made radical changes in scientific nomenclature, concluding that many species had been placed in the wrong genus. He shifted some species of the genus *Nucula* to the genus *Leda*. Thus *Nucula limatula* Say became *Leda limatula* Stimpson. In all, he modified 207 of the scientific names used by Gould just a decade earlier. He justified his decision as follows: "In citing authorities I have referred to the author who gave to the species the name which it now bears;—the whole name, and not a part of it. . . . The practice of citing the author who gave the specific designation—a part of the name—is an *innovation*, which has become frequent among Conchologists during the last thirty years."[47]

Agassiz had long advocated these changes and in following his example Stimpson opened himself up to a storm of criticism.[48] One reviewer doubted that Stimpson's modifications would be adopted, noting that these views of authorship were "not in accordance" with those of most naturalists.[49] Some considered it presumptuous that a teenager had the gall to erase legendary names like Thomas Say and Gould from the scientific literature, and it certainly seemed as if he were promoting himself at the expense of his elders. Stimpson had not intended to denigrate Gould's or anyone else's work; he simply, and probably prudently, followed Agassiz's example. He would later alter his views to conform to the rules of nomenclature that exist today: when a species is transferred to a new genus, the name of the original species describer is retained in parentheses.

In mid-December 1851 Stimpson accompanied Agassiz and his family to Charleston, South Carolina, where Agassiz rented a small cottage on Sullivan's Island that became a base for seaside researches. Despite unseasonably cold weather at the outset, he became fascinated by the new fauna he found here, on what was probably his first extended trip away from home. A month after their arrival Agassiz complained of "an over-excited" nervous system that left him feverish and bedridden. In a January 1852 letter written during his period of illness he referred to Stimpson as a "very promising young naturalist, who has been connected with me for some time . . . draws the crustacea and bryozoa, of which there are a good many new ones here."[50]

With Agassiz confined to his room Stimpson began dredging in the company of Lieutenant John D. Kurtz of the Army Corps of Engineers. Ten years older than Stimpson, Kurtz graduated from West Point in 1842 and worked on the coastal fortifications in Charleston Harbor, giving Stimpson a tour of the still unfinished Fort Sumter. One of their most productive collecting locales was at Fort Johnson on James Island, where nine years later the first shot of the Civil War would be fired. For two months Stimpson and Kurtz scoured the coastline as far as North Carolina. Stimpson wrote home assuring his father that he was in a "happy situation." He bragged that he had discovered at least twenty species here that neither Agassiz nor anyone else had found.[51] In the only example of Stimpson coauthoring a scientific paper, he and Kurtz jointly described six new mollusks from the southern coast.[52]

Stimpson had found many remarkable animals during his few months in and around Charleston. Among them were a sea anemone and a polychaete lugworm, "a large and fine worm" up to sixteen inches long and an inch thick. He expressed surprise that so large an animal had been overlooked as it was quite common in certain localities. He made detailed notes on the habits and station of each species, data not usually found in papers devoted solely to descriptions of new species.[53]

Returning to Cambridge in April 1852, Stimpson's delight over his success soon crumbled in the face of a confrontation with Agassiz. We have only one account of the incident. "It seems that Stimpson went out very early in the morning, at an exceptionally low tide, and dug up several new species of marine invertebrates which Agassiz immediately claimed for his own, but did not succeed in obtaining. The aftermath appears in the laconic phrase in Stimpson's journal: 'May–June 1852. Hard times with Agassiz.'"[54]

The trouble began when Stimpson began working up his Charleston finds. He had uncovered two new echinoderms and promptly wrote out descriptions.[55] These animals were particular favorites of Agassiz's and they had not hitherto been a primary focus of Stimpson's research. Agassiz probably argued that Stimpson would not have found them if Agassiz had not invited him to South Carolina and that therefore Agassiz had a right to any specimens that Stimpson found. We also do not know the exact arrangements that governed Stimpson's studies with Agassiz, as one source states that Stimpson paid his own expenses on the trip.[56]

A weaker man might not have had the courage to stand up to America's most famously imperious naturalist, but Stimpson, now accustomed to seeing his labors in the field turned into publications, refused to give in to what he saw as Agassiz's unjust demands. The incident illustrates Stimpson's desire for independence and Agassiz's insistence on full control over his students. Agassiz's Old World expectations were that students always deferred to their teacher, but part of the problem also stemmed from emerging philosophical differences in the two men's approaches. Stimpson would go on to devote a significant portion of his scientific career to fieldwork and taxonomy, including describing new species, while Agassiz condemned the latter practice as a "pernicious habit" and morally dangerous.[57] Agassiz urged his students to focus instead on embryology or physiology, but Stimpson preferred the freedom of fieldwork over the confines of the lab.

Stimpson became the first in what would be a long line of American students to rebel against Agassiz. His principal biographer frankly noted Agassiz's authoritarianism, suggesting he had "an inability to work with others on terms of equality or near equality."[58] Mary P. Winsor rightly argues that we should think of those who studied under Agassiz in the 1850s not as his students but rather as apprentices.[59] Agassiz's innovative methods allowed his protégés to enter "the sanctuary of science" in a relatively brief period, and he himself admitted that "after a few years' study they are generally as far advanced in one special department of Zoology as the most eminent naturalists."[60]

One of his most devoted students best summed up Agassiz's trouble in dealing with others as peers. Nathaniel Shaler noted that Agassiz used his students to explore certain topics for him and before long Shaler had begun teaching Agassiz new facts, but "what I supplied went into his memory as his own discoveries, which in a way they were, for the direction of the work came from his mind." Tellingly, Shaler also commented on Stimpson's "fierce independence of spirit which did not allow him to profit by mastery."[61]

Stimpson had now faced conflicts with both his father and Agassiz. In an effort to get away from unpleasant matters he returned in July of 1852 to the cold northern waters of Grand Manan, the "Island in the Mist." The largest and most remote of the three major islands at the entrance to the Bay of Fundy, Grand Manan represented virgin territory for devotees of marine invertebrates. Exploring these waters meant taking risks. Shipwrecks in the area were common and John James Audubon called it "that worst of all dreadful bays."[62] The tides at the mouth of the bay are among the highest in the world, reaching fifty feet or more, and dense fogs often blanket the area, making it a dangerous place to prospect for marine life.

Awed by the beauty of this magnificently desolate place, Stimpson described the view. "The watery scene of the region is very striking and beautiful. Long and narrow inlets of the sea run far into the land and seem more like fresh-water lakes and streams, resembling in this, as well as in their great depth and in other particulars, the lochs of Scotland. These inlets are studded with islands, the rocky sides of which are almost perpendicular. The islands are thickly wooded, the trees and boughs often overhanging the water; so that it is possible to glide along under the shade of these trees, while looking down into a depth of twenty or thirty fathoms."[63]

It was at Grand Manan that Stimpson cemented his reputation as one of the best field naturalists of his generation. A typical day's dredging was an ener-vating task and served as a constant reminder that being a field naturalist is physically demanding. The sea gives up its secrets reluctantly, and there were hard lessons to absorb. In heavy seas waves often swamped the small boat, and many valuable specimens were swept overboard and returned to the ocean bottom. The rough seas and cold winds, coupled with the constant work of hauling up the dredge from depths as great as sixty fathoms taxed even young muscles. He taught himself how to pilot a ship no matter what the conditions and he kept a sharp watch for changing weather conditions. He found that the fog frequently came "rolling in with the speed of a race-horse," so he always had with him a large shell with the top knocked off for use as a horn in case he became lost in the fog.[64]

When the weather was too stormy he walked along the shore. Sea cucum-bers were especially plentiful, and at low tide he found it impossible to take a step without crushing these delicate creatures. Limited funds might have been one reason he boiled and ate some of them, which he found to be as tasty as lobsters.

His curiosity extended beyond invertebrates. Some whales approached close enough to the boat that they practically spouted in his face. His dredge brought up many fish from the depths, none stranger than the hagfish, a blind bottom dweller with the remarkable habit of pouring out a gallon or more of slime from its mucous sacs when caught. Stimpson was the first to substanti-ate the existence of this family (*Myxinidae*) from the North American coast.[65]

His observations supported the fact that many supposed deepwater species were found at very slight depths or even on shore. As was common in north-ern regions, there were relatively few species, but the number of individuals was very great. In all, he spent three productive months in and around Grand Manan, securing a rich harvest. His knowledge of America's marine fauna was growing, as within a span of six months he had now spent significant time dredging near the northern and southern extremes of the Atlantic Coast.

Returning to Cambridge in October, Stimpson pondered his options. He did not relish the thought of returning to Agassiz although his father now insisted that he do just that.[66] Agassiz probably wanted him back, as he had sixty monographs on the embryology of marine invertebrates on hand, and Stimpson no doubt had been responsible for making drawings and some of these observations.[67] Stimpson chose to stay and it gave him the opportunity to meet scientific royalty. In 1852 the British geologist Charles Lyell traveled across New England to examine geological formations, often accompanied by Agassiz or Augustus Gould.[68] Lyell later recounted some explorations around Boston on October 30, 1852 "with Strepson, a young engineer and conchologist."[69] There is no doubt that Lyell's "Strepson" was Stimpson, and it was no small thing that Stimpson now had the opportunity to show Lyell the fossil deposits he had uncovered at Point Shirley.[70]

Now wary of his mentor, Stimpson had already begun making scientific plans that did not include Agassiz. Earlier in the year he had begun a correspondence with Joseph Leidy of Philadelphia. Although just twenty-nine years old, Leidy had established himself as one of the central figures in American natural history. Considered by one historian as "the last man who knew everything," Leidy would soon describe the first complete dinosaur skeleton from the United States.[71] He and Stimpson were both interested in small, segmented annelids known as bristle worms, and they discussed a joint work on the marine worms of the Atlantic coast. In his letters Stimpson confided in the quiet, unassuming Leidy. Stimpson wrote that he had been "silently collecting materials" from Massachusetts Bay, boasting that at Grand Manan alone he had found forty-four species of worms. He also related his plan, which he asked Leidy to keep private, to make an excursion to Florida.[72]

A solution to Stimpson's predicament with Agassiz came in the form of an unexpected opportunity, one that most young naturalists could only dream about. Gould had returned from a visit to Spencer Fullerton Baird at the Smithsonian, from whom he learned of plans for a government exploring expedition to the North Pacific Ocean. Gould asked whether Stimpson might be interested in going and he fairly leapt at the chance, with Gould telling Baird that Stimpson was "ripe for it." Gould enclosed a glowing recommendation letter for Stimpson, stating that he "possesses so many desirable qualifications—viz., a competent amount of knowledge—practical skill in collecting both by land and sea—an excellent draughtsman & microscopist—skill in preserving objects—a good seaman—inoffensive manners—enthusiasm."[73]

Magnanimously, Agassiz penned a letter on Stimpson's behalf. He referred to Stimpson as "a young friend of mine" who had been his "special pupil" for nearly three years. Agassiz believed him to be not just "fully competent for the task, but even particularly qualified ... since he has devoted his attention more particularly to the animals inhabiting the northern seas. His great ability in dredging would enable him to make valuable collections even while at sea; and since I do not think that there ever was a dredge let down in those parts of the ocean to which the expedition is chiefly to sail, I do not doubt that the harvest thus obtained would be particularly rich." Moreover, added Agassiz, "Mr. Stimpson is so familiar with all the manipulations of a practical naturalist that he would do every thing easily and with little expense. I should add that his little figure is by no means expressive of his abilities and that from a long intercourse with him, I know him to be a most efficient man."[74]

James D. Dana of Yale also weighed in: "I believe you could not find a naturalist that would be more efficient or more acceptable [in] every way." Among his other qualifications, the one that perhaps meant the most to Dana was that Stimpson was an American. One of the most respected naturalists and geologists of his time and a veteran of the US Exploring Expedition, Dana believed that the United States had erred in giving positions to Europeans on government expeditions.[75]

Given these testimonials, and with his father's consent because he had not yet turned twenty-one, Stimpson secured the appointment of zoologist on the North Pacific Exploring Expedition on November 12, 1852.[76] Rarely if ever in American natural history had such an important position gone to one so young. His connections to Agassiz, Gould, and others were certainly factors but he had also proven himself, having already published a monograph and fifteen scientific papers. His fellow naturalists had even named two species and a genus in his honor, unusual recognition for someone barely out of his teens. If he felt some trepidation about leaving behind family and friends, he also realized that he had been given an exceptionally rare chance to get paid for doing what he loved. Stimpson now became one of a handful of men in 1850s America who could be called a professional zoologist.

CHAPTER 2

NAVAL SERVITUDE

> I have always felt that I owe to the voyage the first real training or educa-
> tion of my mind; I was led to attend closely to several branches of natural
> history, and thus my powers of observation were improved. . . . I feel sure
> that it was this training which has enabled me to do whatever I have done
> in science.[1]
> —Charles Darwin, on his voyage on the *Beagle*

There were hundreds of tasks large and small to attend to before leaving on a three-year cruise around the world, but Stimpson had one major scientific goal to accomplish. He intended to publish a catalog of every marine invertebrate species he had found at Grand Manan, over three hundred in all. This presented an enormous challenge that would require him to iden-tify and classify animals from at least a dozen different phyla. In addition to taxonomy he also addressed questions relating to the geographical distribu-tion of species. This subfield had gained many adherents, particularly among marine zoologists, and Stimpson's Grand Manan paper was nothing less than an attempt to give an accurate assessment of the entire invertebrate fauna of a specific locale, data that could then be compared to those from other regions.[2]

Stimpson arrived in Washington, DC, in late November 1852 and spent much of the next three months preparing for the expedition and writing his Grand Manan report. After the initial sorting and classifying he realized that he had an astounding ninety-two new species to document. He complained to Leidy that the paper kept him "at work all the time—sometimes all day & 3/4 of a night on one species" alone. Every species required a special investigation in the scientific literature to determine its genus and whether it was related to known Arctic or European species.[3]

He traveled to the Academy of Natural Sciences of Philadelphia, which probably had the best scientific library in the country, to locate references and examine the collections. He also initiated a correspondence with for-eign naturalists such as Michael Sars of Norway, who sent him specimens for comparison. Other colleagues generously provided aid, as Dana loaned him unpublished drawings of new genera of crustacea. Charles Girard had left Agassiz to work with Baird at the Smithsonian on a number of projects. In a move sure to upset Agassiz, Stimpson asked Girard to write the section on the flatworms and Girard described five new species, including one named for Stimpson.

One of the groups that Stimpson expended a great deal of time on was the bristle worms, known as polychaetes. These deceptively simple marine animals contain a head, tail, and segmented body but there are over ten thousand different species. Their classification has continued to bedevil taxonomists into the twenty-first century, but more than half the taxa that Stimpson inaugurated are still recognized.

When instituting new species Stimpson exercised caution. Historians have characterized taxonomists as either lumpers or splitters, with the former being cautious in naming new species and the latter more inclined to see specific differences. Field naturalists saw the overwhelming complexity of nature in a way that closet naturalists, or those who did not venture into the field, never did, putting Stimpson squarely in the lumper camp. One of the crinoids, or sea lilies, he had found might have been new but he realized that it could just as easily have been the young of a known species. With dry wit he commented that, "One who is so fortunate as to possess very few specimens, soon becomes perfectly satisfied in his own mind as to the specific distinctions, and finds little difficulty in separating them; while one who has some hundreds, can make but slow progress, the perplexity seeming to increase with the number of specimens."[4]

Certain that many American species were identical with those found off the coast of Europe, Stimpson deferred describing several univalve mollusks because he had been unable to compare them with those from Europe. In a swipe at closet naturalists Stimpson stated that, "every practical naturalist knows how much he is aided in defining species, by seeing them in the beauty of life, in their natural conditions and associations." For many animals he had made drawings from life because the colors quickly faded after death, as if the "soul had pigment" in the words of John Hersey.[5] Other essential characters became indistinguishable as a result of the strong alcohol used to preserve specimens.

Stimpson posed a number of queries relating to the geographical distribution of species. Could a species inhabit two distant localities without occurring in the intermediate space? If so, were geological factors responsible? Were species originally created in several places at once, as Agassiz believed? The answers would help shed light on one of the fundamental questions of the day, namely what constituted a species. Stimpson challenged American naturalists to join him in carrying out further investigations, noting an "increasing taste for pursuits of this kind in our country."[6]

Gould and Dana served as referees and informal advisors for the paper and Stimpson acknowledged Agassiz for allowing him to examine his collection of European books and specimens. He also named a new species of echinoderm for Agassiz but he never told him about the Grand Manan paper, a fact that clearly upset Agassiz when he learned about it from Smithsonian chief Joseph Henry, who planned to publish it.

Agassiz complained that Stimpson simply was not ready, conveniently overlooking Stimpson's numerous publications and the fact that Dana and Gould believed the paper merited publication. Elizabeth Agassiz, writing for her ill husband, made it clear that he had been disappointed with Stimpson's actions. "My husband has usually found Mr. Stimpson too anxious to bring himself into notice by publishing while his knowledge is yet quite immature; but it does not seem an easy matter to repress the vanity of young men in America."[7]

Despite Agassiz's carping, Stimpson's "Synopsis of the Marine Invertebrata of Grand Manan" appeared in the prestigious *Smithsonian Contributions to Knowledge*. This brought Stimpson's name to the international community of science, and it remains one of his major contributions. The British conchologist Philip P. Carpenter later wrote that the article "should be consulted by all who desire to institute a comparison between the sub-boreal faunas on the two sides of the Atlantic."[8]

In the final book of his trilogy on American exploration William H. Goetzmann dubbed the eighteenth and nineteenth centuries the Second Great Age of Discovery. Nationalism has played a major role in scientific exploration, with Goetzmann arguing that the explorer "should be thought of as integral to any history of science."[9] The tradition of the ocean-spanning exploring expedition in helping to make the career of a naturalist had a long history. Two of Stimpson's recent predecessors were Charles Darwin and James D. Dana, who had both set sail in the 1830s. By 1850 the days of epic national oceanic voyages of discovery were coming to a close, with the *Challenger* expedition of 1872–1876 marking the end of the old style of natural history exploring cruises.[10]

An earlier American expedition served as the model for the North Pacific Exploring Expedition (hereafter NPEE). The United States Exploring Expedition (hereafter US Ex. Ex.) of 1838–1842 surveyed the southern and eastern portions of the Pacific Ocean and its naturalists, especially Dana, brought home large collections. The expedition "led to the emergence of the

United States as a naval and scientific power with worldwide interests."[11] The NPEE would be America's second great naval exploring expedition.

Diplomacy and a nascent American imperialism helped shape the NPEE as well. The territory gained in the Mexican-American War fueled the rallying cry of manifest destiny, and once California had been secured, American businessmen and politicians set their sights even farther west, so far west it was East: the huge markets of Asia. Matthew Perry's gunboat diplomacy in 1853–1854 succeeded where others had failed in convincing the Japanese to end their isolation and join the family of nations.[12] The NPEE in effect served as Perry's reserve force, with one American newspaper labeling it "a second edition of the Japan affair, but . . . christened with a more peaceful name."[13]

The stated goals were to survey and accurately chart "for naval and commercial purposes" portions of the North Pacific, the Bering Straits, and the China seas. American clipper ships were already involved in the China trade, and in the 1840s new whaling grounds had been discovered off Kamchatka, the Ohkotsk Sea, and the Arctic Ocean, all areas specifically targeted for reconnaissance by the NPEE. Whaling constituted the fifth largest American industry in the 1850s, and one newspaper explicitly made the linkage between the loss of American whalers and the need for the expedition.[14]

Stimpson's goals were straightforward: to collect as many natural history specimens as possible, especially through dredging for marine invertebrates; to make accurate field notes and drawings of these animals; to carefully document where each specimen was collected, in order to advance the knowledge of geographical distribution; to ensure the preservation and safe packing of collections for delivery home; and finally, to describe and classify the marine invertebrates.

Since arriving in Washington, Stimpson had spent most of his time at the Smithsonian under the watchful eye of twenty-nine-year-old Spencer F. Baird. Appointed assistant secretary of the Smithsonian two years earlier, Baird brought with him a personal collection that filled several boxcars. Baird and Stimpson worked together to acquire a suitable outfit for the voyage, compiling long lists of equipment, including arsenic, dredges, seines, alcohol, jars, corks, insect pins, drawing paper, guns, and ammunition—the tools of the trade for gathering and preserving zoological specimens.[15]

Charles Wright, a seasoned collector, would serve as botanist. Baird insisted that they be supplied with a library and Stimpson purchased many books, with one source claiming that they eventually sailed with one thousand volumes.[16] Stimpson and Wright were to be paid one thousand dollars per year and "one ration per day while attached to a vessel for sea Service."[17]

While one authority has claimed that the expedition sailed with eleven scientists, including six "biologists," in reality, Stimpson, Wright, and Navy officer John M. Brooke were responsible for the overwhelming majority of the scientific work done on the expedition.[18] Baird lobbied in vain for a larger complement of "scientifics," but he believed they were nevertheless much better equipped than the men of the US Ex. Ex.

In a remarkable letter Baird expressed guilt over the likelihood that some of the men he had helped recruit would never return. "I fear I have much to answer for in the way of deluding unsuspecting young (and even old) men to possible destruction from bite of snake, scorpion or centipede, engulfing in caverns while in search of fossil bones, embrace of Krakens, when catching starfish on the seas; or some other undescribed species of calamity, the genus, even, of which is not yet known. The string of scientific expeditions which I have succeeded in starting is perfectly preposterous." Speaking specifically of the NPEE, Baird continued, "I had full authority to prepare at the expense of the appropriation whatever apparatus was necessary to capture all sorts of Sea Devils and Water Kelpies. Getting two enterprising and able naturalists appointed to the expedition, one a zoologist (Mr. Stimpson) the other as botanist (Mr. Charles Wright), we together ransacked our brains, and made out tremendous lists of nets, kettles, dredges, etc. amounting to nearly $2,000, all of which were authorized and paid for without flinching."[19]

At Baird's urging Stimpson wrote to Cadwalader Ringgold, a veteran of the US Ex. Ex. who had been appointed commander of the NPEE, to offer his views on the proposed journey. Given their itinerary new species would be encountered and it was crucial that American naturalists describe them. National pride was at stake, Stimpson explained, as three European naval expeditions were set to sail at the same time as the NPEE. Stimpson spoke of the "mortification" endured by the naturalists of the US Ex. Ex., who through delays in publication saw many of their "discoveries" published by European naturalists.[20]

Stimpson returned home to Cambridge for the holidays in late 1852. His departure from family and friends in early January was gut-wrenching for all involved, and for months to come Stimpson would choke up when he thought about that day. "I did not know I was so much beloved," he wrote in one of his many letters home.[21] For perhaps the first time he realized the depths of his family's love for him, from the thoughtful gifts they sent on his twenty-first birthday to the many expressions of motherly love and sisterly solicitude in their letters. Although they tried to hide it they were worried about him, fearing he might not survive the voyage, a feeling reinforced by the sudden death at age thirty-nine in Bermuda of Stimpson's friend and fellow naturalist Charles Baker Adams. Stimpson fretted that he might face a similar fate of succumbing to disease in a foreign land.[22]

Judging from his letters to his family, Stimpson's health was a major concern, reflected in the statement he made to his parents: "I shall come back to you in much better condition than I left." He felt fortunate to have stayed well in the weeks before reporting for duty, afraid that "rheumatic pains" in his chest would keep him from sailing.[23]

He spent a few days in New York on expedition business before returning to Washington in late January 1853. For the next six weeks he lived in a small room at the Smithsonian, frequently dining with Baird and being drawn into his vast social network. Stimpson took in Washington's lively winter social season, attending numerous parties and balls. One man he met through Baird was William Wadden Turner, librarian at the Patent Office. Turner had married into the Randolph family, a prominent Virginia clan with ties to both Thomas Jefferson and Pocahontas. Stimpson spent a great deal of time with the Randolphs and quickly became smitten with one of the young ladies of the family.

His parents and sister worried about him amidst these "dissipations" and were troubled by his plunge into "the gaieties of the fashionable world." Stimpson justified his actions by telling them that it was necessary for him to know something about the larger world, and he reassured them that he had faithfully attended church since coming to Washington. As if to show his penance, in preparation for the voyage he had five teeth pulled and was vaccinated for smallpox.[24]

By March he was in New York, and on the twenty-first he officially reported for duty on the flagship *Vincennes*. Originally launched in 1826, there was nothing fancy about the *Vincennes*, with one observer noting that the ship had "nothing in her but plain white and mahogany." It was a staunch vessel and the first American navy ship to circumnavigate the globe. This was to be its fourth such voyage.[25] Preparations at the Brooklyn Navy Yard dragged on for nearly two months, and they had a difficult time finding enough officers and crew to man the vessels.

When the *Vincennes* sailed to Norfolk, Virginia many of the crew became seasick but not Stimpson, leading the ships' doctor to call him a "young seadog."[26] Despite the intense heat Stimpson used the time in Virginia profitably, dredging in Norfolk Harbor. Weeks before the voyage officially began he delighted Baird with the news that he had already found and described a number of "new Expedition species."[27] Rumors circulated that their departure was imminent, but according to Charles Wright, Ringgold did nothing for days on end and then pushed the crew hard, threatening to leave before the ship was ready. At the behest of Arctic explorer Elisha Kent Kane, about to embark on a perilous voyage of his own, Stimpson gave a crash course in collecting to Henry Goodfellow, who accompanied Kane.[28]

They finally departed Hampton Roads, Virginia on June 11, 1853. The squadron consisted of five vessels: the sloop-of-war and flagship *Vincennes*, which displaced seven hundred tons, held eleven guns, and accommodated nearly two hundred men and boys; the brig *Porpoise*, displacing two hundred twenty-four tons, with ten guns and seventy men; the steamship *John Hancock*, three hundred twenty-eight tons, three guns and seventy men; the schooner *James Fenimore Cooper*, ninety-five tons, one gun, and twenty men; and the supply ship *John Pendelton Kennedy*, five hundred twenty tons, three guns, and forty men.[29] Both the *Vincennes* and the *Porpoise* were veterans of the US Ex. Ex.

Commander Ringgold had earned a reputation as a tough captain; his chief clerk called him "a law and order man," and Stimpson saw men clapped in double irons months before they officially set sail.[30] While he would have his share of troubles with Ringgold, Stimpson spent most of his time in the company of the officers and enlisted men and they made life extremely difficult for him. There were a variety of reasons for this ill treatment. For years the navy had relied on its own officers to make natural history collections and many navy men resented civilian naturalists. The presence of naturalists, especially one focused on marine animals and dredging, also meant more work for them. Ringgold's order that "Fish caught must be preserved until opened in the presence of the Zoologist" seemed guaranteed to foster resentment.[31]

It did not help that Stimpson was absurdly young to hold the coveted position of zoologist. His relatively affluent background and rather slight build and glasses may also have worked against him. He quickly became a target of abuse and derision from Jack Tar, as the common sailors were called. At the time a life at sea was considered a refuge for the "poverty-stricken, orphaned, criminal, and insane," and many crewmen were indeed crude and illiterate. Despite reform efforts, naval vessels of the day were seen as cesspools of immorality and depravity. The sole preoccupation of many, according to Wright, consisted of sneaking a snort of whiskey whenever possible.[32]

Stimpson received a taste of what was to come only two days out of port. One of the officers wrote in his journal, "Great amusement, among officers and men, in consequence of the excitement and enthusiasm exhibited by the Naturalist when a lot of [sea] weed is picked up." Stimpson knew that many tiny invertebrates were to be found in these patches of seaweed, yet for his exuberance he suffered taunts and jeers from many aboard. This incident set the tone for the rest of the voyage, and Stimpson quickly acquired the dismissive nickname Bugs.[33]

In the face of such open hostility Stimpson's high spirits quickly faded. For much of the next three years the ship served as both home and jail: there was no place to escape aboard a cramped sailing ship. While constantly in the company of others, he felt isolated and alone.

Charles Wright felt that Stimpson went about his duties so zealously that he inevitably became an object of ridicule. He explained that Stimpson "rode his hobby too hard. Hence all our messmates delight in picking at him. He appears to think (and appears amazed that others do not see through his *specs)* that no branch of Nat. Hist. is so worthy of study as *marine* worms and almost goes into ecstasies at times over a nearly shapeless mass of jelly or some ugly worm that anyone else would shudder to touch."[34]

A few officers were sympathetic to the aims of science, notably John Rodgers and Brooke, but Stimpson later admitted that the experience had been painful. Years afterward he counseled a fellow explorer, William H. Dall, who was encountering similar problems: "I am familiar with the nature of the stumbling blocks which brainless officials delight to cast in the way of men of Science. You must keep a stiff upper lip, mind your own business, and resent any interference in it by others unauthorized. Commence by keeping men at a proper distance, and you will rarely have any difficulty with them afterwards."[35] These words, written with the benefit of hindsight, encapsulated lessons that Stimpson was just beginning to learn.

Like the other officers, Stimpson kept a journal during the cruise. In fact he kept two, an official version and a private one.[36] In the official version Stimpson's writing is workmanlike and generally lacking in deep reflection. The focus is squarely on the task at hand, namely documenting the day's collections, along with occasional observations on the distribution of species or geology. Stimpson had good reason to be somewhat guarded in what he wrote. All officers' journals were the property of the expedition's commander, who would use them to write his report of the voyage. Stimpson had heard from Joseph Couthouy about the perils of excessive candor in journal writing on a government expedition.[37]

On the fourth of July, while others were celebrating Independence Day, Stimpson took advantage of a calm day to gather various pelagic animals, devoting two full pages in his journal to their description. Most fascinating was a crustacean that lived in "holes which it seemed to have excavated ... in dead masses of the [me]dusa-jelly, three or four times larger than it in size. This novel kind of house is carried about with it in swimming." A shipmate and good friend of Stimpson's, Edward Kern, also noted Stimpson's discovery of a tiny crab exhibiting "all the colors of the rainbow but no larger than a pea."[38]

Five days later the *Vincennes* reached Madeira, "a green spot on the ocean" 340 miles off the coast of Morocco. Stimpson expressed delight at anchoring for the first time in a foreign port, and although this was supposed to be a brief layover, major repairs to the *Porpoise* took nine days. A grape blight had caused privation and famine on the island, dramatically cutting production of the island's chief export, Madeira wine, esteemed for its potency. Stimpson and the other officers each gave five dollars towards relief efforts, and the five thousand dollars that the crew spent here was "a God-send for the inhabitants."[39]

With its rugged cliffs and cultivated terraces, the island, under Portuguese rule, presented a picturesque scene. Stimpson and several officers rode on horseback through the mountains along a rough and dangerous road to the peak of the Curral das Freiras. Down at sea level he dredged in Funchal Harbor, procuring an immense quantity of animals. Several locals offered to collect specimens for him, with Stimpson noting that they were "miserably poor" and happy for the chance to make a bit of money. Stimpson loved everything about the island, with his only disappointment being not able to swim in the harbor due to an abundance of sharks.[40]

The ships experienced a very slow, tedious passage, enlivened only by the sight of numerous flying fish, to the next stop at St. Jago in the Cape Verde Islands. Despite stifling heat and humidity Stimpson dredged with good success during an eleven-hour stopover in the harbor at Porto Praya, discovering at twenty fathoms a new genus of crab that he later dubbed *Micropisa*. While dredging he was thrilled to see the USS *Constitution* anchored nearby, as "Old Ironsides" brought back memories of home. He spent part of the day collecting on land, finding a large land crab that ran swiftly and killed chickens. He also recorded seeing a kingfisher, which he recognized as a bird that Darwin had noted when he had visited the island twenty-one years earlier. He enjoyed walking through a grove of coconut trees and tasting the fresh milk.[41]

For the next forty-five days the *Vincennes* sailed for southern Africa, settling into the regular routine of a long sailing voyage. Breakfast was at eight and when the weather was good Stimpson spent the rest of the morning preparing, labeling, drawing, and describing his collections. There was no time for dredging but Stimpson used his tow net to gather pelagic animals. In the afternoons he and two others studied German, while the evenings were spent reading or writing letters or journal entries.

On August seventh they were treated to a stunning sight, the appearance of large colonies of *Pyrosoma*, a sea squirt noted for its phosphorescence. Stimpson's pen captured the scene: "a milky way of mellow light, containing bright stars which increased in numbers or thinned out as the ship approached or receded ... the darting of these animals ... being often simultaneous ... gave a singular appearance, something [of] a flight of small rockets."[42]

One officer who quickly recognized Stimpson's ability was Ringgold's personal secretary, Frederic D. Stuart, another veteran of the US Ex. Ex. After his initial amusement at Stimpson's youthful ardor, Stuart came to see him as an industrious and careful observer. Stuart was especially impressed when Stimpson captured a living paper nautilus with a delicate egg case, a rare and beautiful mollusk not often seen alive. When the ship crossed the equator Ringgold refused the men's request to celebrate the occasion by honoring Neptune, arguing that, "much evil frequently attended the ceremony," known for its drunken revelry and horseplay.[43] As they neared the coast of Africa Stimpson noted the different species of birds flying about the ship. For sport some of the officers shot albatross and other oceanic birds. Stimpson, perhaps influenced by Samuel Taylor Coleridge's injunction against killing these birds in his epic poem the "The Rime of the Ancient Mariner," showed either inexperience or reluctance in killing these well-known animals. Stuart recorded in his journal, "poor thing it [an albatross] had a hard time of it, such bungling in taking its life; certainly our Naturalist requires some experience in this part of his business." Anxious to prove his worth, the very next day Stimpson deftly caught a petrel with a bit of line and a hook baited with salt pork.[44]

Early in the voyage Stimpson instituted a series of meteorological measurements, each day noting the longitude and latitude, wind speed and direction, and later in the voyage, the air temperature. On selected days he recorded the water temperature at various depths, looking for a correlation between the water temperature and the vertical distribution of animals in the sea. He found that many species came to the surface only after nightfall, when the water temperature dropped.[45]

The next port of call, Simon's Town on the Cape of Good Hope in southern Africa, was intended to be a short stop for supplies. However, four of the five ships again needed repairs, particularly the *Porpoise*, leading to a two-month delay. Lieutenant Alexander Habersham of the store ship *Kennedy* later complained that the squadron had been sent out in a "miserably unseaworthy condition."[46]

Stimpson's stay began on a down note when he failed to receive letters from home. Ships from America arrived every few days for the next eight weeks but none brought him news from home. Receiving letters was one of the most important and cherished events for those on long cruises. Increasingly despondent, he wrote several times to his parents, wondering whether something had happened to his loved ones.

He poured out some of his frustration in a letter to Baird, revealing his growing difficulties with negotiating the complexities of shipboard politics. Francis H. Storer filled the dual positions of chemist and taxidermist. The son of David Humphreys Storer, a noted authority on the fishes of Massachusetts, Francis was the same age as Stimpson and had attended the Lawrence Scientific

School. He would go on to have a long and distinguished career as a chemist, but he showed little inclination for doing taxidermy. Stimpson usually did the skinning and stuffing of birds himself, but one day Ringgold spotted Stimpson on deck working on some birds. Ringgold ordered him to cease doing such work and chided Storer for not doing his job. The incident caused a rift between Stimpson and Storer, one that caused Stimpson "much pain."[47]

Given his problems with the crew it probably never occurred to Stimpson that he might also clash with his fellow "scientifics." One of the people that made the cruise especially trying for him was assistant astronomer Philip Sidney Coolidge. A great-grandson of Thomas Jefferson, Coolidge was two years older than Stimpson and had also been born in Boston. He spent most of his early life in France, however, where he attended a military school. The exact nature of his problem with Stimpson is unclear although at one point Stimpson said something negative about Napolean Bonaparte, Coolidge's idol. Stimpson felt that Coolidge had an "unaccountable hatred" towards him, and at one point he referred to Coolidge as the only "enemy" he had.[48]

One naturalist assigned to the expedition was Alfred A. H. Ames of the *Hancock*. A former student of Baird's, Ames had signed on to collect fish and reptiles. On reaching Africa Ames was sent home after being judged unqualified by Captain John Rodgers. Stimpson conceded that Ames had not worked very hard but argued that "it requires an immense deal of interest in a subject . . . to stand the man-of-war treatment . . . I have had the go home fever more than once, which might have reached its crisis had it not been for the soothing application of a new species or so."[49]

Men of science faced myriad hindrances and indignities. Wright complained "the majority of the mess have a most sovereign contempt for science and no esteem for its devotees." Still, "scientists in America often tolerated the nationalism, navalism and commercialism which so often accompanied exploring expeditions because it was a prudent way of obtaining popular support for scientific observations."[50] Shortly after Ames's departure, Ringgold had Wright transferred from the *Hancock* to the *Vincennes*. Perhaps Ringgold wanted to give Stimpson a scientific companion or, as Stimpson later suggested, "the Commodore wishes to have his Scientific Corps around him constantly." Wright had been treated well by Captain Rodgers of the *Hancock* and was furious over his transfer. "I am afraid I shall not have the same facilities for laboring on board the flagship" he huffed. His fear proved justified as he now found himself in a filthy room near Stimpson's located amidships "with the racket of Babel" surrounding them. Steady, uninterrupted concentration on scientific subjects was near impossible in such conditions. Wright felt that the move limited his collecting, increased his expenses, and lessened his freedom.[51]

Twenty-one years older than Stimpson, Wright had already made a name for himself as a professional botanical collector by supplying Asa Gray and others. One historian referred to him as Gray's "touchy and eccentric, but nonetheless hardy, resourceful, and intelligent legman." Largely content to let others describe the specimens he gathered, Wright had already experienced the pitfalls of being a naturalist on a government expedition. While on the Mexican Boundary Survey he had even been denied the assurance that he would receive meals or that his specimens would be transported back to the United States.[52] Wright felt that his present predicament was even worse. "The officers, care not a fig—any of them—for our labors and never put themselves to any trouble to facilitate them. . . . I met with more sympathy among rude teamsters on the plains of Texas."[53] The "us versus them" mentality on board gave Stimpson and Wright a common bond, but differences in age and outlook probably precluded a close friendship between them. Both preferred to focus on collecting and their respective interests usually took them in different directions.

The long stay in Africa provided an unexpected respite from shipboard life. For the first few weeks heavy seas prevented Stimpson from undertaking his usual dredging so he ventured inland instead. The season was still too early to get many animals but he saw antelope and flushed numerous flocks of quail, making for fine sport. Stimpson and Lieutenant William Van Wyck of the *Porpoise* teamed up to procure twenty-four species of birds, and much to Stimpson's delight Van Wyck had also collected marine invertebrates that Stimpson had not encountered on the cruise to the Cape. The two men quickly became friends.[54]

Dredging in and around Simons Bay, Stimpson's most successful efforts were made on a few patches of rocky bottom where he found large lobsters and a remarkable new genus of crustacea. His most striking find was a large basket star over three-and-a-half feet in diameter.[55] The results of his collecting in and around Simon's Town were impressive considering the unfavorable season. He amassed hundreds of birds, fishes, insects, and reptiles as well as countless marine invertebrates, sending off a first shipment of specimens for home but keeping all of the five hundred species of marine invertebrates for further study.[56]

The burdens of command had begun to weigh heavily on Ringgold in Africa. Some enlisted men deserted and others were convicted of crimes ranging from sodomy to murder.[57] John M. Brooke, who began the expedition with high regard for his superior officer, reported that Ringgold "seems to be totally changed by the anxieties to which he has been subject. . . . I keep out of his way as much as possible. . . . You have no idea how completely show reigns in this expedition. Utility seems to have no place. . . . It makes me despondent, to look around and see the universal dissatisfaction. I never saw anything like it." As Brooke's biographer put it, "morale edged toward a state of collapse."[58]

Up until this time Stimpson had recorded that Ringgold had treated him well and seemed pleased with his work, but Stimpson now had his own run-in with the commodore. With Ringgold's permission he had taken a room ashore where he used a microscope to make detailed drawings of more than one hundred marine invertebrates. When the time came to settle the charges a "tormented and worried" Ringgold, facing unexpected expenses necessitated by repairs to the ships, refused to reimburse Stimpson. Thirty dollars poorer, Stimpson had learned a valuable life lesson and vowed to never spend another dime without a written order.[59]

Finally setting sail in early November, the *Hancock*, *Cooper*, and *Kennedy* headed for Indonesia while the *Vincennes* and *Porpoise* made for Australia. In an extraordinarily rough and difficult passage across the Indian Ocean the latter two ships weathered storms for nearly a month. The men hoped to reach Sydney by Christmas, a desire that was thwarted when the winds deserted them in sight of land. When they finally entered Port Jackson, Stimpson was duly impressed. "The harbor is one of the most beautiful I have ever seen, the verdure descending to the water's edge. It is so land locked and its waters are so smooth, that it presents rather the appearance of a pond of fresh-water than an inlet of the sea. It is a long, crooked bay, with a series of beautiful coves along each shore, and containing many small islands."[60]

It had been only sixty-six years since the English crown had turned the continent into the world's largest penal colony, and Sydney already had a population of sixty thousand.[61] Unlike the experience he had had in Africa, Stimpson found several locals interested in zoology, notably James Fowler Wilcox, a "natural history dealer" with extensive collections. Using Stimpson's little sheet iron boat dubbed the "pollywog," the two men dredged for several days in the harbor.

Wilcox mentioned that two eminent naturalists lived in the region, William Sharp Macleay and William Swainson. Macleay was the originator of the already discredited circular or quinary theory of classification, and Swainson had been one of the most ardent supporters of the theory. It posited that, "Natural groups of the Animal Kingdom were supposed to be related to one another after the analogy of impinging circles."[62] According to Stimpson both men were said to be "still hopelessly moping over zoological subjects" and it seemed to him that "these two men were banished . . . from the Scientific world of the Atlantic shores, for the great crime of burdening zoology with the false though much labored theory which has thrown so much confusion into the subject of its classification and philosophic study." Stimpson later met Macleay but was disappointed that he showed no interest in marine invertebrates.[63]

In the waters around Garden Island, Stimpson was astonished by the activity of the large brittle stars and he marveled at the size of the ascidians and chitons. On his last dredging excursion he brought up a "rich harvest" of shrimps and mollusks. In company with Wright and Storer he stopped at the newly completed Sydney Museum (a forerunner of the Australian Museum), where they met the affable curator, William S. Wall. Stimpson evinced particular interest in the impressive collection of cetacean skeletons.[64]

During this two-week stopover in Australia life was not exclusively devoted to science. At Ringgold's invitation Stimpson and other officers dined with Sir Charles Nicholson, whom Stimpson called "probably the chief literary man in the country."[65] Everywhere they went on the voyage they ran into Americans, with Stimpson meeting a man that lived near him in Cambridge.

Their departure was delayed upon the arrival of the British surveying ship *Herald*, whose captain supplied them with the latest charts of the region. Stimpson reported that he had "a long and exceedingly interesting conversation" with the *Herald*'s naturalist John W. MacGillivray, who encouraged Stimpson to publish his descriptions of new species. Stimpson welcomed this support from a fellow naturalist, but in a letter to Dana revealed his more competitive side, admitting that he wanted to publish his Australian findings before English naturalists could do so.[66]

The *Vincennes* now began a survey of the Coral Sea. Ringgold called it "the most intricate and dangerous of any portion of the Pacific Ocean." Hidden reefs and shoals made sailing hazardous, and all hands kept a sharp watch. Stimpson suffered greatly from the heat and humidity and everything below decks quickly became moldy. He later told his parents, "I thrive only in cold weather." He occupied his time during the slow passage by catching tiger sharks and dredging. One particularly fertile area, dubbed Groper Shoal, yielded 117 species of mollusks.[67]

Dana had encouraged him to use the microscope while at sea, and an occasion to put these skills to good use came on January 31, 1854, when John M. Brooke helped usher in a new era in the study of the Pacific Ocean.[68] A pioneer in the field of deep-sea soundings, Brooke was a brilliant and innovative naval officer, later becoming one of the leaders of the Confederate navy. At a depth of 2,150 fathoms (12,900 feet, roughly 3,900 meters, or about two-and-one-half miles), Brooke's new sounding lead brought up a quantity of mud from the bottom of the Coral Sea. At the time it was the deepest sounding ever made in the Pacific, and Stimpson was the first zoologist to examine samples fresh from the abyssal depths.

He quickly brought the soft, gray ooze back to his small cabin amidships. On observing the specimens under his microscope he found the remains of numerous minute foraminiferans. He later concluded that they closely resembled a fossil genus characteristic of the Cretaceous period and he proposed naming one of the new animals in Brooke's honor. The suggestion incurred the wrath of Ringgold, who refused to allow it on the grounds that Brooke was merely an agent of the expedition.[69]

As the verdant Santa Cruz Islands came into view Stimpson longed to go ashore, and on February 8 he witnessed an eruption on the appropriately named Volcano Island (now Tinakula). They subsequently sailed temptingly close to other islands, but offers to pilot the ship in were refused. The same scene played out when they anchored offshore of Ponape in the Caroline Islands. Stimpson had made extensive plans and preparations for an investigation of the island, and in his journal he lamented his "hard fate" at not being allowed to land. Using much less restrained language, one of the officers noted that "the disappointment gushed forth in torrents of disgust" from the scientific corps.[70] As the flagship approached Guam in the Mariana Islands, Stimpson again was "doomed to disappointment." Nothing could have been more discouraging to Stimpson and Wright than continually bypassing largely unexplored islands. Their anger further intensified when Ringgold sent men ashore for supplies but did not allow them to accompany the shore parties. Wright complained that, "Had our chief started with the predetermination to tantalise us as much as possible he could not have accomplished his end more perfectly."[71]

Ringgold pushed steadily on, trying to make up for the time they had lost. As a result Stimpson did not set foot on land from the time they left Australia until they reached Hong Kong seventy-one days later. One of Ringgold's most ill-conceived ideas was to put his two "scientifics" aboard the Vincennes, which stopped only at well-known ports. Stimpson often wished he had been assigned to the Hancock, which visited many out-of-the-way places and whose skipper permitted his scientific corps a greater degree of freedom.[72]

By early 1854 Brooke recorded that the officers were uneasy over Ringgold's erratic behavior. Stimpson could not help feeling peeved at his commander when he was told that he would no longer be allowed to bring any specimens on board that might smell or cause a mess. This order made his job virtually impossible, as dredging is often a muddy business and sea creatures can emit noisome odors. Chafing at this restriction, he grumbled that he could only collect small species of fish as a result.[73]

They finally anchored at Hong Kong on March 20, 1854. Ringgold had expected to meet Commodore Matthew Perry here, only to find that Perry had sailed two months earlier. Since December of 1853 Perry had waited daily for Ringgold's arrival, but when he learned that Russian naval vessels in the area were also attempting to sign a treaty with the Japanese, he left in mid-January.[74]

Less than a year into their journey the rigors of ocean voyaging had taken a toll on both men and ships. The *Kennedy* was found to be totally unfit for surveying work and the *Porpoise* arrived in a "very rotten condition," again needing extensive repairs. Reaching land did nothing to mollify the growing anger of the crew towards Ringgold, who further alienated them on Easter Sunday when he had an officer suspended and a seaman clapped in irons for failing to assemble the men for church services.[75] Many began to doubt Ringgold's ability to command, and his behavior became a topic of furtive conversation.[76]

With morale sinking to new lows more men bailed on the expedition. Stimpson was thankful and relieved when Coolidge resigned in April 1854. Francis Storer also left, with Stimpson writing that he was "one of the most hard-working students I ever saw" and suggested that he would have developed his talents had circumstances been more favorable.[77] Wright summed up the feelings of many when he observed, "We have no more idea what part of the earth or ocean we are next to visit than if we were brutes." While tempted to leave, Stimpson vowed to press on, hoping that conditions might improve.[78]

CHAPTER 3

THE FIRST AMERICAN NATURALIST
IN JAPAN

For all that has been said of the love that certain natures (on shore) have professed to feel for it, for all the celebrations it has been the object of in prose and song, the sea has never been friendly to man. At most it has been the accomplice of human restlessness, and playing the part of dangerous abetter of world-wide ambitions.[1]
—Joseph Conrad

Stimpson's spirits were given a much-needed boost when he finally received a large packet of letters from home, the first since leaving home nine months earlier, but they also brought word that his paternal grandmother Eleanor had died at age eighty-nine shortly after he set sail. The first few days in Hong Kong were very hot and then the weather turned cool and damp, and within a week Stimpson came down with a case of influenza that left him shivering and bedridden for days. A violent cough lingered for weeks and every deep breath caused him pain. In a letter home he noted that other officers had left for the United States, one due to lung disease and another because he was "somewhat consumptive." Ominously, he added that he hoped that "the difficulty in my lungs will not be permanent."[2] Whether Stimpson had in fact contracted pulmonary tuberculosis is unknown, but he had several serious bouts of illness during the remainder of the voyage.

In addition to regular letters to his family he also confided in James D. Dana and Baird. Stimpson perhaps began to think about his own mortality for the first time, telling Dana, "if God spares my life until the return of the expedition I think you will be pleased with the zoological treasures I can spread before you."[3] All too aware of the deplorable treatment scientists faced on naval expeditions (Dana once referred to his ordeal on the US Ex. Ex. as "Naval servitude"[4]), Dana sent several long encouraging letters.

Just one example of the pettiness that plagued naturalists on government expeditions soon surfaced. In an effort to secure priority for the new species from Australia and Africa, Stimpson labored over a report on some of the new expedition species, but Ringgold found fault with it, as related by Wright. "Poor fellow, he had written out a dozen or more pages . . . and was ordered to rewrite the whole of it because—it was not on paper of the *exact* size to please him [Ringgold]. That is a specimen of Cadwalader Ringgold." Stimpson recopied the descriptions of fifty-one new species, which were published in 1855.[5]

Once he had recovered enough to begin dredging, what impressed Stimpson initially about Hong Kong was the incredible abundance of crustacea, especially the Brachyura, or true crabs. In a short time he had found over one hundred species, many of them new. By comparison, he noted that only twelve species from this group were known from New York Harbor, leading him to exclaim in his journal that, "the development of the family at this point is astonishingly great."[6]

On receiving permission to hire a Chinese "sandpan" in order to test his trawl, Stimpson reported "very satisfactory" results as it allowed him to take specimens that a dredge would have missed.[7] Ringgold excused seaman Salvador Pelkey from other duties to assist him, and while Pelkey had "no great inclination to useful employment," according to Stimpson, he now had a large, strong man to "do the rough work connected with my business," namely hauling in the dredge and turning over large stones in pursuit of invertebrates.[8] On a jaunt to the Kowloon Peninsula, Stimpson became intrigued by a species of goby (known locally as the mud monkey) that ran away across the mud flats when he tried to catch them.[9]

West of Victoria Harbor, Stimpson found an astonishing variety of fishes, and much to his surprise many of the invertebrates here were similar or identical to those he had dredged off Nova Scotia.[10] The Chinese fishermen whom he had hired to assist him were interested in his equipment and techniques and followed his movements closely. In turn, he found it profitable to board their fishing boats to examine their hauls. These vessels were veritable "zoological *thesauri*," and Stimpson commented on the fact that the Chinese ate almost everything that came from the sea.[11]

After six weeks of collecting near Hong Kong, Stimpson proposed a coastal excursion to several outlying islands. Ringgold approved and Stimpson hired a Chinese boat and crew. Two sailors and a somewhat reluctant Charles Wright accompanied him and they went well armed: the presence of pirates in the area prompted them to carry a three-pound gun and plenty of ammunition. The first day Stimpson successfully dredged and trawled at various stations among the islands southeast of the mainland. As night approached the crew, fearing an attack by pirates, insisted on anchoring in a little bay on the mainland. Stimpson barely slept, conscious of the "rascally neighborhood" they were in.[12]

For the next three days they explored at their leisure. They usually dropped Wright off to collect plants on various islands while Stimpson dredged. Temporarily freed from the navy's rigid command structure, they luxuriated in their freedom. On Grand Lema Island (now Dangan Liedao) the crew reported that monkeys frequently came down to the shore to feast on crabs and shells. Stimpson was determined to bag one but in the initial encounter the smaller primates (probably rhesus macaques) made monkeys out of their human pursuers. "We perceived three of the creatures ... sitting on a rock above us.... Soon after we perceived several other groups of four or five, in all cases sitting on rocks which afforded good points of observation. These surly fellows seemed to be holding a consultation upon our appearance, as they sat in a half circle, pointing at us and then turning to one another with some remark, undoubtedly very witty, from the grins which followed it. We wanted very much to punish their impudence by sending in a bullet among them, but the rocks were inaccessible, and even if we had shot one we could not have secured our prize."

A later attempt to secure one also failed and nearly resulted in Stimpson's firing a load of shot at two of his companions.[13]

On the last day of their excursion Stimpson dropped Wright off on an island while he dredged nearby. When Wright did not meet them at the prearranged time Stimpson began to feel "very much troubled," but his fears were allayed when he spotted the dedicated collector staggering down a hill under an immense bundle of shrubbery.[14]

The mood changed abruptly on their return. Stimpson continued to face ridicule and personal insults from his messmates, leaving him angry and hurt but powerless to do much about it. Ringgold ordered him to discharge his Chinese boatman, a move that effectively ended Stimpson's investigations in "this interesting locale."[15] On another occasion, upon returning from a particularly long day of dredging, Stimpson found that the *Vincennes* had sailed off without him, leaving him to track down its new anchorage. Wright summed up the dismal situation when he complained that, "very few facilities are offered us naturalists."[16]

During the stay in Hong Kong Stimpson did meet one man interested in natural history, British merchant John Charles Bowring, son of Hong Kong's governor. Stimpson was impressed by Bowring's large collection of beetles, and the two explored together, Bowring showing him several large, land-based flatworms that Stimpson later described as new.[17]

After two months in Hong Kong the crew were anxious to begin the primary work of the expedition, namely the surveying of the North Pacific, but the *Porpoise* was in the midst of a major overhaul. Hong Kong's "villainous" climate, especially now that the rainy season had arrived, began to take its toll. Stimpson again became sick, suffering for several days from an attack of what he called cholera-morbus (acute gastroenteritis), characterized by severe cramps, vomiting, and diarrhea.

When the steamer *John Hancock* arrived in late May after four months exploring little-known islands off Borneo and Sumatra, Stimpson could only listen with envy when Lewis M. Squires of the *Hancock* talked of collecting at unknown islands. Stimpson "growled" in his private journal that "we scientifics are here merely for display . . . to form vertebrae of the Commodore's tail."[18] With the *Hancock* in port all hands anticipated a speedy departure. It had been over a year since they left the United States, and the best time to explore the North Pacific was from June to September. Instead Ringgold embroiled the expedition in China's ongoing civil war. The Taiping Rebellion (1851–1864) pitted Christian-inspired rebels against the ruling Manchu dynasty, and one historian has recently suggested that this was the bloodiest civil war ever, resulting in the deaths of twenty million people.[19] Matthew Perry had twice refused pleas to intervene in the conflict, and Ringgold's orders made it clear that he was not to interfere in the internal affairs of any country. Ultimately, Ringgold acquiesced to the pleas of American merchants ("pusillanimous puppies" to some) who sought protection from the rebels.

In early June the *Vincennes* began to shuttle back and forth between Hong Kong, the Whampoa (Huang-pu) anchorage below Canton (Guangzhou), and Macao, hoping to dissuade attacks on American and European interests.[20] Outraged and appalled at defending a "few purse-proud effeminate denationalized dime-grasping Anglo-Saxons who are too abominably lazy to shoulder a musket in their own defence," Wright and others were left stunned by Ringgold's decision.[21] The situation in and around Canton beggared description. One American called it an "arena of fire and slaughter." During a brief visit Stimpson could only watch as the horrors of war unfolded around him, looking on as starving families in rickety boats collected slops from the ship's head. Every night he could see the fires from burning towns and hear the cannons booming.[22] These were sights that Stimpson would never be able to forget.

The crew increasingly engaged in drunken brawls and some "became demoralized by disease and liquor."[23] Stimpson began to write less and less in his journal, instead wandering the streets of Hong Kong and rolling ten-pins in order to get some exercise. He had grown "disgusted and thoroughly convinced of the folly of seeking pleasure in the *abstract*. Pleasure is like a coquette: take no notice of her and go on with your investigations, and she will come to you begging, but run after her and she gets puffed up, soon begins to flirt and ends in laughing at you in your disappointment."[24]

Stimpson and Wright were stymied in their efforts to secure boats to go ashore and collect, and Ringgold refused to pay for them to rent a boat. If they wanted to do their jobs they would have to pay for it themselves.[25] Stimpson made one such trip while the ship was anchored at Whampoa, securing a number of large, pale green shrimp.

With collecting opportunities largely curtailed Stimpson read, including a favorite from his childhood that particularly resonated with him just now, John Bunyan's *Pilgrim's Progress*. In late June, he wrote that he and Wright were essentially "unemployed." Four weeks later they remained in a "quiescent state." Frustrated and angry, Stimpson penned a dispirited letter to Baird, telling him that Ringgold would "not allow anything to be preserved on board the ship which will make any dirt or create the slightest smell. . . . The ship is kept exceedingly neat, bright and shining."[26] Ringgold had come down with malaria and Stimpson remarked on his deteriorating condition in his official journal, candidly writing that the Commodore was "still quite unwell, and is evidently much out of his head. . . . [He] gives very strange orders, which of course are obeyed with many reservations." In a letter home Wright was more blunt, insisting that Ringgold was "as crazy as a bedbug."[27]

By this point John Rodgers and the other officers were desperately awaiting Commodore Matthew Perry's return from Japan, which had been expected for weeks. Fresh from signing the Treaty of Kanagawa, the situation awaiting him in Hong Kong when he arrived in late July surely took some of the glow off his success. Rodgers succinctly captured the main complaint against Ringgold. "While it is impossible to obey his orders it is, while he retains the command very difficult to refuse them. Who shall draw the line between an order lucid enough to be obeyed and one mad enough to be neglected?"[28]

Admitting to nothing more than bouts of delirium from fever, Ringgold sought to have Rodgers and other officers court-martialed. After separate meetings with both men Perry ordered the fleet surgeon to conduct a medical survey on Ringgold, and in short order the Commodore was judged "incapacitated by reason of insanity."[29] Bitter and angry, Ringgold eventually returned to the United States in the company of Frederic Stuart.

Moving quickly, Perry appointed Rodgers to take command of the expedition. The forty-two-year-old Rodgers had gained respect as one of the navy's most scientifically minded officers. The son of a former navy commissioner, he had previously worked for the US Coast Survey assisting Agassiz in his explorations of the Gulf Stream. He now faced a formidable reorganization, with men asking for new assignments and others resigning, further exacerbating a shortage of officers. The skipper of the *Porpoise* was sent home for being drunk on duty, and acting lieutenants were given command of the *Porpoise* and *Cooper*, while another had day-to-day control of the *Vincennes*. For the rest of the expedition they would be dangerously short-handed.[30]

Rodgers promised Stimpson and Wright that he would facilitate their work and would allow them to land at islands to collect whenever practicable. Pleased with the appointment of the energetic new commander, Stimpson later called him "my best friend on board."[31] Stimpson posted a few letters to his family before leaving Hong Kong. Thoughts of home sustained him, and he reiterated his determination to finish the voyage. "I may come home sick at the end of the Expedition, but not before it," he informed them.[32]

By early September, Rodgers had all in readiness for their departure. Now too late in the year to venture into the North Pacific, the *Vincennes* and the *Porpoise* would instead survey the islands south of Japan, including the Bonin Islands and Loo Choo (Okinawa). The *Hancock* and the *Cooper* went north to chart the Yellow Sea. With typhoon season in full swing Rodgers knew that the *Vincennes* and the *Porpoise* might face rough weather, but he decided to risk it. The honor of the expedition had been tarnished, with the *New York Times* calling it "a miserable farce" due to gross mismanagement.[33] Staying in Hong Kong was simply no longer an option.

Rodgers's bold move led to disaster. Shortly after setting out the *Vincennes* and the *Porpoise* ran into severe weather. Stimpson, who rarely got seasick, called it "a most uncomfortable passage." On September 21, the ninth day out of port, the crew of the *Vincennes* lost sight of the *Porpoise* during a terrific storm near the Pescadore (Penghu) Islands. It took the *Vincennes* seventeen days to finally reach the placid waters of the Pacific instead of the normal three days. Following the standard protocol, the flagship continued on to the prearranged meeting place, the Bonin Islands (known in Japan as the Ogasawara Islands). Arriving at Peel Island (Chichi-jima), their fears for the *Porpoise* increased when a typhoon battered them as they lay anchored in the sheltered harbor at Port Lloyd.[34] No trace of the *Porpoise* or its crew was ever found. Over fifty men perished, one of the heaviest losses ever in the name of American science and exploration, but their deaths have been all but forgotten today.[35]

Located approximately six hundred miles south of Japan, the Bonins had been uninhabited until 1830, when a small group of runaway whalers and their Hawaiian wives established a colony. In 1854 there were about forty people living a life of drunken idleness and subsisting on green turtles, fish, sweet potatoes, and rum.[36] More than one sailor felt that they had found paradise and dreamed of settling there.[37]

During the turbulent passage Stimpson's face had swelled up and painful boils erupted on his face and forehead, symptoms associated with stress. The ship's doctor bound him up in poultices so that his head resembled "an enormous cabbage" and seeing became difficult.[38] The symptoms subsided once the ship reached land, and despite his anxiety over the fate of his friends on the *Porpoise*, Stimpson came to enjoy his time at the Bonins. For the first time he explored on little-known islands, and he now had the cooperation of a commander that valued his work.

Dredging for several days in the harbor, Stimpson obtained a new species of crab at a place he dubbed Ten-Fathom Hole. It seemed that everywhere he went he found something new, and some of the most valuable specimens on the voyage came at this stop. Venturing inland he marveled at the luxuriant subtropical forests and mountain streams and killed a bat (the now critically endangered Bonin Fruit Bat) with a five-foot wingspan. He found a new genus of land crab and watched as they fought successfully against attacks from introduced wild dogs.

As an avid angler Stimpson had fine sport here and the extra specimens furnished the mess with many a good meal. On the coral reefs he caught over fifty species of brilliantly colored and strangely shaped fish. He could clearly see down to a depth of four fathoms, where he counted at least twenty additional species that remained out of reach, thereby causing him to suffer "the torments of Tantalus."[39]

After two and a half weeks the *Vincennes* reluctantly continued on to Okinawa (then called Loo Choo), largest of the Ryukyu chain, arriving on November 17.[40] Naha Harbor is enclosed by a series of coral reefs but high winds hampered Stimpson's dredging. He instead trudged through the paddy fields near Tumai where he bagged water snakes, frogs, and land crabs. Along the beach he shot birds by the dozens and the men enjoyed "many a good breakfast" of snipes and sandpipers. Stimpson wryly noted that these birds were easy prey as they were "engaged . . . in a similar investigation of invertebrate zoology."[41]

The beach was also the scene of a lively competition for gathering shells, with Wright and two navy men, Robert Carter and Anton Schönborn, always in on the action. Despite the unfavorable season, during their three-and-a-half-week stay in Okinawa they found nearly four hundred species of marine mollusks, many of them at low tide on the reefs a mile from shore.[42]

When Rodgers invited the regent of the island and his retinue to dinner aboard the *Vincennes*, Stimpson was seated between two "Mandarins" who between them spoke perhaps a dozen words of English. Despite the language barrier the locals evinced "great vivacity and hilarity" thanks to repeated toasts of champagne. Relations quickly soured, however, as Rodgers faced the first test of Perry's diplomatic efforts. Like the Treaty of Kanagawa, the Treaty of Naha stated that Okinawans would provision ships in distress. Rodgers asked for a large quantity of wood and water before the *Vincennes* departed and feeling that the regent was slow to deliver he ordered a "military demonstration." The sight of one hundred armed sailors and Marines marching on Shuri quickly resulted in the delivery of the requested supplies.[43]

Five days out from Okinawa the *Vincennes* communicated with the *Nightingale,* one of the most fabled of all the American clipper ships, but it carried no news of the *Porpoise.* For the second year in a row they held a muted Christmas celebration at sea. Excitement rose as the ship approached the Japanese coast, evidenced by Stimpson's description of Kyushu. "More beautiful scenery can scarce exist, and our enjoyment of it is scarce equalled by anything which has yet occurred in this cruise."[44]

While the Perry treaty was certainly momentous, it had been a cautious first step, as only the ports of Shimoda and Hakodate were officially opened. Rodgers was determined to stretch the interpretation of the treaty to the breaking point, and his opening gambit could not have been more provocative.[45] On December 28, 1854, he brazenly sailed the *Vincennes* into Kagoshima Bay. Kagoshima was strategically important as Japan's southernmost major port, and the Japanese clearly did not welcome this incursion, insisting, with perfect justice, that the Americans had no business being there. For nine tense days Rodgers ignored daily demands that he depart, claiming that the Perry treaty allowed American ships to stop at any Japanese port if distressed. It is doubtful that the *Vincennes* was in need of supplies just fifteen days after being provisioned ("in quantities even greater than required," according to Stimpson) at Okinawa.[46]

Rodgers instituted a survey of the bay and established an observatory on shore to take astronomical measurements. Stimpson came ashore twice in the company of armed parties, where he faced a series of hostile encounters with local officials. He wrote that the Americans were soon "surrounded by crowds of Japanese among whom were several officers, who were distinguished by their fine dress and their two swords. Mr. Wright and myself immediately rushed off in different directions,—he for plants,—I for shells.... I was continually elbowed and pushed by the be-sworded gentry, who tried every means to urge me back to the boat: but knowing they would not proceed to extremities I persevered.... Some of the Japanese even helped me in collecting." Rodgers helped to intimidate the hostile crowd through a less-than-subtle display of marksmanship with his revolver and insisted that a show of arms was his "passport to observation" in Japan.[47]

Stimpson characterized his dredging at Kagoshima as "highly successful" and noted that many of the invertebrates "presented unusual peculiarities of form or habit." He uncovered dozens of new species as well as the bizarre animal called the amphioxus, "that rare and dubious fish." Stimpson was the first professionally trained Western naturalist to explore in Japan and his time at Kagoshima "passed away very pleasantly" despite cold, snowy weather and the equally frosty reception from the Japanese.[48]

With "no further excuse or occasion to remain," in Stimpson's laconic phrase, the *Vincennes* briefly explored other nearby islands. With Rodgers's blessing Stimpson dredged from the flagship, rather than the customary method of employing a small boat. According to Helen Rozwadowski, Stimpson was among the first naturalists to apply dredging techniques on such a large vessel.[49] He suffered a setback, however, when his large dredge was caught on the bottom and torn asunder off Tanega-shima.

As it was too late to proceed farther north, the *Vincennes* returned to Hong Kong on January 30, 1855. There had been no word of the *Porpoise*, extinguishing any faint hopes that it had somehow survived. Several other officers and crew had died and others were sent home sick, including Robert R. Carter, leading Wright to observe, "death has been somewhat busy among us." Stimpson admitted to his parents that the voyage had done nothing to improve his health.[50]

Shortly after their arrival in Hong Kong he came down with a violent cold that lasted for a week, immediately followed by the cryptic entry in his private journal: "The blackest day of my life. . . . How many blacker are to come, the Almighty only knows." Given his frequent respiratory problems it is tempting to speculate that Stimpson may have received a diagnosis of pulmonary tuberculosis, but it is unclear exactly what he meant by the statement.[51] Letters from home had also brought sad news, as he learned that his friend Dr. Waldo Burnett had died from tuberculosis at the age of twenty-five. Stimpson also probably read in the newspapers of the sudden death of one of his scientific heroes, Edward Forbes.

By now Hong Kong retained overwhelmingly negative associations for Stimpson. Knowing that they would be stuck there for the next ten weeks must have been utterly depressing. He wrote Baird that he often wished he had never signed on for the voyage. Brooding over the death of Lieutenant William Van Wyck of the *Porpoise*, he reiterated his complaint that most of the officers and crew considered natural history a complete "humbug." Desperately missing his family and friends, he was also stung to learn that a lady friend that he had left behind in Washington had moved on.[52]

Once he worked through bouts of self-pity and doubt he realized that his sufferings only strengthened his resolve. Natural history served as therapy for both mental and physical ailments. He now had time to reflect on a number of questions raised by his work, particularly regarding the geographic distribution of marine life. He queried Dana, "What must one do when he finds species apparently the same in extremely distant localities?" He had a freshwater shrimp from Java that seemed identical with an oceanic form he had found in the Bonin Islands. He knew it would take an examination of "very numerous individuals of all ages" to distinguish between them, so he floated the idea of adopting a trinomial nomenclature to mark geographic subspecies, as a few European naturalists had begun doing in the 1840s. One historian has noted that field naturalists "with firsthand experience of the variability of animals in nature" were more inclined to see the need for trinomials than museum men who worked only from dead specimens.[53]

With one hundred eighty species of true crabs from the Chinese seas alone, Stimpson was stumped in his attempts to classify them, even at the generic level. His thoughts on diversity led him to share his newly developed views with Dana. "I think I have data to show that many species formerly supposed to be identical (particularly among the shells) are really distinct," he said. "Nature seems to have distributed animals, as she did the stars, in clusters and nebulae, crowded at some points, scattered at others; we resolve nebulae with our telescopes, and see clearly between closely approximated *species* by minute examination."[54] While not original, Stimpson's analogy shows that he was becoming increasingly confident in his ability to generalize from his field observations.

He spent several days packing and labeling crates of specimens to be sent home and with them went descriptions of eighty-one more new species, including corals, gastropods, and amphipods.[55] On April 6, 1855, amidst a farewell of firecrackers, the squadron, now down to just three ships, finally set sail to begin surveying the North Pacific (usually delineated as above 35 degrees north latitude). Between Hong Kong and the Ryukyus Stimpson had a lucky break when he recovered a crab clinging to the sounding lead from ninety fathoms.

While visiting the island of Amami-Oshima, Stimpson had begun to recover from his Hong Kong blues. In his journal he wrote, "The weather being most pleasant, the breeze light but sufficient, our powers of enjoyment were fully occupied, and to most of us this was certainly the most agreeable part of the cruise."[56] For six days Stimpson came as close to zoological nirvana as he ever had, roaming the island unhindered and searching paddy fields and hillsides for land snails, frogs, and birds. A plentiful supply of wild raspberries provided an agreeable addition to his diet.

Dredging in a sheltered cove, he found a small *Lingula*, the first of this ancient genus of brachiopods he had ever seen alive. As the tide receded it exposed "the finest beach for zoological purposes that I have ever met with," and he soon gathered "the finest basket of spoils that it has ever been my fortune to collect in one day.... I am satisfied that four or five times the number of species might have been collected in this locality with further opportunity," he exulted in his journal.[57]

On May 13, 1855, the *Vincennes* reached the treaty port of Shimoda on the southern coast of Honshu. A small fishing hamlet, Shimoda had suffered extensive damage in a devastating earthquake and tsunami five months earlier. During the first part of the two-week stay Stimpson did not collect with his usual tenacity, having contracted a severe cold. On recovering he learned that the boat he used for dredging had been appropriated for another purpose. Wright, encountering similar hurdles in carrying out his work, noted sharply that, "in supreme disgust our Zoologist hitherto indefatigable lounged about most of the time." At a bazaar that had been established for the Americans, Stimpson purchased a number of "splendidly ornamented nothings" and a lithograph of Mt. Fujiyama.[58]

Unlike most of the crew, who viewed the Japanese as savages, Stimpson made at least one Japanese friend, as he later named a new genus in honor of the interpreter at Shimoda, Hori Tatsunosuke.[59] He found other genial companions here, meeting several Russian officers from the frigate *Diana*, stranded when the earthquake damaged their ship. Favorably impressed, Stimpson called the Russians "intelligent men, extremely well educated."[60]

Back at sea, he whiled away some of the tedious hours learning to play backgammon. After fighting rough seas in the strait between Hokkaido and Honshu, the *Vincennes* anchored in the harbor of Hakodate or, as the crew called it, Hack-your-daddy, on Hokkaido, the northernmost of Japan's islands. Following the same routine for the next three weeks, Stimpson collected at the shore in the morning and walked the countryside in the afternoon. One morning he and a few others ventured out in a small boat to see a cave that the Ainu people used as a place to make offerings to their sea-god.[61] On another occasion Stimpson and Wright climbed the hill behind the town, topped by a twelve-foot-tall statue of Buddha. An old man welcomed their visit and showed them his flower garden, but he could not fathom why the two Americans were so intent on chasing dragonflies.[62]

The Perry treaty stated that the Americans were not allowed to shoot animals in Japan, frustrating Stimpson's attempts to collect vertebrates. As they were wont to do, the Americans bent the rules and Stimpson killed a gull with a "lucky shot" from his pistol at a distance of two hundred yards.[63] His quest for marine forms proved successful, leading him to sing Hakodate's praises. "The field is a rich, and almost untrodden one, and probably more new species occurred to us here than at any other point of our cruise."[64] The marine fauna here bore little resemblance to that of southern Japan and instead reminded Stimpson of New England's. He found and consumed great numbers of a mussel identical to the blue mussel (*Mytilus edulis*) of the New England coast. He speculated that the two regions owed this convergence of specific forms to the fact that they had nearly the same latitude.[65]

A historic discovery was made as the ship sailed farther north into the Bering Sea. Brooke had continued improving his deep-sea sounding equipment, and on July 19, 1855 he succeeded in reaching the bottom approximately 150 miles north of the Komandorski Islands. The goose quills on the end of the sounding lead contained mud from the ocean floor at a depth of 2,700 fathoms (over three miles). No one had ever recovered a sample of the sea bottom from such a depth.[66] Brooke quickly turned the sample of greenish mud over to Stimpson, who cut the quills in half and examined the middle section under a microscope. He recognized forams as well as diatoms, and from their fresh appearance Brooke concluded that the animals recovered had lived on the sea bottom. He solemnly recorded that he and Stimpson were the first to see living creatures from such depths. More cautious in his initial appraisal, Stimpson nonetheless supported the idea that the diatoms had recently been alive and had been killed by the change in pressure as they were brought to the surface.[67]

Along the Siberian coast they groped their way through dense fog, entering the Bering Strait on August 1 and anchoring in Glazenapa Harbor off the Chukotski Peninsula. Rodgers decided to send eleven men to stay on Arakamchechen Island (Kayne Island on American maps) to make astronomical observations, collect specimens, and study the manners and customs of the native Chukchi people. Meanwhile the *Vincennes* would sail north to chart the relatively unexplored Arctic Ocean.

Although he was far from well at the time, Stimpson asked for and received permission to join the shore party, a curious decision given that he rarely missed an opportunity to sail and dredge in uncharted waters. But he was tired of the taunts and sarcasm from the crew and the constant noise on board. His choice to remain on terra firma just below the Arctic Circle shows that he had had his fill of shipboard life. When he had begun this journey he had no idea of the hardships that he would face, but he now wrote his parents that he had "condemned myself to three year's confinement in this dungeon of pestilent vapors."[68] He reasoned that he would benefit from the fresh air and it helped that two of his allies, Charles Wright and Edward Kern, were part of the group. Brooke took command, also accompanied by three sailors, three Marines, and a Russian interpreter.

Rodgers considered the Chukchi to be potentially dangerous, and they were indeed an unconquered people living in this remote corner of Siberia.[69] They survived by herding reindeer, fishing, and hunting whales, walrus, and seals, and their homes consisted of large whalebones stuck in the ground and covered with skins. For security the Americans brought with them a twelve-pound field piece and assorted arms and ammunition. A breastwork of barrels filled with dirt and gravel served as a protective cordon around the tent-house made of sails and spars that would be home for the next six weeks. In the event that the *Vincennes* failed to return they were also provided with a whaleboat large enough to carry them to American shores. Fortunately, the Chukchi and their chief Attila maintained a peaceful coexistence with the Americans, even though Stimpson took a human skull from a local graveyard.[70]

The campsite near the village of Gar-rang-gar at first glance seemed to contain little other than damp rocks and stunted vegetation. Pleasantly surprised by the weather, with temperatures rising to 60°, Stimpson found fields covered with a carpet of purple, yellow, and pink flowers. Despite swarms of mosquitoes and frequent fog and rain he stalked the tundra in pursuit of arctic foxes, ground squirrels, hares, and ermine. He sampled reindeer as well as bear meat and gave a detailed description of a successful native whale hunt.[71]

Curious about their uninvited guests, the Chukchi swarmed around the crew and virtually surrounded the small encampment. The men were short and powerfully built, with black hair shaved at the top of their head, while the women were tattooed on their chin, cheeks, and forehead. Stimpson failed to specifically mention females in his journal but some of the other men were surprised and then smitten by their beauty, tempered somewhat by the fact that they seldom bathed. Some of the crew even commented on the native custom of lending wives to friends.[72] Their hosts staged dances and wrestling matches for the amusement of their guests, and Stimpson spent many hours recording their language and compiling a vocabulary.[73] Amazed to see the natives consume small fish that were still alive, he learned they would do most anything for a chew of tobacco.

As usual Stimpson dredged in the bay amidst seals cavorting about his boat. Salvador Pelkey, the seaman assigned to haul in the dredge, began to grumble about Stimpson's pertinacity in pursuing his work. Pelkey had "never exactly suited" Stimpson, and on a day that Pelkey had other duties Stimpson dredged alone. On his first cast he lost the dredge and was convinced that Pelkey had intentionally sabotaged it, thus abruptly ending these investigations.[74]

Stimpson instead turned his attention to ornithology. The region is still known today for the tremendous numbers of seabirds that breed there. Watching the natives hunt seals Stimpson became fascinated by their small, frail-looking boats called "kyaks." He bought one and soon learned to manage these one-man boats, cruising about alone and shooting ducks and other waterfowl.

One incident marred Stimpson's relations with the natives, an episode that he failed to mention in his journal. An inebriated native tried to give Stimpson a hug only to be rebuffed. In anger he shoved Stimpson to the ground and the furious naturalist wanted the enlisted men accompanying him to shoot the man. They wisely refused to do so, and Brooke also rejected Stimpson's demand that the man be barred from the camp.[75]

When the *Vincennes* returned in early September the men ashore celebrated the event with several booming cannon shots. They were surprised to find Stimpson in good health, as they had received a report from the natives that one of the shore party had been ill and they assumed it had been Stimpson. Scurvy, long a scourge among seafarers, had afflicted at least one-third of the men of the *Vincennes*, who suffered the agonies of swollen joints, lesions, bleeding gums, and loosened teeth. Stimpson and the others stayed on the island until the last possible moment, gathering berries and claytonia to aid their stricken comrades. Delighted to learn that Rodgers had ordered the dredge used almost daily, Stimpson inspected a small but valuable collection of invertebrates that he would never have seen if not for Rodgers.[76]

With all hands now anxious to return to American shores, the *Vincennes* sailed through the Aleutians to San Francisco, arriving on October 13, 1855. Rodgers wanted to keep the expedition in the field for one more season but his request for additional funding was denied. Stimpson spent his first two weeks in San Francisco in bed on shore recovering from an unspecified illness, and a week after their arrival he experienced the sharp shock of an earthquake.[77]

A boomtown, San Francisco had been incorporated just five years earlier when California became the thirty-first state. Regular passenger service from New York, via Cape Horn, brought hordes of gold seekers while others arrived via the dangerous overland route. The Gold Rush attracted a motley assortment of fortune seekers, and gambling dens, saloons, billiard halls, and brothels lined the steep, narrow streets.

By early November, Stimpson had recovered sufficiently to begin searching the seaside market for specimens. The marine invertebrates of California had not received much attention from American zoologists, and while several

Europeans had published on the topic, their papers dealt with specimens from southern California. Stimpson now rushed to fill the void in the knowledge of northern California's invertebrates. Hiring a boat for three dollars a day, he dredged in San Francisco Bay, navigating through a maze of ships abandoned by their crews in the mad rush for gold. While noting that the northern California coast was not as rich in marine invertebrates as other coastlines in similar latitudes, he soon amassed twenty-two species of crustacea, fifteen of which were new to science. By April 1856 he had completed the descriptions and his paper was published in the first volume of the *Proceedings of the California Academy of Natural Sciences.*[78]

Stimpson had ceased writing in his journal upon their arrival in California so we know relatively little about his activities there, other than the fact that he also visited Petaluma and Sonoma County.[79] As Rodgers made plans for the long cruise back to New York, both Stimpson and Wright expressed a strong desire to return via the steamer *Cortes,* bound for New York via Nicaragua. Wright alluded to "the extreme unpleasantness of my position in the mess," and after some hesitation Rodgers gave him permission to join the *Cortes.* Perhaps concerned about Stimpson's illnesses, Rodgers refused his request, even after a desperate Stimpson offered to pay his own expenses and forfeit his pay from the expedition. As it turned out, Rodgers did Stimpson a favor. Wright stayed in Nicaragua but a short time due to the war which resulted from American adventurer William Walker's brief takeover of the country.[80]

Tremendously disappointed, Stimpson knew that there would be few opportunities for finding novelties on the well-traveled return route, as both the Hawaiian and Tahitian Islands had for years been rendezvous points for whalers. The *Vincennes* sailed alone from San Francisco on February 2, 1856, the other ships having been reassigned or decommissioned. Stimpson merely recorded in his journal the longitude and latitude each day during this homeward leg, an indication that for him the expedition was effectively over. But there were still new discoveries to be made. During his two weeks in Hawaii (then referred to as the Sandwich Islands), Stimpson's dredging among the coral reefs turned up four specimens, all female, of a crustacean new to science. He recorded that the tiny crab was found "clinging to the branches of living madrepores" and he correctly surmised that it probably fed upon the coral. After several years of study he formally described the first known gall crab, an unusual family whose members force corals to form galls to shelter them.[81] At Hilo he collected a most remarkable little freshwater fish, a goby that climbs up waterfalls using suckers in its mouth and belly. New to science, it was later named in Stimpson's honor.[82] These new finds, coupled with the sheer beauty of the islands, must have cheered his spirits.

After a layover in Tahiti they made a mad dash for home, rounding the treacherous waters off Cape Horn in a snowstorm. The *Vincennes* passed through the South Atlantic, weathered the doldrums and finally anchored in New York Harbor on July 11, 1856, having made a quick passage (seventy-four days) from Tahiti. It was three years and one month to the day since they had set sail from Norfolk, Virginia. After a brief stay at the Quarantine Ground on Staten Island, Stimpson treated himself to a few days at the plush Delmonico Hotel.

Intrepid explorers are often hailed as heroes on returning home but the arrival of the NPEE attracted little notice. The country was preoccupied with the increasingly divisive debates over slavery, symbolized two months earlier when Senator Charles Sumner of Massachusetts had been savagely beaten in the Senate Chamber after delivering an antislavery speech. The small-scale civil war known as Bloody Kansas had taken center stage, and the hotly contested presidential campaign also served to deflect attention away from the return of the NPEE.

At the age of twenty-four Stimpson's life seemed to be moving on fast forward. In just six years he had gone from a smooth-cheeked high school boy to a seasoned veteran of a difficult, three-year exploring expedition. Despite periods of ill health he had joined the ranks of a small but distinguished group of naturalists who had sailed around the world. He had collected and studied marine invertebrates in places that no zoologist ever had, and the collections he brought home would occupy him for years and provide the basis for many of his most influential scientific papers.

CHAPTER 4

DEFENDING AMERICAN SCIENCE

There is a pleasure in the pathless woods,
There is a rapture on the lonely shore,
There is society, where none intrudes,
By the deep Sea, and music in its roar;
I love not Man the less, but Nature more.
—George Gordon (Lord) Byron, *Childe Harold's Pilgrimage*[1]

On returning to Cambridge in mid-July Stimpson still suffered from the lingering effects of what he called rheumatism, leaving him bedridden and in pain for a week. As Darwin and others had learned, around-the-world voyages could exact a toll on one's health and Stimpson did not fully regain his strength for many months. After being gone so long it took time to readjust, but family and friends were eager to hear his stories about far-off lands and people.

John Rodgers had allowed Stimpson to bring some specimens to Cambridge for identification. Rodgers felt sure that an initial sum for permanently curating the collection would soon be allocated but Baird had his doubts, telling Stimpson that he hoped Navy Secretary Dobbin would "give all the necessary aid, but I fear he will not. He don't appreciate these things."[2] Rodgers's attempts to buy alcohol and glass jars were rebuffed.

In late October, Rodgers gave Stimpson permission to come to Washington, as a small sum had been procured to pay for maintaining the collections. Stimpson began the laborious work of unpacking and organizing the collection, putting in twelve-hour days filling jars with alcohol and recording accession numbers into ledger books. Getting Congress to appropriate funding to pay for publishing the expedition reports proved to be challenging, as many politicians considered such reports to be utterly worthless.[3] Stimpson moved to enlist important allies and called on several institutions for aid "in a matter concerning the advancement of our science."[4] The Boston Society of Natural History and the Academy of Natural Sciences of Philadelphia sent testimonials, and in March 1857 Congress approved fifteen thousand dollars for publication of the natural history portion of the work.[5]

Writing nine months after the expedition reached home, Rodgers urged that "whatever is done should be done quickly," proclaiming that "to discover second is not to discover." He formally recommended that Stimpson superintend the collections, telling the new navy secretary, Isaac Toucey, that Stimpson "unites zeal, industry, [and] knowledge to great interest." Stimpson received $1,500 per year, a decent wage for a twenty-five-year-old naturalist.[6]

Stimpson's preliminary report on the expedition's zoological results appeared in the *American Journal of Science* in early 1857. The size and scope of the collections were impressive: some fifty-three hundred species and over twelve thousand lots, or groups of specimens. Given Stimpson's predilections, fish and invertebrates comprised the majority of the collection, although most other phyla were represented. Some two thousand species of mollusks and nearly one thousand of crustacea testified to his industriousness. Stimpson had made colored drawings of many species and had amassed voluminous notes and drawings on some three thousand living animals.[7] The conchologist Philip P. Carpenter later praised Stimpson's efforts: "It is safe to say that in this short period he collected more trustworthy species of shells, with localities, than were received at the Smiths. Inst. from the united labours of the naturalists of Captain Wilkes's celebrated expedition."[8]

While Rodgers had overall responsibility for the publication efforts, he relied heavily on Stimpson to help supervise the writing of the final zoology report. The scope of the undertaking was unlike anything that Stimpson had ever attempted and required a great deal of organizational skill. The distribution of specimens to specialists would seem to have been a relatively easy task, since the number of qualified American naturalists in any given field "could be counted on the fingers."[9] Those originally recruited to write the reports included some of the best men of the era, including John Cassin, Augustus Gould, Asa Gray, S. F. Baird, and Louis Agassiz.[10]

Not all of these men completed their work, however, so Theodore Gill and Alexander Agassiz, among others, ultimately published papers on parts of the collection. In the late 1850s and early 1860s these reports began appearing, with over fifty separate papers published, most consisting of descriptions of new species. Stimpson eventually wrote the sections on the crustaceans, annelids, and others. He would later send the corals to a promising student of Agassiz's, Addison E. Verrill.

Between 1856 and 1861 Stimpson devoted much of his time to classifying the marine invertebrates and coordinating the scientific work of the expedition. In 1857 a series of papers began to appear in the *Proceedings of the*

Academy of Natural Sciences of Philadelphia. Eventually published in eight separate parts, all in Latin, they encompassed 138 pages of text and detailed 573 species of worms and crustaceans, 331 of which were new to science.[11] This is commonly referred to as Stimpson's Prodromus, a word meaning a preliminary publication. The use of Latin meant that these works would only be consulted by a handful of specialists, and these would be Stimpson's final papers written entirely in Latin, as by 1860 he had made the decision to use English. Stimpson's was probably among the last generation of Americans to write taxonomic papers exclusively in Latin, another sign of his place as a transitional figure in natural history.[12]

The first paper dealt with marine flatworms (Phylum Platyhelminthes). For several months in late 1856 and early 1857 Stimpson studied nothing but these often brightly colored worms. He and Joseph Leidy, one of the few Americans who had studied them, exchanged half a dozen letters on the subject. Stimpson taxed his eyes straining to make out anatomical details through a microscope, and when he finally completed his work he complained that his predecessors on the group had made it "a most irksome job from the number of badly characterised genera & species." Showcasing his taxonomic skill at creating suitable names for new species, he dubbed one genus *Galeocephala* for the "cat-like appearance of its head when at rest & and its seizing propensities."[13]

Part two of the Prodromus covered the ribbon worms, and over two-thirds of the taxa he introduced are still valid today. Some of these marine worms reach ridiculous lengths of nearly one hundred feet. In choosing names Stimpson often selected Latin terms relating to the color of the animal— griseus for gray, albulus for white, and piperatus for peppery. Names also described morphological characteristics—sinuosus for bending or winding.

The last six articles of the Prodromus described decapod crustacea, an order that includes the shrimp, crabs, and lobsters. Much of Stimpson's research over the next five years would be on this group. Crustaceans are far more complex in structure and behavior than most mollusks and worms and thus represented more of a challenge to classify. On a practical and professional level, there were also fewer carcinologists (those who study crustaceans) than there were people studying mollusks.

Differences between species can be quite subtle, and extraordinary patience and acumen is required to recognize the most important characters for distinguishing between species. A fellow naturalist wisely noted that "to indicate many apparently new species, is the work of an hour; to establish one on a sure foundation, is sometimes the labor of months or years."[14] In many of his papers Stimpson was careful to state exactly how his new species differed from previously known ones. This information was particularly useful to fellow zoologists. As an aid to other researchers Stimpson also listed the names of the museums where the type specimens of each species were housed.

A rigorous knowledge of the anatomical terms used to describe invertebrates was essential in taxonomy, and mastering this terminology was beyond the abilities of more casual naturalists.[15] Providing detailed measurements of key features was also critical in doing accurate taxonomy and provided a more quantitative basis for classification. Nearly all of Stimpson's published work provided measurements, but these became more elaborate as his career progressed. As his focus moved from mollusks to the higher crustaceans the number of external anatomical characters increased, as did the precision of his measurements, to the hundredths of an inch. He sometimes included tables to better illustrate his findings. In one 1857 paper on crustaceans Stimpson gave dimensions including the following: length of carapace; width of carapace; proportion of carapace; length of rostrum from base of praeorbital spine; length of first ambulatory feet; and others. In some cases he also took into account sexual dimorphism by providing measurements of both males and females.[16]

In fact, during the years he labored over the NPEE decapods he created a new term for future generations of carcinologists: chelipeds. In mid-1858 he sought the advice of the Philadelphia entomologist John L. LeConte regarding the creation of this word. "I want to get rid of the Greek terms used by Milne-Edwards, and substitute Latin ones, which are much more appropriate in Latin descriptions. For instance, can I manufacture the word *Chelipedes* from Chelce and pes? You know we have maxillipedes from maxillie and pes—the same will apply in other instances."[17] LeConte gave his approval and today this is a standard term for the pinching leg.

Stimpson's Prodromus described 488 decapod species, 246 of which were new. Significantly, he went beyond describing new species by instituting new genera and families and renovating the overall classification of decapods, thus improving the existing framework of crustacean nomenclature and contributing to a higher order of understanding of the group.

Like Stimpson's Prodromus, other scientific achievements of the NPEE have not been generally recognized, but there are ample reasons why they deserve a more prominent place in both the history of science and of American exploration. Best documented is the fact that Asa Gray's study of the plants gathered by Charles Wright on the expedition led him to further explore the striking similarity between the plants of Japan and those from eastern North America. Gray argued that common ancestry explained the correspondence, and his thesis provided powerful early support for Darwin's theory.[18]

Stimpson's observations on the NPEE, although unpublished, dovetailed nicely with Gray's, as he had noted the parallel between the marine life of

Japan and that of New England.[19] Stimpson's extensive dredging in Japanese waters had uncovered "the interesting fact, that while the southern shores presented a fauna essentially Indo-Pacific in its character, and abounding in the usual Cones, Cowries, and Olives, etc., the northern slopes of the same islands presented an assemblage of forms far more analogous to the fauna of Sitka and Vancouver region, and containing many species common to the American coast."[20]

The NPEE also helped extend knowledge of the ocean floor thanks to Brooke's sounding lead, and the surveys and maps of Japanese waters later proved to be invaluable to the US military during World War II.[21] Ocean historian Helen Rozwadowski nicely summed up the expedition's impact. "The NPEE assembled, and expanded, available expertise at measuring and collecting from the deep sea. Its members made significant contributions to the practice of nineteenth-century American ocean science as well as to the emerging scientific investigation of the depths."[22]

Although not officially a product of the NPEE, Stimpson's time in California led him to write a lengthy paper on the crustaceans and echinoderms of the region. Weighing in at eighty-nine pages, it is the longest individual paper Stimpson published during his lifetime. He noted that while many government expeditions had surveyed the American West over the previous five years, little had been done to document Pacific coast marine invertebrates. He now called attention to the "opening of a rich field" for study.[23] His few months collecting in the region, a thorough search of the literature, and an examination of thousands of museum specimens allowed him to identify over 130 species of crustacea and 25 echinoderms. The article is the starting point for anyone interested in the subject and it has stood the test of time. Over eighty years after its publication the novelist John Steinbeck and the biologist Ed Ricketts lauded it in their book *The Sea of Cortez* as "an interesting, newsy article."[24]

Stimpson's motivations for undertaking the paper were prompted in part by nationalistic motives. Of the two dozen articles he cited dealing with Pacific Coast crustacea, just nine were written by Americans. Nothing provoked American naturalists more than when Europeans announced discoveries of "American" species. Like Dana's, Stimpson's patriotism stemmed in part from serving on a government exploring expedition. His Boston upbringing and his family's role in the American Revolution also helped instill in him a sense of national pride.

Thomas Jefferson had combatted the European notion that American plants and animals were weak and degraded versions of Old World species.[25] Likewise American science had been viewed as inferior. In 1838, Joseph Henry commented on the low opinion Europeans had of American science, an attitude that was still prevalent twenty years later. One historian noted that "the unblinkable fact of European scientific superiority inspired not humility and resignation but appeals to national honor and calls to action."[26]

Nationalism permeated American natural science in the 1850s and 1860s and Stimpson helped lead the charge. When he and Leidy were contemplating a joint work on East Coast annelids, Stimpson urged that they begin immediately, as he had "reason to believe that certain European collectors ... are mousing about our own coasts at this moment." When the Charleston naturalist Edmund Ravenel discovered new shells, Stimpson urged him to write them up quickly. Sounding almost desperate regarding perceived European intrusions, he stated, "we are daily losing our coast shells by the activity of European collectors, and I am so desirous that they should be described by *American* naturalists that I am often willing we should risk adding a synonym." Later he pleaded with Ravenel to write a description despite not having access to the latest references. "Nevertheless we cannot wait to find out, for *we must have names* for our coast shells."[27]

A book published in 1851 went a long way towards explaining Stimpson's strong feelings on the subject. Amos Binney's book *Terrestrial Air-Breathing Mollusks* included an incendiary chapter entitled "Of the ignorance and neglect of American labors in zoology exhibited by European naturalists," in which Binney argued that Europeans had neglected American research in natural history by design, giving dozens of cases in which Europeans had ignored or misrepresented American contributions to invertebrate research.[28] This condescension was offensive to American naturalists, and confrontations with Europeans came most often over issues of priority.

Stimpson's fervor for "American" species must be seen in light of Binney's slashing attack. Increasingly vocal in demanding recognition and respect for their efforts, by around 1860 Americans were justifiably proud of their achievements, with major natural history museums in Washington, Philadelphia, and Boston and several outlets for publishing their findings.[29] In addition, the country's territorial expansion had been coupled with a new confidence in foreign affairs, most notably with Perry's opening of Japan.

At least one prominent European naturalist was surprised and pleased to hear of recent American progress. In early 1857, Dana informed Charles Darwin of the work of the "excellent Naturalist" Stimpson and the nine hundred species of crustacea that he had collected on the North Pacific Expedition. Darwin responded in a way that must have gratified Stimpson in particular and American naturalists in general. "How the U. States are going ahead in Natural History! I had not heard of the late expedition, though I had heard of Mr. Stimpson before; indeed he formerly sent me some cirripedes."[30]

The Englishman Philip P. Carpenter credited Americans with increasing diligence in ferreting out new knowledge, at the same time noting one crucial difference between science in America and Europe. In America "the young are now learning freely in the public schools what in the Old World has long been the property of only the learned few."[31] Privately, Carpenter probably spoke for many Europeans when he stated, "I think the general impression which prevails in other countries about US naturalists is that they are very clever, very hasty and very conceited."[32]

The most accomplished American zoologists saw that it was in their best interests to cultivate ties with their scientific brethren across the sea. "For information and example, American science needed easier access to European publications. For encouragement and criticism, it needed more European comment on its own publications," explained one historian.[33] But keeping abreast of European science remained challenging, since obtaining European books and journals was extremely difficult for many Americans into the 1850s and beyond.[34] The few scientific journals in existence offered limited opportunities to exchange ideas, so personal correspondence played a key role in diffusing information.

However, something as basic as knowing where to send packages hindered efforts at communication. Stimpson wrote Frederic Putnam, who was compiling the first directory of naturalists, "I am constantly bothered to know how to address pamphlets to European naturalists, and packages for me come directed indifferently to Cambridge, Boston, New York, Philadelphia, and Washington. No doubt many are lost."[35]

Stimpson's earliest European correspondent, going back to at least 1852, was the zoologist-theologian Michael Sars of Norway. In early 1857 Sars dispatched a shipment of annelids to Stimpson, who responded by sending a large batch of shells and other marine invertebrates in alcohol, as well as copies of his publications. In return he pleaded for copies of Sars's articles, explaining that specialized books in natural history were hard to come by in America. He closed by affirming Sars's declaration of love for his science. "Well may you call our Science dear! So great a pleasure as we derive from it will never be felt by those who ridicule it because it produces no cotton, sugar, or corn."[36] A few years later, Stimpson again asked Sars for his publications, stating that he had "not been able to get them from the booksellers, tho they have for me extreme importance."[37]

Another of Stimpson's correspondents, Johannes Japetus Steenstrup of Denmark, sent a valuable series of starfish to the Smithsonian. Many had previously been described by Americans under different names, so Stimpson asked for the articles in which they were described in order to determine which names had priority. Steenstrup obliged and Stimpson thanked him effusively, stating that they were "of the greatest value to our zoologists here—we could not in fact go on with our work without them." He also asserted, perhaps not quite convincingly, that the matter of priority was "of no personal consequence to me: it is only that we may know how to designate the species."[38]

Three years later, as Steenstrup labored over a major monograph on cephalopods, he asked to borrow the Smithsonian's holdings of this unique group of mollusks. As no Americans were working on them Stimpson petitioned Joseph Henry for permission, turning the nationalism argument on its head when he wrote, "Prof. S . . . will describe and label the species in a way which will redound greatly to the advantage of students of American zoology."[39]

Stimpson also exchanged specimens with Steenstrup's colleague Christian F. Lütken, who in 1859 had sent a "magnificent collection" of invertebrates to the Smithsonian. In return Stimpson shipped over one hundred species of decapod crustacea, "mostly types of the species of Dana & myself," and later sent a number of starfish, one of which he had named for Lütken. In July of 1863 Lütken sent more specimens to the Smithsonian, including crustacea from Wilhelm Liljeborg of Sweden and shells from Otto A. L. Mörch.[40] It is likely that Stimpson had ties with other European naturalists, but these examples illustrate the importance of international information and exchange networks to American zoologists such as Stimpson.

Through his publications, his fieldwork, and his ties to the Smithsonian, Stimpson had earned a reputation at home and abroad as a serious investigator of marine invertebrates. One other role served to solidify his status within the scientific community. From 1858 to 1863 he served as an editor of *The American Journal of Arts and Sciences* (*AJS*), the country's leading scientific journal. Known popularly as Silliman's Journal, in honor of founder and publisher Benjamin Silliman Sr., the AJS began in 1818.[41] Silliman had consciously established a rallying point for American science, and one historian credited the journal with "having been the greatest single influence in the development of an American scientific community."[42]

James D. Dana (who married Silliman's daughter) had helped edit the journal, taking on an increasing proportion of those duties in the early 1850s. By the late 1850s Dana's health had begun to falter, and in October 1859 he sailed to Europe. Stimpson, who was just twenty-six, took his place as one of the primary zoology editors.[43] It seems likely that Dana recommended him for the post and even groomed him as a replacement, as Stimpson's first review appeared in 1858, a year before Dana's departure. While his name never appeared on the journal's masthead, Stimpson, writing under the initials W. S., penned over two dozen credited reviews and probably other uncredited ones during a time of great upheaval in natural history.

Given a unique platform to voice his views, Stimpson never shirked from the responsibility of criticizing works that did not meet the highest standards of professionalism. Enthusiastically taking up the cudgel that others had wielded, he used his position to defend American science, repeatedly pointing out cases in which European naturalists had failed to cite the publications of Americans. He just as harshly censured sloppy or inadequate research, irrespective of nationality. "Science," he wrote, "is cosmopolitan, not national," but Stimpson made it his mission to point out American contributions to zoology.[44]

His first review established the pattern for many that followed. In appraising a Danish work on the natural history of Greenland, Stimpson acknowledged its "great importance to American naturalists" but found many instances where the labors of Americans had been overlooked, including several species that Stimpson had described.[45]

Stimpson also played a role in a debate over the creation of the self-supporting aquarium. As noted earlier, Stimpson had maintained a number of aquariums around 1849–1850. By the early 1850s aquaria had become immensely popular in England and the craze soon took hold in the United States.[46] The credit for being the first to invent the self-supporting aquarium had been given to Robert Warington of England. The *North American Review* of July 1858 contained an unsigned review of two books on aquaria, arguing that Stimpson should be credited with the discovery. The journalist Robert Carter echoed this claim, writing that Stimpson "had a number of aquariums in successful operation long before anything was heard of the kindred experiments of the Englishman Warrington [*sic*]."[47]

Aquaria were near the height of their popularity in America at the time, and Stimpson had recently prepared a large one for display at the Smithsonian. The front page of the December 19, 1857, issue of *Scientific American* highlighted the new exhibit. As curator of the aquarium Stimpson proudly showed off his handiwork. He had gathered the plants and animals and carefully prepared the miniature ecosystem. Sea horses, crabs, clams, seaweeds, and a solitary flounder were among the three hundred inhabitants, and the tank attracted many visitors.[48]

Given this context, Stimpson's impassioned arguments against European influence in American books took on a decidedly personal note when, just months after the *North American Review* article, he found fault with the book *Life Beneath the Waters, or, the Aquarium in America*, by Arthur M. Edwards, of the New York Lyceum of Natural History. Stimpson felt that the book's title implied that aquaria had been newly introduced to America by England, when "in fact, these self-supporting aqua-vivaria were used here before their invention in Europe. Still nothing had been published in regard to the subject." Stimpson, a strict follower of the laws of priority, admitted that Warington had published on the topic first.[49]

More tellingly, Stimpson castigated the book because "notwithstanding its patriotic title, it is still a *British* book. ... Were this openly avowed, no harm would be done, for the reader would then understand that the zoology of the work applied mostly to Great Britain." Edwards's book would inevitably mislead the American public since he relied on illustrations of English species, many of which were not found on the American coast. Stimpson concluded that Edwards had "but a very slight and imperfect acquaintance with our marine zoology" and he capped his tart remarks by engaging in a bit of humiliating humor at the expense of Edwards, who urged his readers to "take this or any other book that treats of the subject, and see if you can find out the names of your plants and animals." Stimpson recommended that the reader should "by all means try the 'other book.'"[50]

His reviews returned time and again to the theme of Europeans neglecting the work of Americans. He chastised Henri de Saussure of Switzerland for failing to credit work by Thomas Say and Lewis Gibbes. "At the present day it will be found highly desirable for those writing upon subjects connected with American zoology, to consult the works of American naturalists," he thundered.[51]

In a paper clarifying the synonymy of a genus of flatworms, Stimpson found that Ludwig K. Schmarda of Leipzig and E. Percival Wright of Britain had both missed his description of the genus (despite the fact that it had been published in the widely available *Proceedings of the Academy of Natural Sciences of Philadelphia*), leading Stimpson to comment sourly that "both Schmarda and Wright have overlooked previous labors, so that the genus now rejoices in three distinct appellations, all given within four years."[52]

Showing that he was conversant with the literature in other disciplines, Stimpson critiqued a British Museum catalog of fish that in his opinion had not done a good job in categorizing American freshwater fish. "If recent American works had been consulted these errors would not have occurred," he stated flatly.[53] All of these examples are indicative of Stimpson's goal of improving the accuracy and reliability of descriptive natural history.

Claims of priority in naming new species were obviously linked to national pride, but it is important to remember that men such as Stimpson who risked their lives in the field were especially anxious to protect their discoveries. For some the privilege of naming a new species was the only compensation they would ever receive. The sociologist Robert Merton has suggested that the people involved in priority disputes were often "men of ordinarily modest disposition who act in seemingly self-assertive ways only when they come to defend their rights to intellectual property," a description that aptly fits Stimpson.[54]

Not all of Stimpson's comments on European works were negative. He praised Heinrich G. Bronn's introduction to zoology for being written "in a masterly manner." He also liked the Swedish naturalist Wilhelm Liljeborg's work on the parasites of decapod crustaceans and R. M. Bruzelius's contributions on the amphipods.[55] Because he noted the titles and authors of numerous European publications, Stimpson provided a valuable bibliographic tool for Americans striving to stay abreast of the latest research from Europe.

Candid assessments of other naturalists' work characterized Stimpson's writing, and many an American came in for a verbal scalding. When George W. Tryon Jr., a respected Philadelphia conchologist, described a new species of mollusk, Stimpson bluntly wrote, "I have satisfied myself that Mr. Tryon is wrong."[56] Inferior science infuriated Stimpson. Lambasting a paper on worms by William H. Pease of Honolulu, who had established a new genus—named Peasia—based on five supposedly new species, Stimpson noted that all of them had previously been described, adding, "This kind of progress can scarcely benefit science." When John B. Trask published descriptions of a new genus of microscopic "crustacea," Stimpson pointed out that they were in fact annelids.[57]

As a taxonomist, Stimpson maintained an abiding passion for all aspects of classification and nomenclature. Standardization of nomenclature is a prerequisite for professionalizing any discipline, a maxim especially true of biology.[58] He spent a considerable portion of his energies attempting to straighten out nomenclature, in one case complaining to John W. Dawson about a genus of mollusks in which the nomenclature was "in inextricable confusion."[59]

Protesting against the nomenclatural system used by Francis Holmes in a work on southern fossils, Stimpson corrected many of the Latin names, changes that were necessary in order to gain a truer picture of the percentage of extinct species. Some of the mollusks thought by Holmes to be extinct were in fact still living on the coast, and Stimpson also deduced that one of Holmes's new species was in fact only the tip of another known species. Never averse to a pun, he wrote, "a little *tipical* knowledge is quite necessary to an investigator of fossil shells."[60]

A more serious breach of standards led to a withering review of Timothy A. Conrad's catalog of the Miocene shells of the Atlantic Coast. Conrad had adopted generic names used by Jacob T. Klein, an "avowed enemy" of the Linnaean system, who had given species names to what were more akin to Linnaean genera. To Stimpson it seemed clear that the Linnaean system had become "the *law*," having proven its value for over one hundred years. Recognition of non-Linnaean names would, insisted Stimpson, constitute a taxonomist's nightmare. "[W]e shall have all writers ... searching through every forgotten and almost extinct work at all relating to natural history, which has appeared since the invention of printing, seizing upon every case in which an author *happened* to designate a group of animals by a single name, and adding this to our already overburdened synonymy, to the suppression of the last name in vogue."[61]

His interest in classification led Stimpson to write an important, if little remembered paper on the topic. He began with a stern appeal: "A more careful attention to the subject of nomenclature is urgently demanded of the followers of all branches of Natural History." The British Association had adopted a comprehensive code of rules in 1842 and Americans ratified it three years later, but Stimpson saw that this document needed "much extension and many additions." Stimpson was among the first to explicitly call for an international congress where the "leading lights" of science would mediate taxonomic disputes. "Let a convention meet at Paris or some other central point, composed of delegates from all the scientific societies of the earth, and representing at least all the departments of zoology and botany."[62] This illustrates the belief, common among those working to professionalize science, that a small group of specialists should set the standards for those working in descriptive natural history.

As we have seen, the great burden for all taxonomists of this period was trying to sort out the synonymy of each species. A synonymy is the often-lengthy list of all the scientific names ever assigned to a particular species. The plethora of names caused confusion among professionals and amateurs alike, and a good deal of Stimpson's writing consisted of the thankless but essential task of sorting it all out and coming to some "final" determination.[63]

Stimpson despaired that the greater part of many new zoological works, including his own, consisted of synonymy. "One author, after six pages of historical and synonymical matter, evincing great critical acumen and much bibliographical research, will arrive at what appears to him to be a certain and final conclusion. ... The next writer who succeeds him in the same field will triumphantly prove in *ten* pages that it is not that species at all. ... All this discussion, let us bear in mind, is merely preliminary, and for the purpose of indicating with certainty an object about which the author has perhaps not a dozen words to say."

Stressing the need to disavow all personal considerations, Stimpson expressed his belief in a community of science that crossed all borders. "The smallest interest or convenience to the science in general, followed as it is by a republic of thousands, is of more importance than any compliment to the feelings of a living, or the memory of a deceased naturalist. In fact our mere recognition of an author's names is not of such vast importance to his reputation. His fame must rest upon a securer foundation than this. . . . Does the fame of the great Linnaeus depend upon the number of species he described?"

Above all, he pleaded for uniformity. The Germans, the English, and the Americans all had their own way of spelling Latin names; some capitalized nouns while others used lowercase letters. His proposed tribunal would settle all of these vexatious questions. Men of science would have to respect the decisions as final, and for those who might ignore such decisions, Stimpson had a warning. "Let the names of those who will not abide by them be placed upon a new edition of that black-book which Linnaeus kept of old,—the list of *Damnati!*"[64]

The nomenclature paper appeared in the March 1860 issue of the *American Journal of Science*. Whatever its merits, the article was almost guaranteed to be overlooked, as Charles Darwin's book *On the Origin of Species* had just begun to generate debate in America. Stimpson's article was in the same issue that contained Asa Gray's first long defense of Darwin and natural selection. For all of its significance to zoologists, the arguments over nomenclature and priority meant nothing to the general public. Evolution was another matter entirely and Louis Agassiz, the leading American opponent of Darwin's book, quickly responded to Gray.[65]

Following his mentor Cuvier, Agassiz never accepted the notion that species changed and insisted that there was no connection between animals and plants of the past and those of the present day. He believed that species did not change and were literally created by the "thought of God." Ronald Numbers has described Agassiz as a scientific creationist while asserting that Agassiz was no "Bible-thumping creationist" since he rejected the story of Adam and Eve and ignored the biblical record. Agassiz did, however, believe in repeated acts of special creation to explain the geographic distribution of species.[66] This decidedly nonscientific explanation did not satisfy some. Ralph Waldo Emerson counseled a friend, "Don't set out to teach theism from your Nat. History, like [William] Paley and Agassiz. You spoil both."[67]

Darwin knew that many older naturalists would reject his theory, so he pinned his hopes on "young and rising naturalists who will be able to view both sides of the question with impartiality."[68] Some have suggested that field naturalists like Stimpson found Darwin's theory less controversial than closet naturalists, while others assert that later-borns were more likely than first-borns (like Stimpson) to accept the theory.[69] During the arguments over evolution Stimpson, like many of his American colleagues, remained on the

sidelines and "locked themselves up in their studies and continued their specialized work without giving public support to either side." It seems clear from
his writings that Stimpson was not interested in developing grand theories.
Perhaps he would be better known today if he had publicly addressed the
greatest scientific question of the day.[70]

However, during that first contentious summer of Darwinian evolution,
Stimpson and Agassiz played out a very public grudge match in the pages
of the *American Journal of Science*, with evolution as the subtext. Years earlier Agassiz had found a single specimen of the brachiopod genus *Lingula* in
South Carolina. Stimpson had asked for a description only to be rebuffed.[71]
This particular genus held great importance for Agassiz, who argued that the
existence of the *Lingula*, which had remained virtually unchanged for over
400 million years, refuted Darwin's claim that lower forms became extinct in
favor of their more advanced descendants. To Agassiz the *Lingula* showed the
immutability of species.[72]

In the spring of 1860 Stimpson embarked on a collecting trip to Beaufort,
North Carolina, one object of which was to find the elusive *Lingula*. Although
known from the Pacific (Stimpson had found them on the NPEE) it had not
been officially documented on the Atlantic coast. Stimpson had blustered to
Baird and Philip P. Carpenter that he would find one, and he fairly crowed
when he did, firing off a telegram announcing the discovery. "We have at
last accomplished the object of the Expedition, and performed our promise,
(which must have seemed a vain boast to you and Carpenter.) At 3 p.m. this
day [March 23, 1860] we found the *Lingula*!!!!! Yea verily the unapproachable
Lingula! for all knowledge of which, forsooth, American Naturalists were to be
dependant [*sic*] upon [ripped page]. . . . he may keep [ripped page]. . . . in his
cabinet till it rots for all I care—our catalogue of U.S. shells is no longer dependant [*sic*] upon him for completeness. Tell Carpenter not to let the Cambri[d]
gers know anything about it or they will rush out a description of theirs tho'
so tardy when we requested it and they thought the game was in their hands."
Stimpson's anger is as palpable as his joy. To his way of thinking Agassiz's
unwillingness to share information ran counter to the interests of science.[73]

Within months Stimpson published a typically thorough description
of *Lingula pyramidata*, and in a footnote he got in a barb that infuriated
Agassiz. "We understand that there is a specimen of *Lingula* from the coast
of South Carolina in the possession of Prof. Agassiz. Not having access to
this specimen, we are unable to say whether it be identical with ours or
not."[74] Stimpson clearly implied that Agassiz had been unduly possessive in
not providing access to the specimen. The fiery Swiss had been shown up
by a former student and in America's most prestigious scientific journal to

boot. When the two men met later that year one observer noted that they did not seem to like each other.[75]

While Stimpson made no definitive public statement on evolution, his writings provide clues to his leanings. One important underpinning to Darwin's arguments came from the geographic distribution of species, and one of Stimpson's earliest allusions to the debate came in this context. While classifying crabs from California, Stimpson came across one that resembled a species known from the "Gallapagos." Not having access to the original type specimen, Stimpson decided to describe a new species, risking what no respectable naturalist ever wanted to do, namely adding a synonym. He felt the risk was warranted, as he did not want to refer a known species to a locality where it did not exist. To his mind it would be a far graver crime to distort data relating to the geographic distribution of species, "knowledge of which constitutes one of the most important aims in our investigations of species," as he put it in the spring of 1860.[76]

Even if Stimpson was not ready to defend Darwin publicly, he and another Agassiz student, Nathaniel Shaler, debated the hypothesis privately, for as Shaler later put it, "to be caught at it was as it is for the faithful to be detected in a careful study of heresy." Both men had read Robert Chambers's *Vestiges of the Natural History of Creation*, Lamarck's *Philosophie Zoologique*, the first Darwin-Wallace letters, and finally *On the Origin of Species* itself.[77]

A perceived conflict between science and religion has been a hallmark of the evolution wars from the beginning, and Agassiz's objections to evolution were in part religious in nature. As one historian has written, Agassiz's "professional approach was an expression of his commitment to religious and scientific certainties."[78] Stimpson's letters give us glimpses of his attitudes towards both God and nature. In 1863 he wrote a fellow naturalist, "For there is something beyond this world, and shall we not better understand the grandeur of the greater worlds, by the study of the Great Plan of this one?" In another letter he remarked that John Newberry's religious beliefs forbade him from going for a nature walk on Sunday after a night's drinking. Stimpson felt that Newberry's "mistaken religious ideas would not permit to go out on Sunday mornings to recuperate and walk with us, breathing God's own free air and seeing his glorious works yet uncontaminated by man—the true *Church* for sedentary men. Meeting hours are well enough for farmers who are out all the week taking physical exercise."[79]

In another instance Stimpson's religious views are seen through the prism of his irreverent sense of humor. Writing on a Sunday in the spring of 1860 from North Carolina he jested, "I went to church this morning and got nearly froze to death, and caught cold—whence I conclude that improper indulgence in piety is unhealthy." Two years later, he informed a correspondent that in order to collect on Sundays he had aligned himself with the "Broad Church," a faction of the Church of England that believed that the Bible and church authority could be subjected to historical criticism. Raised as an Episcopalian by a father who served as a warden of Christ Church in Cambridge, Stimpson's letters home from the NPEE contain frequent references to God and attending church, but there are few references in his letters after 1863 concerning religious matters.[80]

Stimpson addressed the question of human evolution in an 1862 review in the *American Journal of Science*. John Lubbock, a Darwin supporter, had discovered stone tools in association with extinct mammals, a finding that pointed to a much earlier appearance of man on the geological stage than had been supposed. Stimpson noted that such finds gave a powerful impetus to further research on the origins of humanity. "We may ascertain whether any intermediate types have been created, and if so, what has been the range of their development,—matters about which we can now only speculate." He continued, "To those who object to such speculations we may observe that, in a religious point of view, it matters little by what method the Creator made physical man, when we know that his creation as a spiritual and accountable being dates with the time when He 'breathed into his nostrils the breath of life, and man *became* a living soul.'"[81] To some the thought of discussing mankind in evolutionary terms was profoundly disturbing, but Stimpson supported research that might shed light on the question. Like Gray and others, his beliefs were consistent with the tenets of natural theology, which made room for unfettered scientific inquiry as well as a faith in a creator.

Many zoologists did not make formal statements of their conversion to evolutionary doctrines but instead incorporated the new views into their work.[82] There is confirmation of this in Stimpson's writings. Unlike Agassiz, he rejected the notion that today's species had no connection with those from the past. In 1865 he thought it not unlikely that the fossil record would show European and American whelks to have been derived from a common ancestor. He also stated that if mollusks from the Miocene period had survived to the present day they might have "undergone some slight change during so great a lapse of time," and he suggested that fossils from intervening periods might settle the question.[83]

In his correspondence with the paleontologist Fielding B. Meek, Stimpson wrote of his belief that classificatory terms such as isopod and amphipod were not "necessarily and trenchantly distinct," predicting that connecting links would be found between these groups. He thought it possible that one of Meek's fossil echinoderms might represent a transitional stage between the

Echini and the Holothurians.[84] Connecting links and transitional forms were terms often used to indicate an acceptance of evolution.

William H. Dall discussed "development theory" with Stimpson in 1865 and concluded that Stimpson was a Darwinian. Dall later stated that Stimpson "stood ready to welcome the theory of evolution with all the light it shed in dark places," and an early Stimpson biographer wrote that Stimpson "readily accepted Darwin's views." A slightly different take is given by another friend of Stimpson's, who, in rejecting Agassiz's statement that a species was a "thought of God," wrote, "[T]he logic of these views bothered Stimpson less than it did me, because he was a man of facts and not fancies."[85] In any event, there is the supreme irony that an entire generation trained by Agassiz came to accept evolution, while he rejected the theory to the end of his life.[86]

The year 1860 marked the peak of Stimpson's connection with the *American Journal of Science*. In the two volumes for that year he contributed fifteen reviews or notices, three papers of his own, and one obituary. He also used his editorial privilege to highlight the work of American naturalists, especially the southern naturalists John McCrady and Francis S. Holmes. By the end of Stimpson's tenure as editor in 1863, the initials W. S. had become familiar to serious zoologists around the world. When Dana returned he resumed, in a more limited way, his former duties at the *Journal*. Stimpson had tried to raise the standards for those working in zoology, thus contributing to the professionalization of science, and his unswerving defense of American science was part of a long struggle to gain international respect. As late as 1876 men like Dana and the paleontologist Edward Drinker Cope still resented European "condescension and disparagement" towards American science. Stimpson had assumed the responsibility of speaking out when American science was slighted, and his own research had done even more to win the respect of his peers worldwide.[87]

CHAPTER 5

THE STIMPSONIAN INSTITUTION

What good men most biologists are, the tenors of the scientific world—
temperamental, moody, lecherous, loud-laughing, and healthy. Once in
a while one comes on the other kind—what used in the university to be
called a "dry-ball"—but such men are not really biologists. They are the
embalmers of the field, the picklers who see only the preserved form of
life without any of its principle.... The true biologist deals with life, with
teeming boisterous life, and learns something from it, learns that the first
rule of life is living.
—John Steinbeck and Ed Ricketts, *The Log From the Sea of Cortez*[1]

One trenchant observer described Washington, DC in the late 1850s and
early 1860s as "a mud-puddle in winter, a dust-heap in summer, a cow-
pen and pig-sty all year round ... and the country all about [is] as primitive as the most enthusiastic naturalist could desire."[2] Many visitors to the
nation's capital in the mid-nineteenth century dismissed it as a provincial
southern town and tainted by the stain of slavery. While the Smithsonian is
today an integral part of the Mall, in Stimpson's day it was in a section known
as the Island, cut off from the rest of the city by the open sewer known as
the Washington Canal. Just getting to the Smithsonian Castle took a certain
amount of resolve, necessitating an "unpleasant walk" through often-muddy
fields that surrounded the building. Even worse, the smell of rotting animals
and the sight of their corpses floating in the canal accosted the senses of even
the most hardened visitor. The building's isolation also attracted criminals,
with gangs of toughs sometimes robbing hapless sojourners on the solitary
Smithsonian grounds.[3]

Thus it was all the more remarkable that a mere ten years after its founding
in 1846, the Smithsonian had become a "fulcrum of scientific power in antebellum America."[4] In the early years of the republic Boston and Philadelphia
had been dominant in American science, but the locus of power had shifted
by the late 1850s. Smithsonian secretary Joseph Henry had impressive

credentials, having conducted important research on electromagnetism that paved the way for the invention of the telegraph, and he used his political connections to obtain government funding for the institution.

From the very beginning there was debate over what the Smithsonian's role should be. Henry's vision was that it should promote original research.[5] He steadfastly opposed using Smithsonian funds to pay for the upkeep of a natural history museum, but the museum operations flourished, thanks to Assistant Secretary Spencer Fullerton Baird. With a single-minded devotion to advancing American science, Baird combined zeal with unparalleled organizational and executive abilities. While an excellent naturalist in his own right, he chose to largely forgo his own research to nurture a legion of naturalists and collectors, leading one historian to dub him a collector of collectors.[6]

Baird began to assiduously cultivate military leaders, politicians, diplomats, and others in order to foster the growth of the American natural history community. Nearly every government exploring expedition had at least one man who collected for the Smithsonian, and with the country expanding westward at a phenomenal rate there were literally dozens of such expeditions. Due in part to Baird's omnivorous appetite, the Smithsonian eventually became the National Museum.[7] Further cementing the Smithsonian's place at the epicenter of American natural history, in 1857 Henry reluctantly agreed to accept the large Patent Office natural history collections (including the material collected on the US Ex. Ex.) on the condition that Congress appropriate funds for their maintenance.[8] The four thousand dollars per year they allocated allowed Baird to hire men to organize and classify the collections.

Thus Stimpson's timing could not have been better when he began work on the North Pacific collections in late 1856. Fresh from three years of isolation he craved scientific companionship, and at the Smithsonian he had ample opportunity to forge relationships with men who had served on government exploring expeditions to the western United States. Among them were the geologists John Newberry, William P. Blake, and Ferdinand Hayden, as well as the naturalist James G. Cooper, whom Stimpson had met in San Francisco. At Stimpson's urging they decided to room and board together at the Rugby House, a small boarding house at Fourteenth and K Streets. Cooper informed his family of the accommodations. "Stimpson . . . & several others keep a kind of private Hotel for themselves, families and friends. I have very pleasant quarters for only $7 a week including *gas lights*."[9]

With the money he had earned from his work arranging the NPEE specimens and the Patent Office collections, in June 1857 Stimpson and the others rented a large brick house just behind the War Department building on G Street and Seventeenth, less than three hundred yards from the White

House.[10] The following year they moved about half a mile east to a cottage at 299 G Street, near the Episcopal Church of the Epiphany. This building, about three-quarters of a mile from the Smithsonian, became a temporary way station to a host of resident and visiting naturalists. They took to calling it the Stimpsonian Institution.[11]

One visitor recalled his impressions of a hot evening in June 1858. He had "sought refuge at the Professor's" not because the temperature there was any cooler but because it "was intellectually and imaginatively cooler. It abounded in objects suggestive of refreshing ideas. There were crabs and shells that had been dragged from the sunless depths of the Arctic Ocean; fishing-lines and dredges that had explored the cool abysses of Kamatchatkan and Siberian seas ... and, what particularly was wont to give an agreeable chill to my fancy, a picture of the prodigious snowy cone of the great Japanese volcano, Fujiyama, made by a native artist at Simoda, where the Professor himself purchased it."[12]

Their dinners were raucous affairs, with endless toasts fueling laughter and singing, the revelries sometimes lasting well into the night. They originally referred to their gatherings as the wigwam of the Skraelingers, a derogatory Viking term for Native Americans. Cooper, with the artist Wilhlem Heine in tow, described a typical dinner. "We have the highest kind of times at dinner every evening ... we six are all naturalists & Geologists, and after working all day spend an hour in eating, drinking, & laughing immensely ... we had a very jolly dinner last P.M. and a grand inauguration of a barrel of Ale, each one solemnly assisting to drive the spigot to the sound of martial music and a great speech from Prof. Heine. The eating club is named the 'Skraelingers' or Redmen (Norse) probably, as suggested by one of us, because we are *well read*. So you perceive we continue as dissipated as usual."[13]

Raw oysters, fresh cod, roasted chicken, and roast beef were consumed in quantity, washed down with whiskey and Hungarian wine. Stimpson particularly enjoyed a good meal and considered himself something of an expert on the edible qualities of marine invertebrates. Extolling the virtues of the American oyster, his nationalism again surfaced when he scoffed at his "ancient and honorable cousins on the other side of the Atlantic" who considered American oysters inferior. Stimpson also praised American lobsters as superior to their counterparts in "the foggy and fogy islands over the water."[14]

Given his love of animals it is not surprising that Stimpson had several dogs, including a big Newfoundland named Carlo. He and his housemates also kept chickens to maintain a steady supply of eggs. On nights that they stayed in they roasted oysters in the fireplace and battled cockroaches and fleas. Eventually their gatherings came to be known as the Megatherium Club, named after an extinct genus of sloth. Inspiration for the name may have come from Joseph Leidy, who in 1855 published a memoir on fossil sloths in the *Smithsonian Contributions to Knowledge* series. *Megatherium* is Greek for large, wild beast, and Darwin and others had commented on these ungainly, almost comical-looking animals.[15] The members of the Megatherium Club were united by youth, burning ambition, a sense of adventure, intelligence, and a deep and abiding love of Nature.

The Stimpsonian bore the imprint of its namesake's personality, as borne out by the following passage from one of his closest friends, who characterized him as "one of those chaps ... who make you like them whether you will or no. He is a perfect gentleman, very well educated and ranks with the first zoologists of the age. ... Stimpson is preeminently conducive and good hearted and withal is a man of refined feelings and tastes—He has the strange weakness of wishing to appear idle, thoughtless, and even dissipated. From merely hearing him talk one would suppose he was a hard-drinking idle fellow with no ambition beyond being considered a good fellow. Yet he is extremely modest and does not in reality care at all for praise. He is very sensitive, warm hearted and a staunch friend."[16]

Whatever his reason for cultivating an image as a drunken layabout, Stimpson's exuberance held full sway among friends. A bon vivant, lover of parties, dances, and other social gatherings, his charisma made him "the life of the party," according to one lady friend.[17] Slender and handsome, in a boyish sort of way, he stood 5 feet 8 inches tall, with a broad forehead and aquiline nose. He let his wavy brown hair grow long and he had a full, thick mustache. His gold-rimmed glasses did not fully obscure his bright blue eyes.

Stimpson clearly enjoyed the social scene, usually with his closest friend during the early days of the club, Ferdinand Hayden. Three years older than Stimpson, Hayden was born out of wedlock to an alcoholic father and an indifferent mother. Friends considered him eccentric and excitable but he was also a shameless self-promoter who took credit for work done by his collaborators. Hayden earned an MD and had begun to carve out a reputation as one of the foremost interpreters of the geology of the American West. He and Stimpson were close for several years, certainly as drinking companions, but later seem to have drifted apart.[18]

Late in 1857 twenty-two-year-old Robert Kennicott came to study at the Smithsonian. Raised in a beautiful prairie grove north of Chicago, Kennicott had been sickly as a child and his father John, a country doctor and frustrated naturalist, encouraged him to spend time outdoors. Robert began collecting animals for Baird, who really took notice when Kennicott discovered what turned out to be a new snake. Shy, insecure, and very rough around the edges, he was, according to Cooper, a "wild Illinois naturalist ...

quite original in looks + manners," and "certainly a genius though unfortunately little educated."[19]

Not without some trepidation, Baird turned the somewhat naive Kennicott over to Stimpson and the Megatheria. Surprised to find that many of the Smithsonian naturalists were scarcely older than himself, Kennicott adjusted to his new surroundings by taking up cigars and scuppernong wine. He and Stimpson formed a close friendship, leading Stimpson to later say, "Kennicott's voice was ever the most cheery, his tale the freshest, and his song the blithest." They were something of an odd couple, with Stimpson's New England polish and scientific credentials standing in sharp contrast to Kennicott's frontier looks and manner.[20]

Another eccentric character arrived soon after. While in New York, Stimpson met a young law student who kept a horse's skull under his desk. Only twenty, Theodore Gill had an encyclopedic memory and had already published a paper on the fish of New York. By August 1858 he began rooming with Stimpson. Kennicott quipped that he considered the pop-eyed Gill, originally an ichthyologist, "about the oddest fish I've come across." Gill rarely ventured into the field and his reputation as a closet naturalist, along with his extreme vanity, made him a frequent target of ribbing. A master at classifying large groups, no less an authority than William H. Dall later called Gill "the most eminent American taxonomist."[21]

Considerably older than the other naturalists at the Smithsonian, Fielding B. Meek was one of the best paleontologists of the era. Arriving in Washington in 1858 at the age of forty-one, he spent many of his subsequent years living in the Smithsonian building. Increasingly deaf as the years passed, he rarely went out socializing, and one observer referred to him as the "old bachelor par excellence." Meek and Stimpson shared an affinity for invertebrate fossils and spent many a day comparing recent and fossil shells.[22]

Baird regularly invited these men to his home for Sunday dinners, and one would later remember the positive influence the Baird household had on poor, "half Bohemian" naturalists living "with little restraint in that great disjointed village." They in turn were fiercely loyal, recognizing Baird as a man who worked himself "half to death" for science.[23]

Due to the peregrinations of its members the Megatherium Club was transitory in nature. With the exception of Stimpson, tied to the Smithsonian by a mountain of specimens and his work for the NPEE, most club members did not stay in Washington for long before heading out on the next expedition. Stimpson watched as his friends left one after another. "I am living a terribly lonely life," he cried to a Cambridge naturalist. The good times seemed to be all over.[24]

In his letters Stimpson did his best to entertain his far-flung comrades, knowing what it was like to endure the rigors of government exploring expeditions. In an effort to cheer up his friend Hayden, Stimpson conjured up a picture of the day when they would be reunited in Washington: "Depend upon it I will have a warm snug place for you a good fire in the grate and a barrel of beer in the cellar and a warm welcome on the lips and then we will go somewhere and see some little bodies in the evening, whose smiles will melt the ice in our hearts which may have been engendered by the cold contact with the world."[25]

One reason that Stimpson lamented Hayden's absences was the fact that the two men pursued eligible young females together, including Joseph Henry's three daughters. Mary, Caroline, and Helen Henry were described by one naturalist as "very intelligent and quite good looking," and one of the most attractive things about them was that they cared little for the fashions and frivolities of Washington's social scene, although they did amuse themselves by roller skating through the halls of the Smithsonian.[26]

Stimpson could not make up his mind which one he liked the most: "They are the best girls in the world, and all so equally fascinating that one always thinks the one he happens to be talking with to be the most agreeable until he gets beside another—You must come home out of that savage wilderness as soon as can come and yet bring the requisite laurels with you." Stimpson bought a black horse that followed him around like a dog, riding him every day for exercise. He later added a buggy, excitedly telling Hayden that they could now "take the girls to ride!!!"[27] Stimpson finally settled on twenty-three-year-old Mary, the oldest daughter, taking her for buggy rides and picnics along the Potomac.[28]

One wonders what Joseph Henry thought of these young bucks courting his daughters. Stimpson hinted at their relationship when he recorded that Henry was "one of the best of men," although "his outside behavior may be occasionally impatient or preoccupied." With Hayden gone Stimpson saw less of the girls. "You may well call them "gentle spirits"! ... I have not seen them now for two months but their images are still fresh in my heart." Later he wrote, "No moments I live through are so pleasant as those spent in their society."[29]

Money, or the lack of it, seemed to hinder his courting. Acknowledging his uncertain financial prospects in natural history he wrote, "dear beings! if they were not so expensive I would go in for one of them sure!" He averred that the Henry girls "seem far beyond my humble aspirations." After several years he admitted that he "really didn't understand them," and in the end all three girls remained amiable but ultimately aloof. None of them ever married.[30]

By the time he reached his late twenties Stimpson declared that he was "right anxious to get married."[31] In one instance he became totally smitten, as recounted in a wonderfully gossipy letter to Baird from William Turner. "Poor Stimpson! he has not slipped, slidden, or even fallen, but has perpendicularly plunged into love again! Just three weeks ago he spent an evening at Dr. Foreman's, and there he met his fate in the shape of a pretty, petite, young, lively, in every way interesting Being. Miss Louisa Sickley by name, of Philadelphia." His letter continues: "Since then he has been her shadow at parties, tea-drinkings, sides, concerts etc.; and finally to conclude he has ended by politely attending her per railway to her home, in the bosom of which terrestrial heaven he is probably repining at this present writing. Remember that this delicate morsel of news is intended not for your own rude consumption but as sustenance to the sympathetic sensibilities of the ladies, to whom be all honor, love, and admiration. Amen."[32]

Stimpson articulated his predicament in a letter to Frederic Putnam, who had recently become engaged: "You have no reason to complain since the lady has said 'yes'—I only wish I was sure *mine* would & blamed if I would not ask her p.d.q. [pretty damned quick]." No other letters mention Ms. Sickley, and Stimpson responded by immersing himself deeper into his work. "Science now absorbs me more than ever before, hard as it is to tear myself from the arms of the Sirens, on whose gentle bosoms I would give my ears (not my science) to recline," he sighed.[33]

Months later he had found a new object of affection in Annie Gordon. In one of his most expressive, jubilant letters, Stimpson reveled in yet another intoxicating infatuation. "Spring is coming fast, glorious season which gives us all new life while nature lures us to her arms. I shall now have more time and take more out-door recreation especially in the form of picnics with the girls, the dear angels some of whom I should certainly try to marry were it not for the pain of leaving the others." He went on to describe a picnic with "the Henries, Miss Annie Gordon, Prof. Mayer, Gill, and myself. We went in 4 teams to the Little Falls over a romantic road, put up our horses in an old mill and had a jolly stroll in the woods, ornamenting the fair heads and brows of the particular fair ones with a profusion of wildflowers." The picnickers "had a splendid lunch with champagne juleps (for I had laid in a great supply of wine, ice, and mint) which however, we could not induce the dear creatures to drink, although they would touch their lips to our glasses to add new zest for us, while they only went through the forms."[34]

The references to alcohol are typical of Stimpson's letters of the period. Nearly all of the club members drank, and Stimpson seems to have consumed more than his share. "The club are well supplied with grog just now—the scuppernong wine having come a week ago 7 galls. [gallons] rapidly disappearing. Last evening we drank 8 bottles ale and 2 scuppernong besides brandies to fill chinks and all went to bed sober. Exercise does it and long walks in the woods." In an apparently alcohol-fueled missive, Stimpson, preparing to go to a wedding, rambled on somewhat cryptically to Hayden: "Am as lonesome as a flea in a water bucket . . . The Kokrochys are xterminated. Phleas skerse. Good bye. Be virtuous among the Cheyennes and see that Meek don't get ----."[35]

Stimpson may have played up his drinking exploits in his letters but he did have some rules. "We never get tight until after dinner which is at 5 1/2 P. M" he chortled to a prospective visitor. In 1859 he vowed to remain "rectilinear" during his father's upcoming month-long stay in Washington.[36] That summer when he went home to Cambridge, he wallowed in "non-conducting misery" and allowed that he was having as much fun as a man could "upon temperance principles."[37]

Baird did his best to discourage drinking but the club's fondness for "demon rum" eventually led to trouble. Neighbors were infuriated by the Indian war whoops and drunken singing but Stimpson wanted no part of these "serious organs." "Did not the great Creator place us on this earth to be happy? The robins sing and we say they are praising God,—are we then the less sending homage to him when we shout in the elasticity of youth and joy?"[38]

William H. Dall later recounted the growing tension between the club and the local residents. "Rumors spread, and multiplied as they spread, of awful doings in the Megatherium Club. Well-meaning neighbors carried the tales to the Baird household. Miss Lucy, [Baird's daughter] despite her youth, assumed, like her mother, a maternal attitude towards these young gentlemen whom she knew so well, and took occasion to inform them of the reports, doubtless with a gentle intimation that the reputation of the Smithsonian coterie was in danger. At a later meeting of the club it was decided to sacrifice their refreshments to the general good."[39]

Stimpson feared losing the respect and good will of Baird and Henry, whose patronage could make or break a career. Access to the Smithsonian museum and library were essential to carry out his work. Henry, sensitive to political matters given the Smithsonian's financial support from the federal government, would not tolerate any shenanigans that might besmirch the reputation of the Smithsonian. Washington was already an inhospitable place for naturalists, as many politicians had little respect for or understanding of their work, with one congressman deriding it as "utterly valueless." Attacks from Congress reached a peak in early 1858 and, tongue firmly in cheek, Kennicott commented humorously on the slur, writing one of his siblings "Naturalists are going up! We have even gotten to be abused in Congress *and if that ain't encouraging I dont know what is.*"[40]

With such pressures bearing down, Stimpson decided to disband the Megatherium Club. In fact, the club did not vanish but instead transformed itself into a more formal scientific society. The first meeting of the new Potomac Side Naturalists' Club (hereafter PSNC) took place on January 29, 1858. Stimpson, Kennicott, Cooper, and Hayden, the core Megatheria, were all founders, but they wisely recruited older hands, many of whom worked at the US Patent Office, in order to gain an air of respectability.[41]

One of the elder statesmen of American naturalists was fifty-nine-year-old Titian R. Peale, whose father Charles had established a popular museum in Philadelphia in 1786. Titian assisted Thomas Say in his explorations along the Missouri River in 1819 and later served on the U.S. Ex. Ex. Since 1849 he had worked as an examiner at the Patent Office.[42]

The physician Edward Foreman assisted at the Smithsonian and also worked at the Patent Office. Other founding members included George C. Schaeffer, a chemist and examiner for the Patent Office, and Charles F. Girard.[43] W. R. Smith, a botanist, and William W. Turner, librarian of the Patent Office, did not attend the first meeting but were recognized in spirit as original members. Baird carefully kept his distance from the club and was not one of the originators.

The constitution of the new society was kept simple. A secretary, the only officer, was elected yearly. Meetings were held every Monday night, and each member took a turn as chairman. There would be no permanent meeting place, no collections, and the members hoped, no expenditures. One nay vote was sufficient to exclude a potential applicant. Every member was expected to give a lecture on a scientific topic or to provide a substitute, followed by a discussion and refreshments that often lasted well past midnight.[44]

Unfortunately, we know very little about the early years of the PSNC. Geology, zoology, and the relatively new art of photography were frequently discussed, and average attendance hovered between ten and fifteen. In just over eight years, between 1858 and 1866, they met 146 times, an average of about eighteen times a year.[45] As field naturalists they often met outdoors and had plans to catalog the flora and fauna around Washington. Stimpson found rare shells not far from the city limits, and several of his subsequent publications were based on these collections. He probably wrote the section on conchology for a popular guidebook to Washington.[46]

Reflecting both his leadership role and an adroit social touch, Stimpson proposed at least eight men for membership, more than any other member. Nurturing ties with potential patrons made sense, and his choices illustrate the close links between government expeditions and descriptive natural history. Among them were: Gouverneur K. Warren, an army engineer who produced one of the first detailed maps of the American West and was later a hero at Gettysburg; Albert H. Campbell, an artist on the Pacific Railroad Expedition who later became chief topographical engineer of the Confederate armies; and James M. Gilliss, who played a key role in the establishment of the Naval Observatory.[47]

Stimpson also nominated the physician Nathan Smith Lincoln, a distant relative of Abraham Lincoln. Dr. Lincoln served on the faculty of Columbian College (now George Washington University), also home to the National Medical College. Lincoln may have had a hand in Stimpson's being awarded an honorary doctor of medicine degree from the college in 1860, a ceremony held appropriately enough at the Smithsonian. This is the only college degree that Stimpson ever acquired.[48] He may have been self-conscious about his lack of a college education, and he used the title of "Dr." for the rest of his life.

Aside from Baird, Stimpson's oldest and best Washington friend was the philologist William W. Turner. Born in England, Turner possessed a brilliant intellect, mastering both ancient and modern Oriental languages and others as diverse as Hebrew, Sioux, and Yoruba. A pioneer in the field of comparative linguistics, his death at age forty-nine in November 1859 hit Stimpson hard. He felt his friend had "worked himself into the grave" through the "brain-exhausting labor" of translating Phoenician inscriptions, leading Stimpson to write a touching obituary for him. He privately lamented Turner's departure for a "higher sphere," telling Hayden that "There are few relatives whose loss I would have felt more than his."[49]

The premature passing of a close friend or family member often compels one to look inward, and Stimpson evidently did not like what he saw. Nearly twenty-eight, he told Hayden that he had given up drinking associates. He worried about Turner's sisters, especially Jane, whom he had come to know through her work helping her brother catalog the Smithsonian library's holdings. They had been left with very little money and, eager to help them, he suggested a plan to Hayden. "When you return, I propose that the sisters take a house and we all board quietly with them. I believe the rest will agree, and where there are ladies we can keep out worthless whiskey-drinking and be taken care of. The club can go on in an improved basis." Jane Turner stayed on at the Smithsonian for the next thirty years.[50]

By July of 1860, Stimpson and Gill had moved to a large apartment building or hotel in Washington's Second Ward containing thirty-one people, including six African American servants and several army and navy officers. The building's owner, Zalman Richards, had two years earlier been elected the first president of the National Teacher's Association (now the National Education Association.)[51]

The PSNC meetings constituted a "perfect galaxy of science," welcoming anyone with a genuine interest in the subject.[52] This included foreign-born naturalists and artists, significant at a time when the anti-immigrant Know-Nothing Party had become a force in American politics. Wilhelm Heine, August Schönborn, and Henry and Julius Ulke had all come to America from Germany after the turmoil of 1848. Heine had toiled as artist on Perry's expedition to Japan, where he met Stimpson, and while in Washington he wrote a three-volume work on the NPEE.[53] Schönborn, best known for helping design the celebrated dome of the Capitol, drew beautiful silverpoint drawings of Stimpson's North Pacific crustacea.[54] Henry Ulke had fought with the revolutionaries in the uprising of 1848 when he was wounded and captured. Fleeing to the United States the following year, he soon amassed a large collection of beetles. Along with his brother Julius he ran a photography studio in Washington and they served as the club's unofficial photographers. Henry Ulke is perhaps most famous today for his paintings of political figures, and his portraits are displayed at the National Portrait Gallery and the White House.[55]

Stimpson also enjoyed the company of a Russian diplomat and entomologist, Carl Robert Romanovich Osten Sacken, known to friends as Baron. A St. Petersburg native, he served as secretary to the Russian legation at Washington from 1856 to 1862 and in his free time studied flies and mosquitoes. Despite his foreign birth and amateur status he rightly considered himself the grandfather of American dipterology (the study of flies).[56]

In May and June of 1860, Stimpson and other PSNC members met publicly with a group of doctors from Japan, part of the first official Japanese delegation to the United States. Their arrival triggered a perfect mania for all things Japanese, and Stimpson's first-hand knowledge of Japan probably made him a man in demand at the time.[57]

Stimpson often invited naturalists visiting Washington to stay with him. To Frederic Putnam he wrote, "I am happy to hear that you are coming on & freely repeat my invite—we are ... pretty full just now but we can yet stow you away comfortably ... it will cost you but little as we board at Beguin's where the living is very good indeed & only $3.25 per week and only scientifics at the table, which is conducive. There are no non-conductors among us, but we never get tight until after dinner. ... I think you will have a jolly time with us." Putnam seems to have enjoyed his visit and it is probably not a coincidence that just weeks after returning to Cambridge he helped found the Agassiz Zoological Club, which also decided to hold meetings on Monday nights.[58]

Two Philadelphia naturalists, the ornithologist John Cassin and the entomologist John L. LeConte, paid visits to the club. Kennicott called Cassin "a glorious man ... a little on Prof. Baird's style though not so *perfect* a man in all respects." Others were not as complimentary; one unnamed naturalist referred to Cassin as "a vulgar looking, rough tobacco chewing politician!"[59] His peers considered LeConte to be the best entomologist America had ever produced.[60]

A decidedly less conducive Philadelphian arrived in January 1861. Edward Drinker Cope, a devout Quaker from a well-off family, later became famous for his long-running feud with Othniel Marsh over dinosaur discoveries. Although only twenty at the time, Cope maintained an air of superiority, commenting to his father that Washington was a "decidedly second-rate place ... the habituates of the Smithsonian, though undoubtedly very superior as regards scientific attainments, are not unexceptionable."[61]

About the same time, three of Agassiz's students similarly did not find all to their liking in Washington. Putnam returned in early February 1861, accompanied by Addison E. Verrill and Albert Ordway, having been sent by Agassiz to bring back duplicate specimens for the newly formed Museum of Comparative Zoology (MCZ). Agassiz believed that his would eventually be the leading museum in the country and that it should thus have first pick of the Smithsonian duplicates. Joseph Henry encouraged the distribution of the Smithsonian's holdings, and given the tense political situation just prior to the Civil War it made sense to distribute portions of the collection to other cities. The Cambridge students expected to take back a large haul.

Baird would follow orders and see that specimens were shared, but he would have some say in who got what. He and Agassiz had a frosty relationship at best, and by the time Verrill and the others showed up, Henry Bryant of the Boston Society of Natural History and Cassin of the Academy of Sciences of Philadelphia had taken many of the best duplicate specimens for their respective institutions. Both organizations had been around for decades and possessed superior collections to those in Cambridge. The MCZ did get a fine series of marine invertebrates, especially corals, but Stimpson followed Baird's lead in making sure that they did not get too much.[62]

As always, Stimpson played host to newcomers, showing up at their rooms at the National Hotel before they had even begun unpacking. That night he took them out drinking. Already close to Putnam, he soon developed a similar connection with Verrill. The two had met at the MCZ in the summer of 1860, and Stimpson had come to appreciate Verrill's keen mind and all-consuming interest in marine invertebrates. Over the next six weeks the two men had many a discussion on corals.

Stimpson led the Cambridge trio on a Sunday walk to Georgetown and then down to Rock Creek. At night their amusements were of a different character.

Verrill stayed in one night only to have Stimpson, Hayden, Gill, and Cope drop in on him. After some preliminary drinking ("Hayden became slightly 'tight'") they all headed out to an oyster bar, where "Hayden got worse and all the rest got somewhat lively" much to Verrill's disgust.[63]

On Stimpson's twenty-ninth birthday he gathered with Gill, Verrill, and several others. Someone suggested that they organize "an association of young naturalists," which sounds essentially like a merger of the PSNC with the Agassiz Zoological Club. Apparently nothing came of the suggestion but the idea reflects the camaraderie that existed among the Cambridge and Washington naturalists.[64]

They needed all the encouragement they could get, as naturalists often faced ridicule, with one noting that, "in the public mind such students were regarded as akin to lunatics."[65] During its most fruitful periods the Megatherium Club provided its members with a close-knit support network. Stimpson stood at the center of both the Megatherium Club and the PSNC, and he passionately articulated his views to Hayden on the fellowship of science. "We are both meeting in the Arena of Science with jealous old fogies scowling at us from the high seats." Stimpson and others of his generation wanted nothing less than to attain the front ranks in the "glorious Army of Science." In another letter he summed up the way many of them felt about establishing themselves in the world. "I think you are all right and making your way and mark in the world in a firm and manly manner. You have certainly as good a chance in sensible minds as the papilionaceous [relating to butterflies] squirts of fashionable life who buzz about this place. Keep up a good heart and prove yourself a staunch Megatherium. Stand up to your liquor and face danger when it comes and above all 'never let your evenings amusement be the subject of your mornings reflections' Vid Megather By Laws XI.2."[66]

This macho maxim perfectly encapsulated Stimpson's attitude towards life at the time. The reference to old fogies also reflects the zeitgeist of his era, particularly the Young America movement, which often contained references to the new generation taking over. Ultimately, Stimpson believed that the study of nature was the ideal way to spend one's time in an often confusing and wicked world: "What more noble pursuit for immortal souls? Riches? War and Butchery? Political chicanery? Superstition? Pleasure? What we seek is TRUTH!!"[67] In pursuing scientific truths the members of the Megatherium and Potomac Side Naturalists Clubs were among the most distinguished aggregations of naturalists America has ever produced. Collectively they stand high in the pantheon of American natural history.

CHAPTER 6

FROM NATURAL HISTORY TO ZOOLOGY

> Hurrah for the dredge, with its iron edge,
> And its mystical triangle,
> And its hided net with meshes set,
> Odd fishes to entangle!
> The ship may rove through the waves above,
> Mid scenes exciting wonder;
> But braver sights the dredge delights
> As it roveth the waters under!
>
> (Chorus)
> Then a dredging we will go, wise boys!
> Then a dredging we will go!
> —"The Dredging Song," Edward Forbes[1]

The years preceding the outbreak of the Civil War were arguably the most productive of Stimpson's life. Early in his career he had set himself an ambitious goal, seeking to document all the marine mollusks of the Atlantic coast from Greenland to Georgia. This elaborate monograph, to be published by the Smithsonian, would contain detailed descriptions and illustrations of each species, both of the shell and the animals inhabiting them. Stimpson recorded bathymetrical data and notes on life histories, updated the nomenclature, and made hundreds of drawings, some of which he engraved on wood blocks. He also vowed to include fossil shells, knowing that they needed to be compared with current species in order to establish a more complete understanding of their diversity and distribution over time.[2] Nationalism and ambition alone do not fully explain Stimpson's drive to document marine life. As one historian aptly noted, "an unnamed species pains the tidy mind of a systematist."[3]

A paper published in 1858 marked Stimpson's formal return to the study of mollusks. He described a remarkable new species, a tiny blood-red animal that he had first observed in South Carolina in 1852 living on giant tube worms. He made three detailed drawings of the animal, which was unique enough that he created a new genus for it.[4]

Stimpson had continued to keep a private cabinet of shells and used his duplicates to trade with naturalists and collectors, while duly noting their contributions in his papers. One of the more interesting men that enhanced his research was Samuel Tufts. Simpson had recommended Tufts, a native of Swampscott, Massachusetts, for a position as naturalist on Elisha Kane's 1853 Arctic expedition. Tufts declined the post and instead made a name building, stocking, and selling aquaria.[5]

In seeking to add to the knowledge of American marine mollusks, Stimpson cajoled his well-traveled friends to collect. When Ferdinand Hayden was assigned to Hilton Head, South Carolina, Stimpson urged him to focus on the smaller mollusks that were likely to have been overlooked by previous investigators. "What a magnificent chance you have to collect shells etc. Do pitch in and improve it. *We have nothing from that locality.*"[6]

Stimpson cultivated connections with men from all regions. His relationship with Edmund Ravenel is particularly noteworthy. They had met in 1852 in South Carolina, and Stimpson had never forgotten the older man's kindness on that occasion. Extremely wealthy, thanks to a large plantation and over one hundred slaves, Ravenel possessed one of the finest shell collections in the South. The dearth of scholarly books and specimens for comparison frequently forced the sixty-year-old southerner to send specimens elsewhere for identification. In December 1857, he dispatched several lots of shells to Stimpson, who quickly returned them fully labeled. "It is some years since I ceased investigating the shells of our coast, my attention having since been confined almost entirely to other departments of invertebrates ... but I shall be most happy to render you any aid in my power."[7]

Over the next few years Ravenel sent several additional packages of shells, and in 1858 Stimpson told him that three of his specimens were new and urged him to publish descriptions immediately. Ravenel promptly followed this advice. The two men respected each other, with Ravenel nominating Stimpson for membership in the Elliott Society of Natural History.[8]

Two years later Stimpson updated Ravenel on his Atlantic Coast mollusk book. "I am doing all I can to perfect my collection both in the recent and tertiary shells of the Atlantic coast, which, to be studied with advantage, must be studied together." After identifying and labeling more of Ravenel's shells, Stimpson asked for "a specimen of each, as I am extremely anxious to make my already large collection of American shells as complete as possible."[9]

Another southerner, John D. Kurtz, also helped bring Stimpson back into the world of mollusks. In 1860, Kurtz published a list of shells found on the coasts of North and South Carolina, briefly describing four new species. Other supposed new species were mentioned by name only and credited to Stimpson, who disavowed them. In reviewing the paper Stimpson noted that Kurtz had "contributed very largely to our knowledge of the marine animals of the same coasts [N. and S. Carolina], in other departments."[10]

At the opposite end of the geographic spectrum, Stimpson first met Edward S. Morse in 1858 when he traveled to Portland, Maine. Learning of Morse's interest in shells Stimpson sought him out, and much to his delight realized that Morse had found several new species. Eager to work with such an astute collector, he invited Morse to join him on an upcoming dredging trip to Mount Desert Island, but due to his father's opposition Morse reluctantly had to turn down the "opportunity of a lifetime."[11]

Stimpson also received aid from John Robert Willis of Nova Scotia, who had sent a box of shells to the Smithsonian for identification. Stimpson encouraged Willis by telling him, "you have done a great service to our noble science by discovering several shells ... which had escaped us all, thereby enriching our fauna and extending the geographical range of ... species. ... There! if that starts your enthusiasm as it does mine I have no fears that the conchological fauna of Nova Scotia will not be soon and well worked up."[12] Stimpson wrote countless letters of this sort throughout his career to inspire others to contribute to knowledge of American natural history.

Thanks in part to Stimpson's efforts, by the late 1850s the Smithsonian had become a clearinghouse for information on mollusks. Underscoring his commitment to the institution, in 1859 Stimpson donated his entire type series of mollusks. This included not only the shell but also the animal itself, preserved in alcohol. In many cases these animals were almost entirely unknown, leading Baird to characterize the collections as "one of the most important additions to the museum during the year."[13] Combining these with those from the two great American exploring expeditions and various other sources, the Smithsonian now possessed the best collection of North American marine shells in the world. Stimpson had collected a fair percentage of them, so it is no wonder then that he is considered the founder of the Smithsonian's department of invertebrate zoology.

When the Smithsonian began publishing pamphlets aimed at instructing the public on how to collect and preserve specimens, Baird turned to Stimpson to pen the section on marine invertebrates. This was as close to writing for a general audience as Stimpson ever got but he clearly had plans to write a popular book on seashore investigations. The Smithsonian archives contain some of his potential titles, including "Sea-notes of a Naturalist."[14]

Another event in 1859 helped spark research on mollusks, when Joseph Henry arranged for Philip P. Carpenter to organize and classify the Smithsonian's West Coast shells. The forty-year-old Carpenter, a Presbyterian minister from Warrington, England, was an ardent abolitionist, a vegetarian, and a devout Christian. By turns compassionate and combative, during his five months at the Smithsonian he found much to complain about. He did not like the smell of the strong spirits used to preserve marine animals; the "long dreary passages to move about in;" and Stimpson's smoking in the workrooms. But Carpenter held his greatest disdain for the city and its inhabitants. "Physically, morally, and spiritually, Washington stinks," he fumed.[15]

Stimpson managed to overlook Carpenter's gruff exterior and referred to him, perhaps tellingly, as "about the best living conchologist." The epitome of a closet naturalist, Carpenter focused on the shell alone and never saw the living animals that inhabited them.[16] Yet each man learned from the other, and when Carpenter delivered a series of lectures at the Smithsonian he cited Stimpson's help in several instances.[17] What bound them more than anything was their tireless advocacy for increasing professionalism in science. Referring to the writings of those he derisively called "trading naturalists," Carpenter raged, "If they would do *nothing*, it would be a blessing. *We* have got to *undo* their errors, before we can do our work; or else we increase said errors. . . . There is a *chance* of keeping West Coast shells right, as Stimpson with [the] East Coast; and I ought to do it, as I have studied them more than any one else."[18]

With Stimpson and Carpenter organizing the Smithsonian's mollusks, Joseph Henry saw a golden opportunity for the institution to publish in this field. By 1862 Henry listed a dozen titles in preparation relating to mollusks. In addition to Stimpson's Atlantic Coast monograph other works included Carpenter's introduction to mollusks, a massive bibliography of North American conchology by William G. Binney, and papers by Isaac Lea, Temple Prime, and George W. Tryon Jr.[19]

The Smithsonian also published a checklist of the shells of North America. Stimpson contributed the section on East Coast mollusks from the Arctic Seas to Georgia, listing 514 known species and forty-five doubtful ones. The remainder of the East Coast shells, including those from Florida, the Gulf of Mexico, and the West Indies, were so little known that no one had yet attempted a comprehensive survey, although Stimpson already had plans to do just that.[20]

Daniel Goldstein and others have noted the important role that correspondence networks played in supplying specialists with specimens and information in many fields of natural history.[21] Stimpson's position at the Smithsonian put him at the center of one of the largest such networks in existence, and his career clearly benefited as a result.[22] People from all over the country sent natural history specimens to the Smithsonian, and it was Stimpson's good fortune to be the first to see the marine invertebrates. The same held true for the many specimens transmitted by government expeditions in the 1850s and 1860s, boom years for American exploration.

The constant stream of new material impacted the way Stimpson and other Smithsonian naturalists approached science. Robert E. Kohler and others have noted that the 1830s through the 1850s were a period when a new surge of discovery, much of it in the American West, led to a large increase in collecting. Zoologists were inundated by a massive and ongoing influx of new material from disparate localities, and the sorting, classifying, and preserving of these specimens necessarily took up much of their time. As one historian has noted, nineteenth-century American zoologists were "happily drunk on data."[23]

Stimpson and the other members of the "Baird school" at the Smithsonian had no choice but to spend a great deal of time doing taxonomy—it was practically part of the job description. Complicating matters was the fact that Joseph Henry, who had never wanted the Smithsonian to be saddled with a natural history museum, demanded that Baird's assistants make duplicate sets for distribution to other museums and educational organizations. This meant that in classifying they had to work relatively quickly, make their judgments and then break up some of those long series of specimens that had helped them make sense of their taxa. When you consider the amount of fieldwork that Stimpson and others also engaged in, it is no surprise that most of the Smithsonian zoologists followed Baird in largely eschewing theorizing.

Baird led by example. His Pacific Railroad Reports on mammals and birds illustrate his meticulous attention to detail, and he insisted that the men under his supervision use "uniformity and precision" in their descriptions to make taxonomy "a precise, empirical science." Baird and his followers used tables and measurements to quantify and explain relationships among and between animals. It was a sign of the growing sophistication and professionalism of taxonomic zoology.[24]

Stimpson's study of one mollusk group led him to reflect on one of the leading questions of the day, namely, what constituted a species? To Stimpson, determining the differences between species required a statement of the facts. To do so one needed to accumulate a large series of specimens of all ages and from different localities, followed by a close examination and comparison of these materials. When these facts were placed before a man of science, the opinion of the author mattered little. "What we want is a knowledge of things as they exist in Nature, and not as they appear in the minds of men. By such a course those who believe that species are actualities may be satisfied as well as those who believe them to be ideas."[25]

The "Humboldtian" tradition that dominated natural history in the first half of the nineteenth century focused on documenting all known species.[26] As Francis Bacon had put it, science was "not a belief to be held but a work to be done," a call to action which Stimpson and others readily acted on. It took time before "biology became exalted from empiricism into a science," in the words of Theodore Gill, and Stimpson can be seen as a transitional figure in the era when field-based natural history gave way to zoology and eventually to a laboratory-based academic biology.[27]

Stimpson demonstrated more professional and exacting standards in zoology through his work on mollusks. His descriptions included details on the anatomy of the animals and were characterized by his embrace of the term malacology, instead of the older term conchology. Rafinesque had introduced the term malacology in 1814 and S. Peter Dance's excellent 1986 book *A History of Shell Collecting* noted that by the 1820s there were "clear indications ... that the shell was considered a poor relation of the animal which formed it; and by inference those who did not include the molluscan animal in their studies were inferior to those who did. That, to put it plainly, is how things stand today."[28]

Europeans seem to have been quicker to adopt the study of the animals in refining their classification. William H. Dall credited Sven Lovén of Sweden with inaugurating "an era in malacological science" when he used the radula, also known as a lingual ribbon, as a key morphological feature. The radula is made of chitinous denticles, sometimes referred to as teeth, and is used in feeding.[29] It was no easy task to find and analyze these "teeth," as it entailed the use of a microscope and required a detailed knowledge of invertebrate anatomy. Amateurs, who had long made significant contributions by describing the shells alone, were less likely to pursue this more analytical research.[30]

Some continued to cling to the idea that the shell alone held the key to classifying families, genera, and species, and the question of shell versus animal generated a good deal of controversy.[31] Of course it was not an either-or situation, and Stimpson and others recognized that shell and animal were one organism. While there were a fair number of Europeans who by the 1850s had made anatomical descriptions of mollusks, most North American species had yet to be so described. This became a primary focus of Stimpson's efforts.

Amos Binney had presented a clear case for studying the animal when he wrote that "the characteristics of the whole organized being are needed; and the description of the shell alone, ought never to be admitted, except when that of the animal itself cannot be obtained. The naturalist who has it in his power to acquire a knowledge of the animal, as well as of the shell, should be held in every case to do so, and to make them both known together." While the study of the shell alone would always be useful to geologists and paleontologists, "it must give way to the more philosophical investigation of the mollusks as living beings." To Binney, malacology represented the future, and those who continued to rely only on the shell ("external envelopes") were mere collectors, not zoologists.[32]

Charles B. Adams, who spent a great deal of time studying the anatomy of mollusks, nevertheless took issue with the suggestion that shells were of no more consequence in taxonomy "than the fur of quadrupeds" and suggested that the use of the shell in classification was perhaps even undervalued. Adams noted that amateurs, who loved shells in part for their beauty, should not be condemned for ignoring the animals and concluded that shells alone were sufficient for identification in most cases.[33]

This more democratic vision for the study of shells was shared by George W. Tryon, the leading American advocate of conchology over malacology. Tryon's career provides an example of the blurred lines between professional and amateur. Six years younger than Stimpson, Tryon, a wealthy Philadelphian with a passion for shells, had by 1860 amassed one of the best private collections in the United States.[34] In one book Tryon omitted technical terms or mention of anything that required microscopic observation. There would be more naturalists if things were not so complicated, he complained, noting that many people, "whose leisure or tastes" prevented them from becoming a naturalist, could still become the next generation of collectors.[35] "It has become fashionable lately to disparage the value of the *mere* shells as a means of distinguishing generic and family groups, and to rely wholly on such differences as may be found in the animals. Without denying the great importance which should be properly accorded to the latter, we would insist that, in general, the *expression* of these differences may be observed in the shell. ... The study of malacology is yet in its infancy, and those who figure in it are very apt to give undue importance to the characters on which they rely for building up their systems."[36]

Stimpson agreed with this to a point but insisted that "little dependence can be placed upon the shell alone . . . the entire animal must be examined for the discovery of the most important characters." The generic position of many freshwater snails known by the shell alone "must remain undetermined" until the animals had been scrutinized, he maintained. "The question of their true position is now solved by an examination of their soft parts," he declared of one group of snails after studying them.[37] In 1864 he decried the fact that some Americans had named new gastropods from the shell alone, "without any attempt to acquire a knowledge of the structure of these animals." He implied that a more professional or thorough worker would and should have made the effort.[38]

While Tryon doubted the value of the radula in classification, Stimpson's own empirical observations convinced him that this was a critical diagnostic character. It took skill and perseverance in the lab to obtain the radula. Stimpson put the shell in a test tube with a solution of caustic potash and used a spirit lamp to dissolve the fleshy parts. By early 1864 he was using this technique to prepare and draw several of these "teeth" every day, a process that he found extremely useful in classification. An examination of one species led him to create a new family and genus, both of which are still valid today, a result of Stimpson's laboratory methodology and his conviction that the animal held the key to the accurate classification of mollusks.[39]

In addition to the professional versus amateur divide, the views of Stimpson and Tryon must also be seen as a classic conflict between field versus closet naturalist. Stimpson had made his reputation as a field naturalist while Tryon was largely a collector. As Stimpson and others noted, the closet naturalist found it relatively easy to see well-defined species in his cabinet. The field naturalist, observing animals in their natural habitat, realized that the differences between species were not so distinct, putting the field naturalist in a better position to make these determinations.[40]

Despite their differences the two men engaged in a long and cordial correspondence. In one letter Stimpson wrote frankly of Tryon's work that "some of your statements appear to me quite true; others not so. Of the former I must observe that it was very gratifying to me to find that you had arrived at the same conclusions, from the shells, that I had come to from studies of the animals."[41] Even when Stimpson praised Tryon he could not help tweaking him at the same time. On instituting the new genus named *Tryonia*, he apologized that it had been described "in a very imperfect manner, the characters of the shell alone being given."[42] Tryon was magnanimous in return, later dedicating a book to the memory of Thomas Say, Augustus A. Gould, and Stimpson, referring to the latter two as his "personal friends" to whom he had always turned for advice.[43]

As evidence of the increasing specialization of American natural history, in 1865 Tryon established and edited the *American Journal of Conchology*. (The French Journal de Conchyliolgie began publication in 1850.) The name itself shows Tryon's determination to keep the shells at the forefront, although Tryon himself came to make investigations of the lingual dentition of mollusks, and his journal published articles from followers of both camps. Stimpson published two articles in the inaugural issue. The journal ran for seven years, and Tryon later went on to publish his monumental *Manual of Conchology*, which eventually comprised twenty-eight volumes and enshrined his name in the annals of mollusk research in a way that Stimpson's never was. In 1888

William H. Dall, by then the dean of American malacologists, wrote a history of American research on mollusks, dividing it into three time periods. The last he named in Stimpson's honor, concluding that while his friend did not produce "any epoch-making work, . . . gradually the old methods were discarded for the new," helping accelerate the shift from conchology to malacology.[44]

The Civil War altered Stimpson's plans for the big mollusk book. He had long hoped to have color illustrations of some species but these were prohibitively costly. By 1863, the Smithsonian had begun paying for the engraving of cheaper woodcuts, and for the sake of accuracy Stimpson drew them himself. This took time, and he might have been better off farming them out to someone like Edward Morse, who petitioned for the job, but Stimpson relied on the income.[45] By early 1864 he expected that the first part of the book would be published soon. But wartime inflation had raised the price of paper and printing, and as a result Henry suspended publication of scientific works. He alluded to the "unexpected length to which some of the works were tending," probably referring in part to Stimpson's ever-growing tome. The following year Henry noted that numerous woodcuts had been drawn or engraved for both Stimpson's and Carpenter's books, but added that "no definite period can be fixed for their completion."[46]

His magnum opus delayed, Stimpson narrowed his focus. His next project may have been prompted by a remark from Carpenter, who in his usual barbed way implied that American naturalists lacked the "zeal" to study a group of large carnivorous snails from the Atlantic. Stimpson took up the challenge and wrote an article on them for the *American Journal of Conchology*. He hoped that his study would now "remove the stigma" that had been placed on American naturalists, and jabbing right back at Carpenter he noted that Americans, "unlike their European brethren, are surrounded by such an abundance of new materials, that it is hardly surprising that so much lies uninvestigated at their doors. We cannot do everything at once."[47] In an unpublished note Stimpson was less diplomatic, speaking disparagingly of "little Great Britain which might be stowed away and lost in a corner of one of our western territories."[48]

A Canadian colleague, the geologist John W. Dawson of McGill University, provided the impetus for one of Stimpson's longer papers. An expert on Pleistocene deposits, Dawson had trouble distinguishing different species of the mollusk genus *Buccinum*. Geologists were desperate for more information concerning the geographic and bathymetrical distribution of species to aid them in using fossils for dating.[49] Dawson asked Stimpson to compare the fossil whelks with the shells of recent species, making this largely an exercise in good old-fashioned conchology.

The *Buccinums* are restricted geographically to the frigid seas of the northern hemisphere. Zoologists consider the genus a difficult one because the shells are quite variable and their surface features are prone to erosion. Each of Stimpson's fifteen descriptions noted exactly how a particular species differed from its closest relatives and whether they occurred as fossils. He appended a synoptic table to help in identifying the different whelks, similar to the dichotomous keys used today.[50]

Stimpson's work served as a blueprint for up-and-coming malacologists. In 1865 William H. Dall expressed the desire to do for West Coast mollusks what Stimpson had done for the East Coast forms. Dall wanted "to produce work not like the loose descriptions of Mr. Carpenter, of shells which he has never seen living, or collected in person; but work like Stimpson's on the East coast mollusks, going into every detail of every species thoroughly and carefully."[51]

The war led Stimpson to serendipitous new avenues of inquiry. With the Atlantic seaboard a combat zone he instead intensified his explorations in the District of Columbia. On one of their hikes along the Potomac River he and Meek found a new species of freshwater limpet in a small pond, from a genus hitherto found only in Cuba, Honduras, and California. With Smithsonian funds Stimpson drew three woodcuts, including one of the lingual dentition, and made notes on the life cycle of these minute animals. He envisioned this as the first in a series of ongoing "Malacozoological notices," but this was the only one that ever saw the light of day.[52]

Other freshwater invertebrates fascinated him, and once again a comment by Carpenter prompted a rejoinder. Carpenter had taken "American collectors" to task for their failure to study a group of freshwater snails that European naturalists had embraced. Bristling at the use of the word "collectors," Stimpson agreed that American naturalists had been lax in examining the minute, innocuous animals that inhabited these shells. There had been a few published descriptions but Stimpson argued that these had focused on trivial or unimportant characters. On examining two species from the Potomac, Stimpson showed that while the shells were dissimilar, an examination of the animals showed that they were in fact closely related.[53]

In reviewing the literature on the group he came to question the conclusions of Samuel S. Haldeman, who in the 1840s had published extensively on them, including descriptions of the animals. Stimpson reclassified many of Haldeman's genera and species and at one point implied that Haldeman's work was "unscientific."[54] This aspersion evoked an angry response and Stimpson wasted no time in replying to Haldeman. "As to criticism in scientific matters, I don't think that the idea of personal attack should ever be entertained concerning it," he wrote. He pointed out that he had been more severe in critiquing Theodore Gill's work on these species but this had in no way jeopardized their friendship. Tryon, another whose work Stimpson had attacked, called Stimpson's article, "a most excellent paper."[55]

Stimpson continued to delve further. He kept some of these aquatic snails alive in tanks to observe their movements. They are tiny creatures, most having shells barely more than one-quarter of an inch long. It is difficult to imagine a more unprepossessing group of animals, but to Stimpson their structure, habits, and classification were a mystery that needed solving. This was zoology in its purest form—an in-depth study of a largely unknown group in order to correctly classify them and to observe and describe behaviors and anatomical details that no one had ever recorded. It did not matter that a mere handful of people would ever truly understand or appreciate his labors.

Members of the new family that Stimpson inaugurated (Hydrobiidae) are found around the world. Stimpson dismissed the notion that they could be classified on the basis of the shell alone, in his mind once and for all putting to rest the outdated views of the conchologists.[56] His field observations were what set him apart from closet naturalists such as Carpenter and Gill. Stimpson noted that the ova were deposited in April and May, and he described and figured the ova capsules, commenting on the fact that the species "huddled together in groups according to their practice at this season." He also expounded on the stepping mode of progression of one species, illustrating the behavior with drawings. His discerning eye even noted the presence of tiny parasites that had infested one species, a discovery made only after he extracted the animal from its shell.[57]

The most taxing aspect of the endeavor was dissecting these minute animals. Sitting hunched over a microscope for hours in a museum workroom led to serious eyestrain and headaches. Stimpson received advice from his friend Alfred Mayer on the best methods to prepare and observe objects under the microscope.[58] Over the course of his career Stimpson probably studied these mollusks more closely than anything else he ever did, taking him the better part of two years. The final result spanned fifty-nine pages, including twenty-nine woodcuts that he drew to illustrate the mollusks' soft parts, including the radula and sexual organs. In another example of a quantitative approach to taxonomy, he provided a formula for the teeth of each species. It can be argued that Stimpson's Hydrobiinae paper represents his most important original scientific research.

Joseph Henry published it as part of the Smithsonian's Miscellaneous Collections series and lavished praise on the quality of Stimpson's work. "The results of the investigation," said Henry, "were a more exact definition of the family ... and the extension of it to include other forms previously scattered.

... This memoir not only furnishes an interesting addition to descriptive natural history, but a method of investigation which may be advantageously applied to other families of the class." Other naturalists quickly adopted his nomenclatural innovations, and nearly eighty years after Stimpson's death a worker on this group lauded the paper as a "classic" and "the greatest contribution to the study" of these animals in North America.[59]

The Hydrobiinae paper is Stimpson's most complete "life-history" of a group. There were several elements that distinguished it from the work of conchologists or that of many naturalists. His field observations and collecting constituted a huge commitment of time. Such monographs also required a great deal of library research in order to summarize and in some cases correct the work of those who had come before him. The precise and tedious work of making detailed microscopic dissections of internal morphology increasingly came to separate the amateur from the professional, as the microscope became a key tool in the rise of zoology. People who drew all of these lines of evidence together were a new breed of "scientific" zoologists that emerged in the 1850s, and Stimpson was in the vanguard of the movement in the United States.[60]

For Stimpson the whole point of natural history was embodied in the field experience. He had become a naturalist because he loved observing animals in their natural habitat and relished the challenge of finding and collecting them. Those who accompanied him in the field recalled his contagious enthusiasm and joy.

Since many marine invertebrates never approached close enough to the shore to be left exposed by the tide, some species could be obtained only by use of the naturalist's dredge. Between 1857 and 1864, Stimpson escaped the heat and humidity of Washington every summer for dredging excursions along the coast. Each trip added to the Smithsonian's collection as well as to his personal cabinet of mollusks. As one of his peers commented, Stimpson possessed "a profounder practical knowledge of our marine molluscous fauna, and their bathymetrical and geographical distribution than any other naturalist."[61] This knowledge took the form of meticulous records that he kept of each dredge haul, inspired by the example of the British naturalist Edward Forbes. In 1850, the year Stimpson began dredging in earnest, Forbes published dredging papers based on his work along the British coast, listing the "date, location, depth, distance from shore, nature of the sediments, and depth zone being sampled for each collection."[62] By the mid-1860s Stimpson had compiled similar data for portions of the Atlantic Coast.

In the summer of 1857 Stimpson chartered the ship *Mystery* for a two-week cruise in and around Massachusetts Bay. Despite rough weather he boasted to a correspondent that he had "dredged many new and interesting species of deep sea invertebrates as well as some European species not before known here." He offered his collection to the Smithsonian if Baird would agree to pay half the costs; Stimpson agreed to pay the other half "for the fun I have had."[63]

Unlike his earliest dredging trips, when he often set out alone, he now brought along friends. In 1859, accompanied by Albert Ordway and the physician John Hamilton Slack of Philadelphia (described by Stimpson as "a scientific man of property"), they sailed among the islands of southeastern Massachusetts. While becalmed off Martha's Vineyard ("the Isle of Wight of New England," according to Stimpson), Stimpson noted a pretty blue isopod swimming near the surface that turned out to be new to American shores. His report on these "miniature sailors" appeared in early 1862.[64]

It was a trip to the south that brought perhaps the greatest rewards, as Stimpson and Theodore Gill were the first naturalists to intensively explore the waters off Beaufort, North Carolina. Leaving the Smithsonian in March 1860, they had an eventful journey that included a broken-down train and several missed connections before they boarded a small boat to Beaufort. They arrived in a bedraggled state, disappointed to find a room barely big enough for the two of them. The locals informed them that only a few shells could be found in the area and that collectors immediately picked up what washed ashore.

While reflecting on this state of affairs they began to doubt the wonderful things they had heard about Beaufort from John D. Kurtz. They had come to associate Beaufort with a vision of a naturalist's paradise, "a synonym for everything that naturalist[s] could desire & where accumulations of animals were asking for collectors."[65] The next day they learned that Kurtz had not exaggerated. In the shallow waters of Bogue Sound they thoroughly raked the bottom with dredges. Stimpson remarked on the decidedly tropical character of the fauna, recognizing mollusks that were more typical of the West Indies. Two ocean currents collide here, the Gulf Stream and the colder Labrador Current, resulting in a unique mix of animals that delineated the southernmost range for many temperate species and the northernmost for subtropical species. No wonder then that Beaufort has since become home to several major marine research centers.

They happily rented a second room to store their collection, and Stimpson penned an ecstatic letter to Baird about their finds: "Verily we are luxuriating in clover of Ye most exuberant growth + magnificence. The zoological riches of this place exceed our most sanguine expectations and we have had hard work to keep from collecting on this most beloved Sabbath."[66] The twenty-three-year-old Gill was a relative novice when it came to fieldwork, and Stimpson joked to Baird that "I have as much as I can do to take care of Gill & keep him from falling overboard into the jaws of the Cyprino-donts."[67] Sergeant William Alexander of nearby Fort Macon gave them specimens of the rare and highly prized *Pecten nodosus*, a scallop known as the Lion's Paw.

In addition to the discovery of the *Lingula* (see chapter 4), there were other significant finds during their three-week stay. Carpenter had asked him to search for the mollusk genus *Tellidora*, hitherto known only as a fossil, and Stimpson had indeed found one alive, "as per request."[68] In all he found nine mollusks that had only been known as fossils and described a new genus and species of crustacean, known today as the olivepit porcelain crab. They found enough to keep Stimpson busy for months afterward.

Later that summer Stimpson embarked on a memorable trip to Grand Manan, accompanied by Nathaniel Shaler and Alpheus Hyatt. Shaler recalled a humorous example of Stimpson's rough manner in dealing with the "remote and primitive people" of the region.

> Between Grand Manan and Eastport, while we were pulling in a rough sea as for dear life towards the shore, a fisherman sailing in one of the sharp-sterned, high-pooped schooners of the time, found his curiosity too much for him. So he bore down on us, caught up, and called out, "Some kind of fishing?" "No," said Stimpson, "we ain't fishing"; whereupon the skipper pays off, and sails away a mile. When overcome again, he turns about, runs up to us, lays to awhile in silence, then, "Lost your anchor?" to have for brief answer, "No, haven't lost no anchor." Forth once more, but lest he burst with ignorance, he comes about very near and in a pleading tone calls, "Wall, what be ye doin'?" Stimpson, in his favorite attitude of one foot on either gunwhale, explains. "Don't you see skipper, we are turning Grand Manan over to fill up Eastport Harbor." "Shough," roars the skipper like a blowing whale . . . and sails away. When we arrived in Eastport, we found that our fame as "naturals" had preceded us.[69]

Despite his cheekiness, Stimpson's seamanship engendered respect from even the most grizzled seamen. A contemporary later wrote, "there is scarcely a sea-port from Nova Scotia to Florida where the hardest fishermen do not tell of the daring and skill of the delicate, gentlemanly student who would placidly face the gravest dangers of the sea in pursuit of its inhabitants. His skill as a sailor and his profound knowledge of the ocean did a great deal to give science the respectability among seafaring people that Tyndall and Huxley . . . gave it among the mountaineers of the Alps."[70]

A five-week excursion to Grand Manan in 1861 enabled Stimpson to get many duplicate mollusks, which were highly valued by many of his European correspondents. He also sold shells to both the Smithsonian and the Museum of Comparative Zoology. He got off quite cheaply on the trip, as a wealthy gentleman lent him his yacht and crew, with Stimpson paying only for potatoes, the staple supplement to the day's catch.[71]

At Eastport he ran into Verrill, Hyatt, and Shaler, just returned from a summer-long trip of their own. Shaler recalled the encounter. "Our bibulous friend called for something to drink by way of celebration. We then remembered that we had a dozen bottles of old Jamaica rum ... which we intended to give to Skipper Small. Stimpson found it so good that he declared people who would carry such a potable for three months without touching it had no real title to it. Suiting the action to the word, he pocketed two bottles ... and went away to his den with it; returned to repeat the process, and in six journeys lightened us of the lot. When begged that he would leave at least one bottle he seemed indignant at the suggestion."[72]

A Summer Cruise on the Coast of New England, published in 1864, gives us a vivid glimpse of Stimpson's personality. Its author, Robert Carter, had coedited a literary journal that published the first printing of Edgar Allan Poe's "The Tell-Tale Heart," but he is best known today for suggesting the name for the Republican Party in 1854. Carter first chronicled the voyage as a series of articles for Horace Greeley's *New York Tribune*, where he served as the Washington correspondent and presumably where he met Stimpson. Carter eventually turned these dispatches on the cruise of the sloop *Helen* in July 1858 into a delightful book that has been hailed as a saltwater classic. An immediate success, Carter's only book has remained popular, going through at least five subsequent editions.[73]

One of its major characters is a man dubbed the Professor. Although never identified by name there is no doubt that the character is based on Stimpson, and his vast knowledge of marine life and waggish sense of humor are captured in Carter's muscular prose. Over the years Stimpson's role in the book, like much of his life, was largely forgotten.[74]

CHAPTER 7

LIVELY TIMES AT THE SMITHSONIAN

The Southerners fight like forty thousand D---ls for their blessed "institution" of Slavery—But however plucky they are they are fighting in a bad
cause. . . . It is like cutting out a big tumor from a man—tis a terrible and
bloody operation—but it *must* be done.[1]
—Robert Kennicott (1864)

By January 1861 Stimpson and others affiliated with the Smithsonian,
like nearly all Americans, were uneasy about the state of national affairs.
Abraham Lincoln's election had quickly led to the secession of seven states,
and with southern sympathizers abundant in the capital, Lincoln's inauguration took place in the midst of preparations for war. Two thousand soldiers
fanned out through the city on the eve of the swearing-in, with sharpshooters
stationed atop public buildings. Emblematic of his feelings for both the Union
and the city of Washington, which he considered his true home, Stimpson
served as one of thirteen marshals representing the city in the March 4 inaugural parade. Decked out in blue scarves and each carrying a two-foot-long
baton, they provided a colorful escort for the president's ride to the White
House. Afterwards the marshals were personally introduced to Abraham
Lincoln.[2]

The Civil War would affect the country in countless ways, not the least
of which was the exchange of scientific information. Stimpson's close ties
with southern naturalists had been essential to his efforts to map "American"
marine invertebrates, and he hoped to continue these profitable relationships.
Charleston, South Carolina served as the hub of southern science, led by the
physician Lewis R. Gibbes.[3] In 1850, Gibbes published an important paper
on crustacea collections in American museums. With James D. Dana now
more focused on geology, Stimpson and Gibbes were practically the only two
Americans publishing on the higher crustacea in the years prior to the war.

Stimpson had initiated a correspondence with Gibbes in 1859 by asking
for a dozen fresh shrimp from Charleston Harbor for his study on crustacean

muscular systems. A month later, Stimpson petitioned Gibbes for the loan of a type specimen. Gibbes sent his entire type series, and Stimpson took the liberty of retaining several of the "imperfect specimens," justifying this somewhat impertinent move by telling Gibbes of his forthcoming monograph on the decapods of the East Coast, to be published by the Smithsonian.[4]

Stimpson also kept up with the work of John McCrady, another Charlestonian, praising several of his papers in the *American Journal of Science* as important contributions to the study of the hydromedusae.[5] McCrady reciprocated the favor. Pleased that Stimpson had described the *Lingula* that inhabited the southern coast, McCrady followed up with new details on the habits of the animal.[6] A staunch supporter of southern culture and a member of the Elliott Society of Natural History, McCrady would later lose his scientific library and unpublished manuscripts when Columbia, South Carolina burned in February of 1865.

Edmund Ravenel continued to send shells for Stimpson for identification, the latest batch arriving in June 1860. South Carolina's secession from the Union on December 20, 1860, may have spurred Stimpson to belatedly respond, as four days later he wrote Ravenel a long letter. Apologizing for the delay, he thanked him for the shells and offered to send a series of northern shells. Stimpson referred not to politics but to duplicates from his recent sojourn to Maine.[7] He again encouraged Ravenel to forward descriptions of new species of mollusks, and Ravenel promptly complied, giving Stimpson leave to "alter and correct as freely as may be necessary."[8] Stimpson had been about to describe several of these species, having found them at Beaufort, North Carolina, but yielded to Ravenel, "being only what is due to one for whom I have so much respect, as one of our earliest cultivators of our *beloved science in this country.*" Stimpson made a few minor corrections and forwarded the paper to the Academy of Sciences of Philadelphia, which published it in its *Proceedings* in February 1861.[9]

Stimpson also expressed concern for the type specimens in Ravenel's collection. Sounding a cautionary note, he reminded his southern colleague that many of Thomas Say's types were almost "inaccessible or entirely destroyed," and he hoped the sixty-four-year-old Ravenel would donate them to the Smithsonian where they would "rest securely." Ravenel held on to his types; it was unrealistic to expect that he would give up his most valuable materials in the middle of a worsening sectional crisis, but later events would prove that Stimpson had good reason for concern.[10]

The last letter between the two men, dated April 14, 1861, found Stimpson writing on the day Fort Sumter formally surrendered. He sent Ravenel a rare shell and offered some of his North Carolina duplicates in exchange for recent and fossil shells.[11] The next day Lincoln called for seventy-five thousand volunteers and for the next four years Stimpson and the southern naturalists had no contact. The Civil War shattered the spirit of collaboration that had existed between the naturalists of the North and South.

During these uncertain times Stimpson faced increasingly bleak financial prospects. His salary for work on the North Pacific Expedition had run out on January 1, forcing him to look for paying work. He sorted and classified specimens from two other government expeditions, the Northwest Boundary Commission, and the Pacific Railroad surveys, but he also needed to find new lodgings. In a move he would later come to regret, Joseph Henry allowed Stimpson and others (including Ferdinand Hayden, Theodore Gill, Fielding Meek, and Thomas Egleston) to move into the Smithsonian building just days after the war began. They paid fifteen dollars a month for meals, provided by the wife of Smithsonian janitor William McPeake, and ate their meals in the basement right under the north entrance.[12]

This perk from Henry was truly a godsend. Rooms became increasingly scarce in wartime Washington as thousands of people descended on the capital. Henry's benevolence also aided Stimpson's scientific work, as he could now work more efficiently with his books, notes, and specimens all in one place. While never officially on the Smithsonian's payroll, Stimpson received some funding from the Smithsonian, and he found enough scientific side jobs over the next few years (including his reviews for the *American Journal of Science*) to make a living.

Few were able to think of work in the days following the declaration of war, however. The Smithsonian was issued twelve muskets and 240 rounds of ammunition and barely managed to fend off a proposal to quarter troops in the building.[13] After Sumter fell the city became a fearful, isolated place with rumors of an imminent invasion running rampant. For a few tense days the possibility of attack seemed real, given Washington's proximity to the slave states of Maryland and Virginia. From his small room in the northeast tower of the Smithsonian (known as the Campanile Tower), Stimpson could see the Rebel flag waving defiantly in Alexandria, Virginia.

The excitement and confusion that gripped the city in April of 1861 are evident in a letter Stimpson penned to John W. Dawson of Canada. "We are all in a state of excitement here and it is difficult to find a quiet corner for the pursuit of Science," reported Stimpson. On April 18 Washingtonians learned of Virginia's secession, leading to a call for troops to protect the Union's capital. The next day Baltimore erupted in violence when a Massachusetts regiment en route to Washington was attacked by angry civilians. At least four soldiers were killed and dozens wounded.[14]

When the battered men arrived in Washington, Clara Barton, among others, tended to their wounds. A handful of outraged Massachusetts natives, including Stimpson, were also present as the regiment bedded down for the night in the Capitol building. Appalled at the lack of official aid to the troops, they called a meeting to form the Massachusetts Association, "in order to secure ... proper care for the wounded and disabled, and decent interment of the dead." Among the more prominent members were Stimpson, the physician Nathan S. Lincoln, and the journalist Benjamin Perley Poore. Said to be the first of its kind, the Massachusetts Association organized care for wounded Massachusetts soldiers until the federal government took charge.[15] Stimpson was motivated in part by the fact that his youngest brother, twenty-two-year-old Frank, had enlisted in another Massachusetts regiment on April 2, so it could just as easily have been his brother wounded in the attack.

War or no war, there was still science to be done. Stimpson contemplated another summer dredging excursion despite the presence of Confederate marauders on the coast. Relating his plans to Baird, Stimpson revealed his weakness for bad puns. "We propose to arm the schooner with a gun capable of firing shells, in case of meeting the privateer now off the coast. If we meet them we will make them SEA-STARS (N.B.—joke)."[16]

After this trip to Grand Manan (see chapter 6), Stimpson returned to Cambridge to attend his sister Sarah's wedding in mid-October. He had hoped to leave Cambridge immediately afterwards, but his father was ill and asked him to stay to assist with business matters. While eager to return to the Smithsonian he used the time profitably by making drawings of the crustacea from the Northwest Boundary Commission. Nervously following news of skirmishes in Virginia and Maryland, he heard rumors of an attack on Washington. Uneasy over the possible danger to the Smithsonian, in late September he asked Baird, "Do you think our grumpies [slang for crustacea] in any danger? If you want me write & I'll appear forthwith!" The following day Stimpson employed gallows humor in a missive to Meek, asking, "Is Washington not blown up yet?"[17] When he finally returned he found Baird in a "terribly gloomy" mood.[18] Washington had become even more of an armed camp, and a steady stream of casualties flooded the city after each battle.

If Stimpson thought of becoming a soldier we have no record of it. The pressure on young men to enlist could be intense, with some women shunning and ridiculing those who chose to stay out of the conflict. The struggle to save the Union was the defining event for many of his generation, and tens of thousands of men Stimpson's age (twenty-nine when the war began) fought in the war. Estimates vary but one reputable source asserts that around 57 percent of the white men of military age in the North served in the armed forces.[19]

Few of the naturalists connected to the Smithsonian enlisted, as Stimpson, Kennicott, and Gill all sat out the war. James G. Cooper headed west to practice medicine and work for the California Geological Survey. Others found noncombat positions, Ferdinand Hayden as a volunteer surgeon and John S. Newberry with the civilian Sanitary Commission. Perhaps one reason why Smithsonian naturalists did not serve is that Baird consistently advised against it, and he himself hired a black man to go to war in his place. Baird's pragmatic views were spelled out in a letter to a fellow naturalist: "Don't talk of fighting: there are plenty without you to do this, and we must keep up a few Naturals to keep Darwinism from being forgotten."[20]

Instead of serving as targets on a battlefield, Stimpson and others channeled their patriotism and intellect towards the advancement of American science. America had little hope of keeping pace with Europe if scientific inquiry ceased for the duration of the war. Fielding Meek best epitomized the spirit prevailing among many when he wrote, "while the men of the north and south are engaged in the benevolent occupation of blowing out each other's brains, I will work off all my vengeance on the fossils."[21]

Only a handful of Stimpson's letters mention the war, but one of his reviews for the *American Journal of Science* contained an oblique reference to the conflict. In a March 1862 analysis of Francis Holmes's *Post-Pliocene Fossils of South Carolina*, he referred to South Carolina as having more extensive formations of post-Pliocene fossils "than in any other state in the Union." Thus, Stimpson showed that he still considered South Carolina to be a part of the United States and not a member of the Confederacy.

Despite Agassiz's urging them to stay out of the war, more of the Boston/Cambridge naturalists signed on, including Albert Ordway, who went on to become a brigadier general, and Philip Sidney Coolidge, Stimpson's nemesis on the NPEE, later killed at Chickamauga.[22] Stimpson maintained close ties with the Agassiz students who chose to stay on at the Museum of Comparative Zoology. By now most of them had been with Agassiz for several years and were, like Stimpson before them, seeking more independence. The first of this generation of pupils to go through the painful process of breaking with Agassiz was Edward Morse. Stimpson played a key role in the incident, having earlier in the year encouraged Morse's interest in scientific illustration. Just before Christmas 1861 he reinforced the suggestion by asking Morse to execute a drawing of a coral that he needed immediately. "Don't fail," he urged Morse, but fail Morse did, through no fault of his own.[23]

Agassiz had intercepted Stimpson's letter, although it had been clearly addressed to Morse, and held on to it for a week. When he finally handed it over to Morse the latter became livid and told Agassiz he was leaving. Agassiz responded by refusing to reimburse Morse for a shell collection that Morse had deposited in the museum. Morse wrote bitterly of Agassiz: "as long as one will toady to him and be content to live on nothing, without hope of anything, so long will everything go smooth; but when one asserts his independence then is the man vexed and indignant. ... O, he was so wicked and unjust." These harsh words were followed by ones that encapsulated the dynamic between Agassiz and his students. "I have learnt everything there and feel indebted to Prof for such good as he has done me; but sincerely condemn his actions in other matters."[24]

It was around this time that Agassiz discovered just how much respect he had lost from his students when he learned of a secret group called the "Society for the Protection of American Students from Foreign Professors."[25] Whether Stimpson played a role in this shadowy society is unknown, but given his disputes with Agassiz, his love of scientific clubs, and his close relationship with Putnam and Verrill, he may have had a hand in it.

The students soon decided to focus their growing animosity in a more productive way. In February 1862, Stimpson received a letter from Putnam containing an idea for a proposed book on the natural history of New England. According to Stimpson, they were keeping the project "*very quiet.*"[26] The plans were ambitious, encompassing a four-volume work of about 650 pages each, and they asked Stimpson to coauthor (with Morse) the section on mollusks. Stimpson served as an example of an ex-Agassiz student who had made his break and succeeded on his own terms, and he supported and encouraged their work.

Probably amused by their cheek, he suggested ways to proceed. He offered to write a piece on the marine mollusks as well as a general article on the physical geography of the maritime region. Stimpson advised that a sample issue be sent free to all interested parties, a teaser that would contain "some of the jolliest plates," but cautioned that the work should not be fully started until they had secured enough subscribers to pay the printers and engravers.[27]

Coincidentally, it would seem, Agassiz asked Stimpson to come to Cambridge to organize the MCZ's large crustacea collection. With Ordway's departure Agassiz had no one to oversee one of the largest classes of invertebrates. The animosity from the *Lingula* incident had faded and Stimpson, short on funds, accepted the offer. Before departing Washington he wrote a touching letter to Baird, who had left the city on account of his wife's illness. Stimpson's words captured the feelings of love and respect that many had for Baird: "The wheels of the Smithsonian will not move without you."[28] Arriving in Cambridge barely three months after Morse's stormy departure, Stimpson brought a peace offering in the form of a large collection of the Smithsonian's duplicate starfish, sure to please both Agassiz and his son Alexander.

Stimpson also carried with him materials relating to the proposed book on New England natural history. When Stimpson suggested that the book be dedicated to the memory of Thomas Say, he was surprised by their reaction. "They had been taught to regard him [Say] with contempt—at least so I should judge from their reception of my suggestion." Stimpson, like many others, revered Say as one of the country's greatest naturalists. For all of the talk the proposed book never materialized, hampered by wartime financial stringencies.[29]

As Stimpson began sorting the MCZ crustacea he brazenly took samples of each species not found in the Smithsonian collection, "even where there are no duplicates properly so called, for, as I tell Prof. A., even with this arrangement we shall not get more than a fair return for those we sent to him per Ordway in Feb. 1861."[30]

After two months in Cambridge the increasing tension between Agassiz and his collaborators had grown palpable. An example of Agassiz's insistence on strict control is illustrated in the case of Albert Bickmore, who planned a collecting trip to the West Indies. As many naturalists of the time did, Bickmore decided to solicit subscriptions to help defray the costs. Stimpson gave twenty dollars for a share of the crustacea and asked Baird if the Smithsonian wanted to contribute. Before Baird could reply Stimpson informed him that Bickmore's plans had "come to a violent end. . . . Prof. A. refuses to allow other parties to have anything to do with the specimens beyond making pecuniary contributions for their collection! This is of course *entre nous*. I write to you freely, in a different way from that in which I should write to anyone else." Referring to Agassiz, Stimpson punned, in Latin, that a good pair of subjects for zoological lectures would be a morning session called "De caudae *longitudine* felis nostrae" and an evening session dubbed, "De caudae *brevitate* felis alienae"; that is, on the length of our cat's tail, and on the shortness of someone else's cat's tail.[31]

Tired of the "bigotry and illiberality" at the MCZ, Stimpson escaped the oppressive atmosphere to attend his brother James's wedding in Newton, Massachusetts.[32] Like his older brother, James Stimpson had also followed the call of the sea and had shown great bravery in fighting off a mutiny of "coolies" during a harrowing voyage on the ship *Norway* in 1860. James had recently received a commission in the Union Navy and would soon begin serving as an officer aboard a ship manning the federal blockade.[33]

At the wedding, Herbert Stimpson offered to pay William's expenses for a trip to Europe that summer. Such a pilgrimage served as a rite of passage for many American men of science and would allow Stimpson the opportunity to meet with his European counterparts and study the collections in their museums. On returning to Washington he quickly packed a box of starfish for Christian Lütken and other specimens for Michael Sars.[34]

Herbert Stimpson's motives for the trip were pecuniary. He and James Durrell Greene, former mayor of Cambridge, planned to make and sell rifles worth ninety thousand dollars to the Russian government. The venture required the elder Stimpson to put up twelve thousand dollars, a figure that stands as some indication of his wealth. Somewhat suspicious of the great profits that were promised, William sounded out his friend Baron Osten-Sacken to see if the offer was genuine. Osten-Sacken confirmed his government's interest and Stimpson even paid a visit to the arms factory. Before departing from New York's Astor House in late May he tried to coax Ferdinand Hayden into joining him on the trip, but to no avail.[35]

Unfortunately, very little information has come to light regarding Stimpson's only visit to Europe. They departed on the steamship *City of Baltimore* bound for Liverpool, England. The only known letter written by Stimpson during his four months abroad was to Baird, dated July 1, 1862, from Paris. Stimpson noted that Henri Milne Edwards graciously assisted him in examining the collections at the Jardin des Plantes. Stimpson also met the ornithologist Jules Verreaux, who with his brother Edouard ran a thriving shop selling natural history objects.[36]

Stimpson complained that he had not been able to delve into scientific matters much as he had to serve as a translator for his father, who could not "speak a word of any language other than English." Much to their chagrin they were advised to cancel their trip to St. Petersburg. Alexander II's recent emancipation of the serfs had produced considerable turmoil, and it appears that the proposed sale of rifles was never completed. The Stimpsons also planned on visiting Germany, Switzerland, and Denmark, and there is evidence that they at least reached Germany.[37]

When Herbert Stimpson returned home William decided to stay and engage in fieldwork. For two months he dredged off England's west coast from the Hebrides to South Wales, making substantial collections at locales including Oban, Scotland and Milford Haven, Wales. Off the magnificently rugged coast of northern Scotland he secured several perfect specimens of a new coral as well as a valuable series of jellyfish. He spent heavily to obtain these specimens but considered the costs to be offset by the richness of his finds. In his view the specimens were "almost as good as types, being obtained at original localities."[38] Many of the species were labeled and identified for him by unnamed English naturalists in London, making the catch even more valuable.

On the return trip Stimpson stopped at Halifax, Nova Scotia, where he received a valuable lot of shells from John R. Willis. Arriving in North Cambridge in October 1862, he realized just how sharply the war had accelerated, lamenting to Baird, "[w]hat a terrible state our country is in! I find one near relative and many friends have been killed in the late battles, and see mourning everywhere." A first cousin, Isaac Hall Stimpson, had been mortally wounded at Antietam a few weeks earlier.[39]

After taking a few weeks to separate out duplicates from his European collections he returned to the MCZ to take up the cataloging of the crustacea. During his absence Shaler, Bickmore, and Alpheus Hyatt had gone to fight in the war and the museum seemed "quite deserted."[40] Stimpson's presence in Cambridge gave him a bird's eye view of the deteriorating relationship between Agassiz and his remaining assistants. Tempers flared when Agassiz discovered that they were pursuing the publication of articles without his knowledge. Just as he had done with Stimpson years earlier, Agassiz argued that they were not yet ready to publish. Giving lie to that notion, Verrill had just read a long paper on corals at a meeting of the Boston Society of Natural History. In it, he acknowledged Stimpson's aid in providing drawings and field notes.[41]

Agassiz would soon be mired in a public dispute over scientific property with Henry James Clark, who had come to work for him around the same time as Stimpson in 1850. Both men had been in South Carolina with Agassiz in 1852 but their careers had taken very different trajectories since then. Stimpson preferred fieldwork while Clark embraced the laboratory. Clark had stayed loyal to Agassiz for thirteen years, doing hundreds of drawings and microscopic investigations for Agassiz's *Contributions to the Natural History of the United States*. Clark was in fact one of America's leading microscopists and a fine zoologist, but eventually he too came to feel that Agassiz had taken advantage of him and deprived him of proper credit and remuneration.

By early 1863, just weeks after Verrill's paper had caused a stir, Clark published a major paper on jellyfish that he had not mentioned to Agassiz.[42] Stimpson had provided Clark with specimens, drawings, and unpublished notes, assistance that Clark duly noted as having come from "my friend." By March 1863 Clark and Agassiz had quarreled, and a few months later Clark issued a pamphlet entitled "A Claim for Scientific Property" before leaving for a teaching position in Pennsylvania.[43]

One can imagine Agassiz reading Verrill's and then Clark's paper, his anger growing at the perceived betrayals, while seeing Stimpson's name mentioned as having aided both men. While working in Agassiz's museum Stimpson had been actively supporting Agassiz's assistants in advancing their careers. Stimpson knew that by simply doing what professional naturalists did, namely, sharing information freely with colleagues, he would come into conflict with Agassiz.

Stimpson finally completed his arrangement of the museum's crustacea, and the collections were promptly put on exhibit in the museum. He badly wanted to return to Washington but told one friend that he was "rather hard up pecuniarily" and had to husband the fifty dollars a month he received from Agassiz. The latter had financial troubles of his own and just before Christmas, Stimpson told Baird that Agassiz "has not paid me a cent." In Agassiz's defense, Stimpson knew that his former teacher had spent virtually all of his own money to support the museum.[44]

With money becoming an issue, Stimpson decided to sell off sets of his European collections. He had hoped to keep these "in view of sometime getting an appointment as curator of some museum, for if I could bring with me this and other collections I have made I should stand a better chance of getting such an appointment."[45] Most naturalists, Stimpson included, kept private collections in the hope that these would one day help secure a full-time position. While he had given many of his most important specimens to the Smithsonian, Stimpson still retained a sizable personal cabinet. With very few museum jobs available and competition for openings fierce, the enticement of donating a large private collection could make the difference in getting hired.

Historians have argued that owning a natural history collection "possessed intrinsic social value," and formed a valuable article of scientific commerce.[46] This "scientific capital" was one of the few rewards that naturalists of the day could expect to accrue. Scientific capital did not always put food in one's belly, however, so Stimpson sold several thousand specimens from his European collections to the MCZ. Agassiz paid two hundred dollars (the money came from the Gray Fund, a separate account used for collections) and Stimpson also sold a set to the Smithsonian for one hundred dollars. He kept a third set for himself, his only "private property."[47] To generate more cash he also sold an expensive microscope that he had earlier purchased from Dr. Waldo Burnett, who had used the instrument to make pioneering discoveries in spermatology.[48]

On arriving at the Smithsonian in January 1863, Stimpson was delighted to find several old friends in residence. Kennicott had returned a month earlier from three years of exploring in the Hudson's Bay Territory, and as true friends do he and Stimpson picked up right where they had left off. Kennicott's room in the South Tower was directly below Stimpson's, and with the aid of a trap door in the floor the two men rigged a rope ladder to provide ready access from one room to the other.[49] In addition to the five members of the Henry family, others living in the crowded Castle included Gill, Meek, Edward Foreman, Thomas Egleston, William Rogers Hopkins, Henry W. Elliott, and Edward D. Cope.[50]

Stimpson now presided over a newly rejuvenated Megatherium Club, gleefully announcing its resuscitation to Hayden. "Glad to hear you are coming on here. *We want you.* Make your disposition to stay here, for *we shall keep you*—lock you up if you attempt to leave. *Science* wants *you.* Plenty of sawbones can take your present place in Philad.[elphia] but none can take your place *here.* The Megatherium is revived. Kennicott, God bless him, has come back. Barrel of ale in the cellar. Digestion howls in our den at dinner. Jolly conduction. Advancement of Science ... Friendly sodality. This is *Home.*"[51]

This incarnation of the Megatherium Club differed from the earlier one in several important ways. Two years of war had dimmed some of the gaiety and optimism. They were now all eating, sleeping, and working in the dark and dusty confines of the Smithsonian Castle. With Washington's once lively social scene curbed, they sometimes stayed inside the building for days at a time. The sense of isolation brought them closer together, as did the "*atmosphere of industry*" that prevailed throughout the workrooms.[52]

Kennicott's letters capture the excitement and intellectual ferment that pervaded the Smithsonian. "We live so much with one another here, and have such an inexhaustible lot of strange things to see and think about that we always have our minds occupied."[53] Around eight o'clock each morning they left their various "rat holes," startling early visitors to the museum by suddenly appearing from behind display cases and through little-used doors. To heighten the effect they would sometimes hide behind mummies in the main hall and squawk and groan for the benefit of passersby. A Megatherian would greet one of his fellows with the boisterous salutation, "How! How!" After breakfast they met in Baird's room before scattering to their assorted lairs to spend the day sorting and classifying.[54] To get to their rooms in the Smithsonian towers in the evening, they had to climb a long set of dimly lit, rickety stairs.

While the Megatheria enjoyed living in the Castle, the ethnologist George Gibbs took a dimmer view of their accommodations, referring to the building as an "old rookery." Gibbs breezily opined that the building, "from its situation and construction was the deliberate and intentional trap to kill off the new breed of naturalists. ... Such a rattletrap, with the prohibition of smoking tobacco and without the regular issue of quinine cocktails in the morning and of bourbon and whiskey at noon and bedtime, will ultimately be fatal to all except supernaturals and you none of you have gone further than prenatural-ism."[55] Stimpson in fact fell ill for several weeks in late March and early April but recovered sufficiently to collect mollusks along the Potomac with Gill and Louis Francois de Pourtales, a friend of Agassiz's.[56]

One uncultured visitor gave further details on the extraordinary goings-on at the Smithsonian. Hoping to instill a work ethic in his immature brother, Kennicott arranged in April 1863 for seventeen-year-old Ira (known to his family as Bruno) to assist him for several months. Kennicott lovingly but accurately called Bruno an "ignoramus," but he was quickly adopted as a cub member. In turn Bruno enjoyed the antics of his roommates, writing ungrammatically but feelingly of these stalwarts of American science. "They are the darndest set of fellows here I ever saw—they are Doctors and Professors and all that, and they talk about their books and all that just as if they *was somebody* and after all they are just like a parcel of boys—Why tonight we all ran races and hopped and jumped in the big museum hall because it was too rainy to go out for a walk—Darned if I ever did see such men. I suppose Stimpson and some of 'em are big naturalists but they act like mighty small boys."[57] At thirty-one, Stimpson still retained a healthy measure of youthful exuberance.

Robert Kennicott pinpointed more reasons why his friend was so popular. "Stimpson is one of those kind of men that add a charm to any pursuit with which they are connected. His presence produces a most happy effect here, giving a rosy tint to the atmosphere of science which pervades this building, while his pleasant style of hard work helps one to be industrious."[58] A month before Stimpson arrived in Washington, Kennicott had told a friend that the Megatherium Club was extinct—it revived only with Stimpson's arrival on the scene.

The Megatherium's more formal counterpart, the Potomac Side Naturalists Club, also experienced a rebirth. These meetings countered "the general tendency to depression" engendered by the war.[59] Between the sixty-second and eighty-ninth meetings (probably from late 1861 to late 1862), thirteen new members joined, including Lucius Eugene Chittenden, register of the Treasury Department. Kennicott pegged Chittenden, whose signature adorned Union money, as "a man of considerable scientific tastes and acquirements [and one who] studies shells, ornithology etc.—only as an amateur however—He ... gives the Megatheria a standing invitation to dinner every time they can come and gives us *good* dinners when we get there."[60]

Kennicott described a typical Potomac Side Naturalists Club meeting in February 1863: "I managed to get up a good entertainment, using alcohol bottles for drinking cups and a big shell for a ladle.... It went off capitally.... We invite people to about the poorest dinners they perhaps ever ate with only ale to drink and yet so excellent is the sauce of conduction that our poor dinners in our coal hole like dining room go off capitally and seem very well liked. At first I felt a little ashamed to invite people to dinner, but I don't see but conduction goes farther than fine table and good eatables, etc." While some seemed to think that "Naturalists somehow lived on birdskins and snakes," they were only too happy to demonstrate their capacity for consuming bivalve mollusks and other delicacies.[61]

A glittering array of learned men dropped in on the meetings, especially those held at Baird's home. The list included James Hall, the paleontologist (whom Kennicott described as "a kind of skezeeks. . . . Half his thunder he stole from others"[62]); Louis Agassiz, an increasingly controversial but still mighty figure; and Louis Francois de Pourtales, a colleague of Agassiz's and an expert in microscopic marine animals.[63] As Kennicott aptly pointed out, "[Louis] Agassiz, [John] Torrey, and most of the naturalists I've seen never grow old."[64]

Joseph Henry surely witnessed some of the Megatherium Club's wild antics in the Smithsonian and increasingly felt that his liberality in allowing naturalists to live in the building had been taken advantage of. The arrival of country bumpkin Ira Kennicott proved perhaps to be the final straw. Henry feared that teenagers were simply not responsible enough to share his home, much less that of the national collections, and he worried about the fire hazard from the prolific use of gaslights.

By August 1863 Henry decided that he could no longer tolerate the museum being turned into a "caravanserai," and he wrote Baird that "he had made quite a stir in the Institution" by telling all of the naturalists except for Meek to find lodgings elsewhere.[65] He softened his position regarding Stimpson, telling him he could not return to the building until October 1, when Henry himself expected to be back to keep an eye on things. At the same time Henry cut funding for natural history projects and warned Baird that he expected Smithsonian work to be pursued "more vigorously" than it had in the previous twelve months.[66]

With the end of the Megatherium Club, the Potomac Side Naturalists Club soon collapsed as well. Two of the main figures, Kennicott and Stimpson, departed Washington before the war's end and the Potomac Side Naturalists Club last met in 1866. A few members resuscitated the club in 1873, only to see it fold five years later. A permanent scientific body was finally founded in Washington in 1880, The Biological Society of Washington, which recognized the PSNC as a worthy predecessor.[67]

Stimpson made his customary pilgrimage to the Atlantic shore that summer, journeying to Union-occupied Hog Island on the eastern shore of Virginia. Despite being plagued by mosquitoes he found wildlife in abundance among the placid lagoons and rippling salt marshes, informing Baird, "I was quite successful in obtaining new facts in geog. [raphical]. distr. [ibution]." Stimpson called the locale a positive "paradise of sea-birds" and much to Baird's delight sent him bird embryos and chicks preserved in alcohol.[68]

Soon after returning to Washington, Stimpson learned that his sister Sarah had taken ill. She had given birth to a son the previous October but the baby died three months later from malnutrition. Stimpson rushed to North Cambridge but after a "short but severe illness," Sarah died from a fever at the age of twenty-seven. William had been extremely close to his only sister and he grieved her death. She was laid to rest in the family plot at Mount Auburn Cemetery.[69]

His thoughts turned to his brothers. Frank, now twenty-four and a sturdy six-footer, had realized a life-long dream of becoming a professional soldier when he was accepted into the Seventeenth US Infantry. He saw action at Second Bull Run, Antietam, Fredericksburg, and Gettysburg and received a steady series of battlefield promotions including a citation for "coolness and gallantry" at Gettysburg, where his regiment came under heavy fire near Little Round Top. Twenty-six-year-old James Stimpson served as acting master of the *Alabama* on patrol with the South Atlantic Blockading Squadron.[70]

After Sarah's funeral, Stimpson stayed in Cambridge for a few weeks to comfort his parents. Sarah's companionship had always been Stimpson's favorite part of coming back to Cambridge, and his letters make clear that he preferred Washington to Cambridge. He disliked the "cold society" of Boston and disparaged the "cerulean influences" of its people.[71] Agassiz's influence was powerful, so it is perhaps significant that Stimpson was never elected a member of the Cambridge-based American Academy of Arts and Sciences, as many of his peers were. Stimpson had forged his identity in Washington and felt accepted there in a way that he never did in Cambridge.

Upon learning that Baird intended to visit Wood's Hole, Massachusetts, Stimpson ordered a dredge and sent it on along with detailed instructions on the best dredging locales. "May success attend your researches 'in fundo maris!'" he told Baird. Wood's Hole would later become synonymous with Baird through his work on the US Fish Commission.[72]

His prolonged stay in Cambridge allowed Stimpson to keep abreast of happenings at the MCZ, where relations between Agassiz and his assistants had reached a new low. Putnam and Verrill had been with Agassiz for over four years and were unhappy over not being paid enough, but the main issue was that they felt that Agassiz was retarding their careers by not allowing them to publish without his consent.[73]

On November 19, 1863, the same day on which Abraham Lincoln gave a short but now famous speech consecrating the cemetery at Gettysburg, Agassiz, in a move designed to reassert his authority over his rebellious workers, introduced a series of regulations aimed at stemming mutinous behavior. Regulation 5 codified Agassiz's views on the thorny issue of intellectual property. It stated that any work done in the museum during regular hours was owned by the museum. That seemed fair enough, but Verrill, Putnam, and the others took it to mean that Agassiz could claim credit for *any* work that they had done. Samuel Scudder wrote that the regulations protected the museum from abuses by the students, but the students had no protection from their superiors against any "infringement of propriety." An indignant Verrill called the new rules a "Frenchy, Louis Napoleonic-tyrannic-Papalistic set of regulations."[74]

It did not take long before the rest of the natural history community heard of the row. Albert Ordway, on leave from the army, visited the Smithsonian and recounted the blowup to Stimpson and others. "We have the full sympathy of the scientific men in Boston, Yale, and Washington," boasted Packard.[75] Stimpson immediately wrote to Putnam concerning the by-now-infamous regulations, saying, "I don't wonder you found them unendurable. Meek says they may all be reduced to two:—that each assistant shall keep his nose clean and ask the 'Curator' when he wants to go out!" The time had come for these men to gain their "scientific emancipation," as Agassiz himself put it.[76] By the spring of 1864 Verrill, Putnam, Packard, and Scudder had all parted ways with Agassiz.

In the bitter fallout Agassiz vented some of his anger on Stimpson and Baird. In 1861 Verrill had organized the Smithsonian's corals and Stimpson asked him to write the report on those from the North Pacific Expedition. With Baird's blessing Verrill took them and other Smithsonian corals back to Cambridge for further study. On the eve of his departure in April 1864, Verrill wanted to return them but Agassiz refused to allow it, although the work on the NPEE specimens was done on Verrill's own initiative and not under the auspices of the MCZ. Agassiz did not return the specimens for many years and when he did some were missing and others were damaged, much to the annoyance of Stimpson and Baird.[77]

As he had demonstrated time and again, Agassiz simply could not recognize as equals the people whom he had helped to train. The students were convinced that Agassiz acted as if he were a king and they were serfs, and many came to think of him as the "Great Annihilator." Another historian sympathetic to Agassiz allows that Ernst Haeckel's view of him as "the most ingenious and most active swindler who ever worked in the field of natural history" had some foundation.[78] Twenty-three-year-old William James found Agassiz's childlike enthusiasm for nature endearing but also saw Agassiz's darker side, likening him to a charlatan. "He is doubtless a man of some wonderful mental faculties, but such a politician & so self-seeking and illiberal to others that it sadly diminishes one's respect for him."[79]

The departure of Agassiz's assistants provides an opportunity to compare Agassiz and Baird. They were undoubtedly the greatest advocates of natural history in nineteenth-century America, but conflict between them arose because they had the same goal: to assemble the largest and finest collection of natural history specimens in the country. Each shared a strong commitment to advancing American science, and to some extent each man sacrificed his own research to teach and encourage others. However, the two were polar opposites in personality. Quiet and unassuming, Baird avoided the limelight and disliked speaking even at scientific gatherings, while the flamboyant Agassiz gave frequent public lectures and mixed in society. Baird got on well with nearly everyone who ever worked for him, while Agassiz ultimately drove off his protégés. Agassiz recommended Baird for the job at the Smithsonian, but some of his later letters to Baird are breathtakingly condescending.[80] Agassiz disliked the fact that Baird and his followers devoted much of their energy to taxonomy, an endeavor that he considered inferior to embryology or physiology. Stimpson and others relished the challenges of taxonomy and became very good at it, but Agassiz tried to persuade Joseph Henry that this focus was actually harming American science. Agassiz even attempted to prevent Baird's election to the National Academy of Sciences over the issue.[81]

As one of the few to work closely with both men, Stimpson's insights into these two titans of American natural history are worth noting. When Stimpson finally had the opportunity to run his own museum in Chicago, he directly compared the methods of America's two great museum builders. "Our Academy is carried on on the principle of *liberality*. *No Popery* here. I tell everybody all I know and help them all I can, but don't assume *infallibility* or choke down *rising talent*. . . . Depend upon it, Baird's system is better than Agassiz's."[82]

William H. Dall also worked for both Baird and Agassiz. Writing for public consumption he was more diplomatic than Stimpson. "To him [Agassiz] is due the awakening of a popular interest in Science in the United States through his contagious and genial enthusiasm, and worldwide reputation. To Baird we owe the utilization of opportunities and the preparation of publications through varied agencies, which gave a systematic base to this interest and ensured its perpetuity. Together they supplemented each other."[83] And both men inspired a generation of students, many of whom like Stimpson became museum builders.

CHAPTER 8

SCIENTIFIC SKIRMISHES

> During the continuance of the war we must to expect find that more
> attention is given to the collection of facts than to the deduction from
> them of general principles; the latter must be deferred to a period of more
> tranquility, when the mind is in a better condition for continued applica-
> tion to the development of a single idea.[1]
> —Joseph Henry (1862)

The historian Robert V. Bruce has sketched the broad impact of the Civil
War on American science, writing that the war "distracted scientists from
their work, reduced their income and support, diverted them—sometimes
for good—to other activities, and even killed them." Doing science requires
intense motivation and a laser-like focus, and with over five hundred men
dying on average every day from the four-year war, there were certainly times
when the classification of an obscure group of crustaceans just did not seem
to be a priority.[2]

Science may not have had his full attention, but Stimpson did undertake important original research during this period. Displaying his usual
wide-ranging interests, he wrote on crustacea, mollusks, echinoderms, paleontology, and other topics during the years 1861–1865. Many of these papers
were short notes clarifying arcane points in classification and nomenclature.
This "scientific housekeeping," epitomized by his reviews for the *American
Journal of Science*, was time-consuming, and such contributions attracted the
attention of only the most dedicated and professional zoologists.

We have seen that the war delayed publication of his book on Atlantic coast
mollusks. Stimpson had also been working on a similar treatment of Atlantic
Coast decapod crustaceans for the Smithsonian. When wartime cutbacks at
the Smithsonian delayed publication, Stimpson looked elsewhere for an outlet. He found one in the *Annals of the New York Lyceum*, which published the
first two parts of his "Notes on North American Crustacea" in 1862. These had
been completed in February 1859 and April 1860, respectively. A third part,

also completed by 1860, remained unpublished for another decade. Together these three articles span almost 150 pages and describe 267 species, more than half classified as new. Stimpson included locality data and a full synonymy for each species as well as drawings of twenty-two of the species.[3]

Just as with his study of mollusks, Stimpson relied on a network of collectors to provide him with crustaceans from diverse locales. One man mentioned prominently in the New York articles was Albert Heinrich Riise, a Danish pharmacist living on the island of St. Thomas. Stimpson named several species in his honor, calling him "an indefatigable investigator of West Indian Natural History." He sent copies of his articles to Riise, and his letters stressed that he was under great obligations to Riise's collecting efforts.[4]

Another collector who figured prominently in Stimpson's work on crustaceans was John Xantus, described by one historian as "overbearing, inconsiderate, rude, and insulting." In 1859 Baird had arranged for the Hungarian-born Xantus to be stationed as a tidal observer for the US Coast Survey at Cape St. Lucas, California, and he asked Stimpson to send directions for securing marine invertebrates. Not satisfied with the response, Xantus testily informed Baird that "Mr. Stimpson in fact wrote to me, but his directions were as follows: 'crustacea are found at low water mark under stone, or by digging in the sand.' Those are all his instructions, which remind me of an old Hungarian proverb, 'Here—I give you nothing, but take very good hold of it.'"[5]

Despite his peevishness Xantus proved to be a prodigious collector whose work brought up interesting questions in the geographical distribution of species. Stimpson marveled over the fact that this "new and rich Carcinological fauna differs entirely from that of the Upper California Coast, *not a single species being identical.*" He felt that the character of the fauna more closely resembled that of the western coast of Mexico and some species reminded him of those found at the Galapagos Islands.[6] In the spring of 1860 Xantus warned Baird that his latest haul of ten thousand crabs would pose "an inextricable dilemma" for Stimpson. Baird replied that "Stimpson and Mr. Gill are in raptures over the new fishes and crabs. . . . Stimpson vows that he already has 200 species from you, most of them new, and his chief difficulty is to find names for them!"[7]

Stimpson came to the aid of a fellow carcinologist when he edited a paper by Albert Ordway, who had written about the genus *Callinectes*. Stimpson had instituted this genus three years earlier and one of the species, the blue crab, is known to crab lovers throughout the United States. Ordway's subsequent enlistment delayed publication, and Stimpson found the paper while working at the MCZ. Recognizing its value, he penned a brief introduction and edited the manuscript for publication.[8]

Carcinology remained a focus during the war years. When Johann K. Strahl of Germany proposed a revision of the decapods, Stimpson responded with a lengthy critique. "Some of Dr. Strahl's conclusions are so surprising," he warned, "that they may well require the closest scrutiny before acceptation." Stimpson recognized the limitations of doing taxonomy solely from dried museum specimens, and in his view Strahl had relied too much on such specimens instead of fresh specimens or those preserved in alcohol, which gave a truer picture of an animal's morphology. Thus he implicitly rejected Strahl's conclusions as those of a closet, rather than a field, naturalist. Stimpson's response to Strahl is also significant in that it is his most extended treatment of using homologies in classification, in this case the antennary joints in decapod crustacea.[9]

Stimpson's most important contribution to paleontology appeared in 1863. During an 1859 visit to Gay Head on Martha's Vineyard, he and Ordway had found numerous fossils including shark teeth, crabs, and whale bones. Subsequently Alpheus Hyatt and Nathaniel Shaler brought back additional examples. Most of these Miocene fossils were broken into small fragments, but Stimpson meticulously pieced together the puzzle of a previously unknown crab. His thoroughness and attention to detail is illustrated by the fact that he devoted five full pages to describing its morphology. He postulated that the animal had lived in a warm climate since all of the current members of the family lived in the tropics. When the paper was read at a meeting of the Boston Society of Natural History, another paleontological topic trumped Stimpson's contribution, as Louis Agassiz commented on the newly discovered and now-famous Solenhofen *Archaeopteryx* fossil.[10]

One of Stimpson's more unusual papers had him exposing a scientific hoax. Benjamin Silliman Jr. had asked him to examine the so-called "glass coral" of Japan, which is actually a foot-long, cylindrical sponge. Many museum specimens were encrusted by a conglomeration of marine animals, including sea anemones and shark eggs. Stimpson credited Verrill for recognizing that some of these groupings of "parasitic growths" were perpetrated through the ingenuity of enterprising "Japanese curiosity-mongers," who used silken threads to bundle the animals together. Stimpson conceded that the fakes were well done. "The nicety with which this is done is wonderful, and the deception is perfect. We should judge that the Japanese must have considerable knowledge of the lower animals, to be able to produce factitious congeries, so nearly agreeing with nature and so well calculated to deceive even practiced naturalists."[11]

Stimpson's proficiency with the microscope brought him another small job at the Smithsonian. At Baird's behest he drew eighteen woodcuts illustrating bird's tongues and he also contributed a note on the subject. He discussed the bifid nature of the tongues and the fact that the base of the tongue was armed with backward-pointing spines, which he correctly posited as an adaptation to keep food from escaping. Although far removed from his usual invertebrate researches, Stimpson was among the earliest Americans to study this aspect of avian anatomy.[12]

The Civil War had its greatest negative impact on the publications of government-sponsored expeditions. In the summer of 1861 John Rodgers ordered all work on the NPEE natural history report stopped, as the Union's huge military expenditures (reaching $2 million a day) took precedence. Stimpson's 1,100-page manuscript on the crustacea (accompanied by twenty-five beautiful plates drawn by August Schönborn) had been completed some time since. Stimpson estimated that he needed about four thousand dollars to see the project through, not a huge sum considering the fifteen thousand that had already been expended. In an undated draft of a letter to Ernest R. Knorr, who had served on the expedition as Rodgers's clerk and now headed the North Pacific office in Rodgers's absence, Stimpson pleaded for the funds necessary to finish the report. Most of the work had been completed and nearly 1,800 pages of manuscript and fifty-four plates were in hand.[13] Stimpson claimed that "the scientific discoveries of the Expedition are as great in amount and importance as those of any expedition yet sent out by any nation," but it all came to naught. By the time the war ended the will to publish the results of a decade-old expedition had evaporated. Stimpson never stopped working toward that goal but with each passing year the prospects became less and less likely.

In the midst of the war Stimpson became embroiled in a scientific controversy relating to the expedition. He had recently read several accounts of animals living at three thousand fathoms. Many scientists had grave doubts about finding life at such depths. but Stimpson found the new evidence persuasive, particularly George C. Wallich's account of deep-sea starfish. This led him to reevaluate the diatoms he had examined in 1855 fresh from the sea floor at 2,700 fathoms. In 1863 he now argued that the specimens he had examined "were undoubtedly living, judging from their fresh appearance and the colors of their internal cell contents."[14]

In a short rebuke Wallich stated that the answer as to whether diatoms lived at great depths "may with certainty be answered in the negative." A pugnacious figure in the world of natural history, Wallich flatly stated that the diatoms recovered from such depths never showed a trace of motion.[15] Stimpson had used practically the same words in his North Pacific journal: "No movements . . . nor other signs of life were perceptible," but had chosen to omit them in his published account. (In his NPEE journal Stimpson originally wrote the word *living* and crossed it out in favor of the word *recent*.)[16]

Wallich was correct in positing that diatoms found on the sea bottom generally lived near the surface. However, it has recently been documented that these diatoms can sink very rapidly, and some are still alive when they reach the ocean floor, a fact that might account for Stimpson's observations.[17] Stimpson remembered the sting of Wallich's review for years. "Those English are cursedly captious. Some years ago I published an account of having found diatoms in mud from two miles deep in the ocean—Dr. Wallich pitched into me ... ridiculing the idea. Now the same discovery (!) has been made by another naturalist (English) and credited!" he snorted to Meek.[18]

Stimpson worked on collections from other government-sponsored explorations and as with the NPEE he had to settle for publishing abstracts of his findings rather than full reports. For example, he and Verrill teamed up to write an 84-page report on the marine invertebrates collected by the Northwest Boundary Commission under Archibald Campbell, but only a nine-page summary of Stimpson's findings ever appeared.[19] He also wrote descriptions of animals found on the Pacific Railroad Surveys and from a survey for a ship canal across the Isthmus of Darien in Colombia.[20]

His knowledge of the Arctic seas led to a paper on specimens from a privately funded exploration, Isaac Israel Hayes Arctic expedition of 1860–1861. Hayes brought back more species (sixty-nine) than any other Arctic explorer ever had, and Stimpson published a brief synopsis of the marine invertebrates. They were noteworthy in that they had come from localities nearer the Pole than previous expeditions had reached on the American side of the Arctic Circle. Stimpson described two new mollusks, naming one in honor of Hayes.[21] While Stimpson's publications during these years are an important part of his contributions to science, the war delayed the release of several major manuscripts, including the Atlantic Coast mollusk paper and the North Pacific Exploring Expedition Zoology report. The Civil War ultimately altered the scientific legacy that Stimpson would leave.

Throughout 1864 Stimpson continued to work at the Smithsonian, but Baird found it increasingly difficult to procure funds to pay him. Kennicott hired him to cull Smithsonian duplicates for the newly reorganized Chicago Academy of Sciences, and Baird arranged for him to receive twenty dollars a month for three months from the Jessup Fund of the Philadelphia Academy.[22] By now he was getting a bit old to be living such a hand-to-mouth existence, but as the war dragged into its third year, funding for science dried up. When Agassiz's former students inquired as to whether the Smithsonian might offer employment, Stimpson told Putnam, "I wish I could find places for some of you but the fact is, I am myself adrift now, waiting for something to turn up."[23]

While keeping abreast of the latest skirmishes, military and scientific, he spent much of his time energetically courting one Ann Louise Gordon. He had known Annie for several years dating back to their days picnicking along the Potomac before the war. By late 1863 he had become a regular visitor to her home near Ilchester, Maryland, about twelve miles from Baltimore and roughly forty miles from the Smithsonian. Annie was twenty-eight, which in those days might have led some to consider her something of an old maid, but according to family legend she cut quite a swath in Washington society.[24]

Like the Stimpsons, the Gordons were from a prominent Episcopalian family, but the courtship did not come without strife. Annie's father, James Frisby Gordon, hailed from an old aristocratic Maryland family. A fiery Scotsman and expert horseman, Gordon had once owned a huge estate but seems to have fallen on hard times. Like many in Maryland he was a vocal secessionist, and it quickly became apparent that he had no love for his daughter's ardently pro-Union suitor. Pointedly, Gordon always sent his "negro valet" to inform Stimpson of Union victories but personally delivered news of Southern triumphs.[25]

Undaunted, Stimpson announced their engagement in February 1864. He dashed off a letter to Frederic Putnam, who had also been recently smitten by a young lady, telling him that he had been spending a great deal of time away from Washington, "in very pleasant company. ... From recent experience I believe I could do twice as much work in Science as a married man, as I have done in former years as a careless batchelor [sic]."[26]

Some of his carefree bliss vanished when he learned that an old friend had fallen in the war. In 1864 Captain Joseph P. Couthouy had been shot and killed on the deck of the USS *Chillicothe* off Louisiana. Couthouy had been one of many to mentor Stimpson early in his career and the death also touched the extended Stimpson clan, as one of William's first cousins had married Couthouy's daughter.[27]

Amidst the mounting death toll came an event that must have seemed almost ludicrously incongruous to Stimpson. At age sixty-two his father became a new dad when William's stepmother (forty-one herself) gave birth to Herbert Francis Stimpson. William never mentioned his stepbrother, thirty-two years his junior, in any of his extant letters.

As Stimpson made plans for his wedding he surely hoped that his brothers would be granted leave to attend. The Army of the Potomac had just begun a major offensive in Virginia, however, and Frank Stimpson attained the rank of brevet captain for gallantry and meritorious service during the brutal Battle of the Wilderness on May 5. Not given to exaggeration, U. S. Grant wrote in his memoirs, "More desperate fighting has not been witnessed on this continent." On May 11 he made his famous vow to Lincoln to "fight it out on this line if it takes all summer."[28]

Men like Frank Stimpson were the fodder for that resolve, and Frank's luck finally ran out the next day at Laurel Hill. In a heavy rainstorm Grant ordered Stimpson's old friend General G. K. Warren to send his men on a frontal assault against a well-entrenched enemy. Warren hesitated, as four days earlier a similar attack had been repulsed with heavy casualties. An incensed Grant threatened to relieve Warren of command, so he reluctantly gave the order for a suicidal charge against a position that he knew could not be taken.[29] In the slaughter that followed twenty-five-year-old Frank Stimpson suffered a mortal wound, just one of an estimated seventeen thousand killed, wounded, or missing at this engagement on one of the most murderous days of the war.

There were so many casualties that evacuating the wounded to Washington, sixty miles from the battlefield, took longer than usual. Like countless other maimed Union soldiers Frank arrived at the massive Armory Square Hospital near the Smithsonian on May 27, the same hospital where Walt Whitman spent time caring for the sick and the dying. Frank died the next day, sixteen days after being wounded.[30] It must have been devastating for Stimpson, coming less than a year after losing his sister. His fears for his remaining sibling must have increased exponentially, as James continued to serve in the Union navy intercepting blockade-runners.[31] Two months later another of Stimpson's first cousins was killed at the siege of Petersburg.

After burying Frank in the growing family plot at Mount Auburn, Herbert asked William to remain in Cambridge for a time. The elder Stimpson had also endured a tumultuous nine months, having buried two adult children and fathered a child. William stayed for a week but made no secret of his wish to return to Washington.

By mid-June he had made it back to Washington. He and Kennicott traveled to Ilchester to see Annie, who may not have made a good first impression judging by Kennicott's comment that she "is a very charming young lady who improves upon acquaintance."[32] After a short visit Stimpson saw Kennicott off to the train station, promising to return to Washington as soon as he completed a few last-minute arrangements for the wedding. Once again the war intruded. Jubal Early and a small band of desperate Southerners had pushed into Maryland. By menacing Washington they hoped to divert Union troops from other fronts, and while Early had little hope of taking Washington he intended to sack it if he could.

Hearing that the rebels were reported at Illicott's mills near Ilchester, Kennicott became concerned about Stimpson, who was stuck for several days in Baltimore unable to get to Ilchester or Washington. The Confederates fought their way to Silver Spring, Maryland just six miles from the Union capital before they were finally run off.[33] Relieved that the danger had passed, Kennicott twitted Stimpson when he wrote Baird, "If I were going to get married I'll bet I'd hurry up a little more."[34] Stimpson and Annie were finally wed in Ilchester on Thursday, July 28, 1864, exactly two months after his brother's death. The newlyweds left immediately for Philadelphia where they spent two days at the mammoth Continental Hotel, one of the first American buildings with a "vertical railway," or elevator.[35]

They continued on to Somers Point, New Jersey, where Stimpson was set on doing fieldwork during his honeymoon. Fifty years earlier Thomas Say had roamed these same beaches and described many new invertebrates.[36] Over the years his collections had been poorly curated and many specimens had disintegrated. Stimpson planned to make extensive collections in and around Great Egg Harbor, and by comparing them with Say's written descriptions he hoped to redeem some of Say's lost types.[37]

Social creature that he was, Stimpson, presumably with his wife's consent, invited several people to join them. Kennicott and his sister Alice stayed for a week and Stimpson also extended an invitation to the somewhat reclusive Meek. He and Stimpson were very close but the sensitive Meek felt he was losing one of his best friends. Nonsense, countered Stimpson. "My wife says that you must not consider that you have lost one friend, but that you have gained another."[38] Marriage clearly brought a measure of serenity to Stimpson. While ensconced on the beach for the start of the second week of their stay, he gushed to Meek that, "I am as happy now as man can be, or I think, *as Man ever was.*" Three months later he told Putnam, "Marriage is the jolliest thing in the world." It was probably around this time that he learned Annie was pregnant with a honeymoon baby.[39] Unfortunately, we know very little about their relationship. Annie would bear his children but it is unknown to what extent she played a role in furthering his career.[40]

When his collecting equipment arrived, Stimpson commenced dredging, giving the landlubber Kennicott his first lesson in seaside natural history. Kennicott confided to Stimpson that he had been offered a position as naturalist on an expedition to Russian America. Having just been appointed director of the Chicago Academy of Sciences, Kennicott sounded out Stimpson about the possibility of coming to Chicago to oversee the Academy museum in his absence.

Kennicott had long hoped to get Stimpson in Chicago. In a remarkable letter to his family written a year earlier, he envisioned a day when he and Stimpson might work together at the Academy. "I have some slight hopes of seeing Stim[pson] at work in Chicago at some future time when we get a little start in science there—In fact once we get such a man connected with any scientific association in Chicago he alone would put it on a respectable basis among the scientific institutions of the world. . . . If I could only have him with me in Chicago we would grease the wheels of the Academy or some other Natural History institution, that would get up a momentum that would run it well up to the top of American zoology in a short time."[41]

Bold words, but Kennicott's instincts were sound and his vision proved to be prophetic. The scientific expertise of Kennicott and of Stimpson complemented each other; Stimpson knew marine invertebrates whereas Kennicott specialized in vertebrates, particularly mammals, reptiles, and birds. He realized that having a man of Stimpson's standing in Chicago would be highly advantageous to the Academy, and that it was essential that at least one person connected with the Academy be solely interested in the advancement of the natural history museum.

As Kennicott pondered whether to embark on what would become known as the Russian American Telegraph Expedition, he felt awkward about the possibility of being Stimpson's boss, even in absentia, anxiously asking Baird "would Stym swallow a position second to me? Faith it will look funny for a chap like me to have an assistant like him. Guess "I'll be director or what not and he curator."[42] While his scientific reputation far outstripped Kennicott's, Stimpson was not one to obsess over titles. He carefully weighed the pros and cons of leaving the East Coast. Annie's pregnancy complicated matters, as did her family's opposition to her moving halfway across the country. But opportunities for full-time employment in his field, even temporary employment, were about as common as two-headed snakes. With research curtailed at the Smithsonian his options were limited. The responsibilities of being a husband and soon a father necessitated that he find a steadier source of income. By taking the job he could help out a friend, acquire some firsthand experience in running a museum, and get a chance to visit the West and see for himself one of the fastest-growing cities in the world.

While in Cambridge for Christmas, Stimpson received a formal offer to serve for one year as the curator of the Academy at a modest salary of $1,500 a year plus travel expenses. On accepting, he wrote the trustees, "I beg leave to add the assurance of my appreciation of the importance of the undertaking, and of my strong desire to advance the interest of the Academy in this direction, as far as will lie in my power." In order to mollify Annie's family, she and William decided that for the time being she would stay in Ilchester. Resigning himself that he would be separated from his new bride, he took solace in the fact that he would be able to visit her often during the coming year and promised to return before her due date in April. Early in 1865 Stimpson packed his bags and boarded a train for Chicago, where he would begin a new and very different chapter of his life.[43]

CHAPTER 9

WILD CHICAGO

American life blossomed more naturally in Chicago; enormous factories, endless streets, fantastic stores. ... At first sight [it] seemed abominable; upon further reflection I recognized that it was admirable beyond all words. ... I wouldn't want to live there for anything on earth; but I believe those who ignore it do not completely understand our century, of which it is the ultimate expression.[1]
—Giuseppe Giacosa

Chicago's transformation in less than forty years from sleepy frontier outpost to bustling crossroads of the nation stands as one of the most compelling stories in US history. Established as a city in 1837 with barely four thousand people, like an awkward adolescent Chicago experienced a massive growth spurt in the years immediately following the Civil War, with the population nearly doubling in six years to over three hundred thousand. Stimpson had witnessed the sudden expansion of Washington's population during the war, but it paled in comparison to the feverish development that characterized booming Chicago.[2]

It was not a place for the faint of heart. The helter-skelter pace was both energizing and exhausting. One prominent clergyman summed it up when he wrote: "The intense go-ahead activity that distinguishes us here is the activity of young men." A visiting New Englander recognized the mercurial nature of the city. "It seems like a place where the unfortunate financiers of older cities have struck out boldly, risked almost madly, and won."[3] People were consumed with making money and were not too particular how they got it. A later resident summed it up best: Hustlertown.

From the beginning Chicago had been locked in a struggle with St. Louis for economic dominance of the Midwest. By the end of the Civil War, Chicago had emerged victorious. The *Missouri Republican* cast a critical and perceptive eye towards its rival, stating that, "wherever anything is to be done for the good of Chicago, somebody is found to do it. ... Keen, sharp-sighted, and long-sighted, quick and bold to the verge of audacity, persistent and, the censorious say, unscrupulous, they rush on, rejecting doubts and conquering difficulties, to triumphant success and prosperity."[4]

During the long train ride Stimpson had time to review what Kennicott had told him about the organization he would now be overseeing. Like many other nineteenth-century academies of science, the Chicago Academy had struggled to find members and adequate funding. Kennicott had helped found it in 1857 and amassed a large collection, including many specimens from the American subarctic.[5] He dreamt of establishing a "young Smithsonian" in the West, but the idea seemed a far-off dream until Louis Agassiz's visit to Chicago in February 1864. Kennicott schemed to use Agassiz's charisma and influence to help raise money for a Chicago natural history museum, confident that Agassiz's support would result in twenty-five thousand dollars being raised in three days. While he had lost the respect of many of his fellow naturalists, to the public at large Agassiz continued to be "the very embodiment of Natural History itself," according to Kennicott, who later told Baird, "You can have no idea of his power among the populace."[6]

After some initial hesitation Agassiz agreed to give his blessing to the Academy. Speaking to a group of civic-minded Chicagoans and using all of his considerable powers of persuasion, he told them that Kennicott's collections were "of great value" and that they would provide a foundation for a magnificent museum.[7] The testimonial had an immediate impact and in a few weeks 120 men had pledged five hundred dollars each to become life members of the Academy. Without Agassiz's intervention the Chicago Academy might never have become a major force among American scientific societies after the Civil War. The irony that Agassiz had made possible his temporary job in Chicago was certainly not lost on Stimpson.

Thanks to the infusion of cash the Academy had a solid economic foundation, but there were still major obstacles to overcome to create public interest in, and support for, a natural history museum in Chicago. Nearly half the population was foreign-born, led by Germans, Irish, English, and Scandinavians. Like most immigrants they arrived at the bottom of the economic ladder, and the day-to-day pressures of making a living and adapting to a new culture did not allow for more exalted pursuits. For most, their closest encounter with nature came through the carcasses of prairie chickens, passenger pigeons, deer, and fish for sale at the local markets.

Certainly no one mistook the place for an intellectual mecca. "Not all the gophers in Illinois would tempt me to dwell in the famed West, even were I as desperate a naturalist as you," wrote one of Baird's correspondents.[8] Kennicott had long been frustrated in his attempts to stir up interest in natural history in

Chicago. A few years earlier he had penned a plaintive letter to Baird. "What will you do to help us let these 'benighted' skunk city folk know that there are some other things of interest in creation than corner lots and railroad stocks ... other ways of worshipping God beside cheating each other all week and going to (sleep in) church on Sunday. ... I don't mean to say that the Chicago people are all ignorant of natural science and its results, but as a class they care as little and know as little of such things as any folk I ever heard of that called themselves civilized and enlightened."[9]

Stimpson thus had no illusions about the difficulties that lay ahead when he arrived in cold and snowy Chicago in January 1865. His presence came as a great relief to Kennicott, who was frantically trying to put the museum in order before heading north. Huffing their way up three flights of stairs, he gave Stimpson a tour of the Academy's rented rooms in the Metropolitan Block at LaSalle and Randolph Streets. Kennicott apologized for failing to do more than complete the display of birds and eggs, but Stimpson expressed amazement at what had been accomplished. "I don't think a year's work in any other museum ever showed as well," he told a fellow naturalist.[10]

Just a few days after Stimpson's arrival he was stunned to learn via telegraph that a fire had ripped through the Smithsonian building. The first accounts that reached Chicago were grim, with the *Chicago Tribune* lamenting the loss of the collections and referring to the Smithsonian in the past tense.[11] He and Kennicott quickly wired Baird for details and without waiting for a reply, Stimpson caught the next train back to Washington.

Fearing the worst, he arrived as the ashes were still cooling, but to his immense relief he found that the collections had been largely spared. However, Stimpson's workroom in the northeast tower had been damaged, and he lost heavily in unpublished manuscripts, books, and personal effects.[12] Disappointment turned to rage when he learned that greed had played a role in his losses. Either during or immediately after the fire, unknown parties broke into his wardrobe, trunk, and bureau and stole the contents, also helping themselves to some of his whiskey. Unbeknownst to them the bottle contained a mixture used to preserve specimens. Joseph Henry related that the thieves were repaid for this transgression, "by the effect of the sulfate of copper which had been dissolved in it several of them became deadly sick and would have died had not they vomited freely." After assisting on the scene for several days, Stimpson fell ill and spent time recuperating at Baird's.[13]

He made it back to Chicago in time to see Kennicott off, and a week later he attended his first of the Academy's monthly meetings. Underscoring the utilitarian nature of Chicago science, the biggest stir came when minerals from Chicago's lake tunnel project were displayed. This ambitious undertaking was designed to provide a source of contaminant-free drinking water to the expanding city, and it was the engineering marvel of its day.[14]

In reading the newspaper accounts of the meeting the next day, Stimpson received a reminder that science would be a hard sell in Chicago. At the same time as the Academy's meeting a much larger audience had gathered in another part of the Metropolitan Block building to listen to a faith healer. The *Chicago Times* spluttered that "Ignorance and credulity have been rampant" and expressed shock that so many "grossly ignorant and easily deceived" people had attended.[15]

Undaunted, Stimpson fully immersed himself in the day-to-day routine of operating a natural history museum, spending most of his time sorting, organizing, and preserving the collections. He had been doing such work on marine invertebrates his entire adult life, but he now faced the challenge of caring for all the myriad collections—birds, mammals, fossils, plants, insects, fish, reptiles, and more. Something as basic as arranging and labeling a collection of mounted birds for exhibit proved to be more enjoyable than he had expected. Like all true naturalists the more he studied any aspect of nature the more his curiosity was piqued. In his breezy way he informed Baird, the dean of American ornithologists, that he had become "quite ornithological. Birds are, after all, kind o' interesting." One batch of Arctic birds arrived in a "monstrously buggy" condition, infested with dermestid beetles. These insects are the bane of natural history museums since they feast on the dead and can quickly reduce a bird skin or insect to a fine powder. Stimpson learned how to bake birds in the oven to kill these pests.[16]

Stimpson was enthusiastic about being a part of Kennicott's grand enterprise, and his letters to eastern naturalists were optimistic. He bragged to Putnam, "we are progressing in a *wholesale way*," boasting that Chicago would soon possess the best fish, crab, insect, and mammal collections in the country. To Baird he wrote, "There is a most excellent prospect ahead. There will be no want of money, for the trustees are most liberal, and have the still greater virtue of thorough confidence in Scientific men, with whose operations they will not interfere except to furnish the funds!"[17]

A primary goal was to gather as large a collection for display as possible, from bison to bird's eggs and mastodons to prairie flowers, in order to spark an appreciation for nature in city dwellers. Stimpson implored Baird to send "everything large and striking" that the Smithsonian could spare. Realizing that living animals would further attract interest, he got Baird to loan him the Smithsonian's aquarium in order to display several Hellbenders, large aquatic salamanders that have been described as looking "more like bad dreams than live animals."[18]

Aid from the Smithsonian would be crucial to the Academy's success. While Baird and Henry assisted many other educational organizations, they had a special alliance with the Chicago Academy, thanks in part to their personal ties to Kennicott and Stimpson. In his annual report for 1864, Henry reported that Smithsonian collections had helped provide a foundation for the new Chicago museum. At the same time he alluded to the positive impact of natural history studies in shaping society when he wrote that the Academy "is diffusing a taste for the study of nature in that city of unparalleled growth, which cannot be otherwise than highly salutary in ameliorating the sensual effects of great material prosperity."[19]

Henry had put his finger on one of the most salient aspects of Chicago in the 1860s. Many considered it to be the most wicked city in America, and there is plenty of evidence to support such a conclusion. The city's reputation for sin went hand-in-hand with an unrelenting focus on wealth and materialism.[20] Museums like the Academy's were an appeal to the intellect and were meant to instill moral improvement.

Stimpson often turned to Baird when he needed advice. Apologizing for his impertinence, he inundated Baird with requests for information and countless favors large and small whether it be egg drills, blowpipes, fine mesh nets, dredges, specimen jars, or parchment labels. Interested amateurs who might be willing to collect specimens for the Academy needed the proper guidance and tools, so Smithsonian publications on how to collect and preserve specimens were also in demand. The cultivation of local collectors would be crucial to the long-term success of the museum, but making and maintaining these contacts took a great deal of Stimpson's time. Baird sent nearly everything Stimpson requested and a good deal more besides. Stimpson tried to repay his old boss in birds, the currency that he cherished most, by imploring Chicago sportsmen to obtain species that Baird wanted, including requests for warblers and cedar waxwings.

Increasingly preoccupied during his hectic first spring in Chicago, Stimpson made ready to travel east to be with Annie as she neared the end of her pregnancy. He planned on leaving on April 13, which would have put him in Washington the next day, Good Friday. The assassination of Abraham Lincoln that night stunned Americans and plunged the North into grief just days after Lee's surrender at Appomattox. Stimpson no doubt heard about the tragedy first-hand from Henry and Julius Ulke, who lived at the Petersen boarding house where Lincoln died and took the famous photo of Lincoln's deathbed.

Stimpson arrived in Ilchester in time for the birth of his son, William Gordon Stimpson, born on April 24. He stayed with his wife and child for several weeks, getting his first taste of the joys and trials of fatherhood. Spencer and Mary Baird agreed to be godparents to Willie, a responsibility they would take seriously.[21] Reluctant to leave his new family, Stimpson consoled Annie by telling her that he would be back again in two months.

While increasing the size and scope of the museum's collections constituted his main priority, there were numerous other tasks to attend to as he began his second stay in Chicago. Kennicott had asked him to boost the holdings of the Academy library, as budding naturalists in Chicago were badly in need of scientific books and periodicals. Stimpson found it to be "in a great state of confusion" and spent hours making an inventory.[22] While he would donate countless specimens to the Academy, Stimpson did not contribute much in the way of books. Much of his personal library, including many rare and valuable works, had been destroyed in the fire at the Smithsonian. Commenting on his wandering ways he told one friend that "I am knocked about through the world so much that a private library is of little use to me."[23]

As the public face of the Academy, Stimpson encouraged individuals and groups that might provide money or other assistance. His gregarious personality helped him cultivate ties with the Academy's Board of Trustees, the benefactors who paid for his salary and virtually all of the museum operations. Most were motivated more by a deep sense of civic pride than a true devotion to science, but they provided funding that allowed the Academy to become the first midwestern academy of science to hire a full-time naturalist to lead it.[24]

During Stimpson's tenure in Chicago the only other major scientific enterprise in the city was the Chicago Astronomical Society. Founded in 1862, it had privately raised funds to purchase the largest telescope lens in the United States. Academy trustee J. Y. Scammon paid for the construction of the Dearborn Observatory to house it and personally paid the salary of astronomer Truman H. Safford. Just as financial support for the Academy came overwhelmingly from a few individuals, the study of astronomy in Chicago depended largely on Scammon. Beginning in 1867 Safford gave several lectures at the Academy, and he and Stimpson were probably the only full-time men of science doing original research in Chicago in the late 1860s.[25]

The nine trustees were a varied bunch, including lawyers, bank owners, manufacturers, merchants, and members of the Board of Trade. Many had been born in New York or New England, and the men who took the most active roles in the Academy were quite young. George C. Walker was only thirty; Eliphalet W. Blatchford, thirty-nine; Ezra B. McCagg, forty; William E. Doggett, forty-five; and Jonathan Young Scammon, fifty-three. At the height of their powers, these men exemplified the best in cultural philanthropy in Chicago.[26] While Chicago had shown the world how much could be accomplished in a short period of time in trade and commerce, these men realized that the city needed to establish the foundations of culture as well.

While he enjoyed getting to know the trustees, Stimpson felt like something of an outsider in Chicago. He took his meals at a downtown boarding house, but eating seemed joyless without his usual cast of friends. Having grown accustomed to dining with the many brilliant minds gathered at the Smithsonian, he had perhaps not fully understood just how much Chicago was an intellectual backwater. Isolated from the scientific community and from the books, specimens, and field of operations that fueled his career, he keenly felt a lack of companionship, scientific and otherwise. He confided to Meek, "it is confoundly lonesome here in this big building without a soul to speak to in the long evenings." He missed the nightly "hygienic" walk to the Military Hall for a glass of ale with friends and his regular disputes with Gill over issues of priority. Instead he drowned his sorrows with alcohol, having discovered that they served "the best ale in the world for only 5 c. per glass" in the basement of the Metropolitan Block.[27]

He eagerly looked forward to getting mail, hungry for the latest scientific news. When he learned that Meek had turned down an invitation to accompany Agassiz on a trip down the Amazon River he wrote, "I am right glad, my dear friend, that you did not go on that crazy expedition of Agassiz.' And although I differ so much from that man in scientific politics, as it might be called, I should be very sorry to see any harm come to him, and I must say I fear for his safety."[28]

Through Meek, Stimpson met one of the few scientific men working in Illinois, Amos Henry Worthen. Nearly two decades older than Stimpson, Worthen came to science relatively late, not publishing his first paper until age forty-four. As Illinois's state geologist he was a regular visitor to Chicago while editing the *Reports of the Geological Survey of Illinois*. Worthen utilized the services of the newly established Western Engraving Company, a company that happened to be located right next door to the Academy, to produce illustrations of invertebrate fossils. Stimpson and Worthen hoped that between them they could keep an engraver busy full-time.[29]

Meek and Worthen were frequent collaborators, which gave Meek an excuse to travel to Chicago to supervise the engravings for an article. A delighted Stimpson promised his friend a "jolly" time and also inveigled him into writing a long article on fossils for one of the Academy's publications.[30] Meek called Chicago a "*live place*" and expressed admiration for the trustees. "They talk seriously of building a fireproof building before long and they are not the kind of men who *talk* very long about a thing before doing it."[31]

With Meek safely ensconced in the Academy's rooms, Stimpson returned to Ilchester to celebrate his first wedding anniversary. Back at an "infernally hot" Washington he packed more Smithsonian duplicates for the Academy. He kept Meek apprised of the freshest dirt on the personal lives of their friends, getting in friendly gibes at Theodore Gill and Ferdinand Hayden, of the latter writing "he is to be married to a Virginia Virgin ... if he don't back out in the interim."[32] Stimpson's hunch was correct as Hayden did not marry for another six years. While he and Hayden had once been extremely close, Stimpson had come to see him as a shameless self-promoter who did not give Meek enough credit in their collaborations.[33]

In early August, Stimpson turned up in North Cambridge for the first time in nearly two years to pick out shells for the Academy from his private cabinet. While Stimpson renewed family ties, Meek learned just what a "live place" Chicago could be. As he slept one night in the Academy's rooms "some rascal without the fear of the law before his eyes... stole my pocketbook, containing over $200 in paper, and a buckskin purse, containing about one hundred dollars in gold, also my gold watch."[34] Notoriously hard of hearing, Meek suspected the Academy's janitor. Mortified when word of the crime reached him, Stimpson felt responsible. Having been victimized by thieves himself, his anger boiled over and he promised, "after this we will keep fire-arms ready, and any thief trying to enter will have a bloody recompense. I have lost heavily to thieves myself and would give half I am worth for the privilege of killing one, in the act."[35] No one was ever caught and Stimpson asked the trustees to make good Meek's losses.

Back in the fall for his fourth visit to Chicago in 1865, Stimpson took stock of the Academy's progress. His inventory illustrates the extent of the Smithsonian's aid to the fledgling Academy. Of the 812 mammals in the collection, 622 came from the Smithsonian. In addition they contributed nearly 2,500 birds, 800 sets of eggs, 1,100 crustacea, 1,500 mollusks, 600 fish, and over 3,000 fossils for a grand total of over 10,000 specimens.[36] Joseph Henry was delighted to see so many specimens *leaving* the Smithsonian for a change. Another important service rendered by the Smithsonian was as a clearinghouse for sending and receiving packages from Europe, thus expediting the exchange of scientific information.[37] While the Smithsonian helped many individuals and organizations, its support of the Chicago Academy was unprecedented.[38]

The core of the Academy's collection came from Robert Kennicott, who had given some 7,500 specimens, mainly birds, insects, and mollusks. The rest of the Kennicott family contributed another 2,300 specimens, many from Robert's sixteen-year-old brother Flint. Stimpson chipped in with nearly 3,000 animals that he and Kennicott had gathered at Great Egg Harbor in New Jersey as well as an assortment of other marine invertebrates totaling 5,600 specimens. Nearly two-thirds of the museum's 39,000 specimens in October 1865 had come from the Smithsonian, various members of the Kennicott family, and Stimpson; one-quarter were from the Smithsonian alone.[39]

Published in late 1865, Stimpson's report on the museum led to a favorable notice in the *American Journal of Science*. Expressing optimism regarding the advancement of science in America's hinterland, the editors wrote, "The establishment and liberal support of such institutions as this and the Chicago Astronomical Observatory are facts that speak well for the intelligence of the citizens of Chicago. They have already shown us that they possess the energy, the capital, and the skill to build up a great and flourishing city in an astonishingly brief space of time, and now they seem to be equally in earnest when they speak of making their city 'the scientific as it is already the commercial center of the west.'"[40]

Prior to the Academy's ascendance, the most prominent scientific organization in the Midwest had been the Academy of Science of St. Louis, established in 1856. Led by the botanist George Engelmann and the paleontologist Benjamin F. Shumard, the St. Louis Academy did not emphasize building a large natural history collection and instead focused on the creation of a library and a publication series. Contemporary accounts suggest that the Smithsonian originally meant to give surplus specimens to the St. Louis Academy, but the Civil War shelved that plan and "the subject was utterly neglected" in St. Louis. By 1864 the Chicago Academy instead had become the favored beneficiary of Smithsonian largesse.[41]

Promoters of science in St. Louis faced many of the same problems as did those in Chicago, especially the "cool apathy of the great public." There was, however, one critical difference. The president of the St. Louis Academy, George Engelmann, bemoaned the lack of support from the "heavy men" of his city. This usually limited them to annual expenditures of less than one thousand dollars, with the publication of a volume of the *Transactions* every two or three years.[42] In sharp contrast, the trustees of the Chicago Academy funded budgets to the tune of six thousand to ten thousand dollars per year between 1865 and 1871, much of which went to the museum and the collections.

Still, the Academy faced a significant hurdle in a lack of public support. By 1865, the population of Chicago had reached 178,000, but the Academy had a mere thirty-four resident members. Given the historic ties between medicine and natural history, it is not surprising that physicians constituted just over one-quarter of the membership and were among the most active members.[43] There were 120 life members but (with the exception of the

Academy Board) most of these men never again contributed after initially paying for a membership.

Outside of Chicago, the Academy boasted thirty-four corresponding members including Agassiz, Baird, Meek, James Hall, and Jared Kirtland. (One curious name on the list is Robert Chambers of Edinburgh, Scotland, later revealed as the author of the controversial book *The Vestiges of the Natural History of Creation.*) By the end of 1865, Stimpson had heard from local men such as John Wesley Powell of Bloomington, Illinois, Dr. Jacob Velie of Rock Island, and Lucian B. Case of Richmond, Indiana. A year later the paleontologist Edward Drinker Cope and Illinois entomologist and Darwin correspondent Benjamin D. Walsh were added to the fold. Many of these men sent specimens to Chicago.

Stimpson exhorted potential donors to the collections to act quickly, "A donation now will tell far more than one two or three years hence, as we want to make a show to excite additional pecuniary interest." Thanks in part to Stimpson's many friends, the volume of specimens increased dramatically in late 1865 and early 1866. Highlights included another 2,000 shells and dozens of birds from the Smithsonian, while Thomas Bland added 2,300 marine invertebrates from the Gulf of Mexico and Robert E. C. Stearns of the California Academy of Sciences shipped over 500 shells as well as samples of gold, silver, and copper ores. Stimpson also donated 1,300 more invertebrates.[44]

One addition in particular elicited a positive response from Stimpson. Trustee George C. Walker purchased from famed English collector Hugh Cuming a perfect specimen of a chambered nautilus (*Nautilus pompilius*) complete with the preserved animal. Expounding on this prize, Stimpson pointed out that cephalopods had been far more abundant in the Silurian period, a fact amply demonstrated by the exquisite fossils of these animals that turned up regularly in Chicago area limestone quarries.[45]

By the end of 1865, the Academy's rooms were nearly full, and the trustees made "another grand attempt to get funds" to secure a permanent home for the museum. They purchased a lot on what was then the far south side of the city at Thirtieth and Prairie Avenue. Weeks later they received a tempting proposal from the executors of the estate of Stephen A. Douglas, offering a tract of land near the old University of Chicago at Thirty-Fifth and Cottage Grove and fronting on the University Square. The sole condition was that the Academy erect a building within two years; after ten years the land would belong free and clear to the Academy.[46]

The trustees favored the proposal but the majority of members vehemently objected, arguing that the site, about three miles south of the downtown business district, was too far from the city's center. With several members threatening to sever all ties with the organization if the move was approved, in April 1866 the members voted to reject the Douglas offer and the trustees yielded.[47]

Stimpson's one-year commitment technically expired at the end of 1865, but the trustees asked him to stay until Kennicott's return. The telegraph expedition had devolved into a nightmare for Kennicott as jealousy, poor management, and internal power plays led to infighting and countless delays. William H. Dall wrote that officers had conspired against the trusting and somewhat naïve Kennicott. Stimpson replied that Kennicott would have been in his rights "in dissolving entirely his connection with the Teleg. Exp. and returning home, and his course would have been endorsed by his friends here. But there is no such word as 'fail' in his vocabulary."[48]

A part of Stimpson might have selfishly wanted Kennicott to come home, as prospects for natural history had improved at the Smithsonian since the end of the war. During the darkest days of a typically brutal Chicago winter Stimpson thought "a great deal" about Baird's promise to secure him a position in the National Museum when it was officially organized.[49] With the date of Kennicott's return appearing more and more distant, Stimpson decided to bring Annie and little William to Chicago. Pleased to be reunited with his family, his joy was tempered by the fact that for the first few months all of them were constantly sick.

Stimpson would have liked the opportunity to run the Academy alongside Kennicott. "Together we could run this institution capitally," he told Baird. "I think I have got more system than he has, but I must have him to rush in the materials and workers and sinews of war." Stimpson admitted that he had no talent for making others work, and he also admired Kennicott's "energy and brass" in asking people to donate money and time to the Academy.[50]

While they paled in comparison to what Kennicott faced, Stimpson had frustrations of his own in early 1866. Academy meetings had increasingly veered away from traditional natural history topics to matters of public health. Despite its swelling population, Chicago lacked an independent Board of Health, and since physicians were such an important part of the Academy's constituency it is not surprising that they looked to the organization as a forum for acting on these issues.

Cholera posed the most immediate threat. Previous outbreaks in 1832 and 1849 had killed thousands in the city and across the nation. The disease appeared in New York in the spring of 1866, and Chicagoans awaited the inevitable arrival of the dreaded scourge. It struck with a vengeance in July and ultimately killed nearly one thousand people.[51] Some had theorized that high levels of ozone in the atmosphere instigated the outbreaks, so in May the

Academy formed a committee that included Stimpson and attempted, with little success, to measure ozone levels.[52]

The Academy also became embroiled in a controversy with the meatpacking industry, one of the most powerful interest groups in the city. Nicknamed Porkopolis, Chicago had overtaken Cincinnati as the center of the pork industry. The Union Stock Yards opened on Christmas Day in 1865, and during peak seasons as many as twenty-five thousand hogs were slaughtered daily with grim efficiency. When nearly two hundred people in two small German towns died from trichinosis it made headlines around the world and the reverberations were deeply felt in Chicago, where German Americans were the second largest ethnic group. Fears of a similar tragedy occurring in Chicago prompted calls for action.[53] At the March 1866 meeting, Stimpson and three of Chicago's leading physicians (Edmund Andrews, Hosmer Johnson, and James Van Zandt Blaney) were named to a committee charged with determining whether the parasite responsible for the deaths, *Trichina spiralis*, existed in the United States and if so to assess the extent of the danger and possible remedies.[54]

The creation of the committee did not go unnoticed. The city's leading pork merchants, Gustavus Swift and Philip Armour, were bitter rivals and cutthroat businessmen. They both knew that the Academy's report had the potential to cost them hundreds of thousands of dollars. According to Stimpson, unnamed "Pork men" made it known that if the Academy proved that no trichinae worms existed they would "build up our Acad[emy]."[55] Science plays by its own rules, however, based not on power or influence but on objective observation. Thorough as always, Stimpson borrowed books from the Smithsonian to research the subject. Approaching the problem from a zoologist's perspective, he detailed what was known about the parasite's life history. He also dutifully examined muscle tissue from 94 hogs while others looked at samples from an additional 1,300 animals. What they found startled them. In one case, Stimpson found as many as 18,000 trichinae worms in a cubic inch of spinal muscle. In all, they discovered that twenty-eight animals had worms, about one in fifty, a rate far in excess of what had been found in European hogs. However, only three pigs were found to be densely infected. The solution to the problem was already well known: cooking pork to a temperature of at least 160 degrees kills the parasite.

Stimpson took the lead in writing the final report. Where some had asserted that trichina had been found in earthworms, Stimpson noted that the "practised zoologist easily distinguishes between" trichina and similar nematode worms. The committee also noted that very few Americans became infected because they cooked their meat, unlike some Germans that ate it raw. The committee concluded that the uproar had been caused by "interested persons, for speculative purposes" and that the "panic which now prevails is unfounded in reason, senseless, and greatly injurious." It was estimated that 90 percent of Americans were largely dependent on pork as a source of food, and it would be folly to discard it when such simple measures existed to make it safe.[56]

The Academy's investigations were widely reprinted and discussed around the country. Stimpson noted that the report had been subjected to attacks and ridicule by the pork interests despite its conclusion that "few sensible persons will refuse pork."[57] The "pork men" never did donate to the Academy.

Stimpson was named to yet another committee in early 1866, this one charged with investigating the city's artesian wells as a viable water supply.[58] This only added to his concern that Kennicott's vision for the Academy as a natural history museum had been eclipsed in favor of purely utilitarian needs. He knew that the Academy could not thrive without the active involvement of physicians, but he also did not want to see traditional natural history neglected. He did his best to guide the Academy back to more familiar if still practical waters. Kennicott's expedition had shipped back a wharf piling from San Francisco that had been perforated throughout by the teredo, or shipworm, leading Stimpson to address a meeting with a brief summary of these animals.[59]

Fortunately Stimpson's budding friendship with trustee Eliphalet Blatchford presented him with an opportunity to return to the field. One of the few men in Chicago with a lively and enduring interest in natural history, Blatchford had made a fortune producing bullets for the Union army. He had donated specimens to the museum for years and amassed a wide array of fine-mesh seines and nets for fishing. Even more crucial was his willingness to spend large sums in the name of science. Stimpson later told Baird that Blatchford could "do for us as much or more than any of our patrons."[60]

After a preliminary seining trip with Stimpson, Blatchford proposed a more extensive field excursion and agreed to bankroll and participate in a canoe trip down the Fox River in Wisconsin to Ottawa, Illinois. The region fits the description of what one historian has called an inner frontier, a landscape of largely undisturbed nature that was still easily accessible to people living in cities.[61]

Stimpson knew that they would need to take precautions lest they be rebuked or worse for hunting and fishing out of season. He opened a correspondence with Increase A. Lapham, Wisconsin's most experienced naturalist, asking for a testimonial letter on their behalf in order to fend off possibly pugnacious citizens by showing that Blatchford and Simpson's operations were of a purely scientific character.[62] Escaping the city in the spring of 1866 made sense with cholera edging ever closer, so in early May they traveled roughly eighty-five miles north to Waukesha, Wisconsin. For much of its length the Fox River is shallow enough to walk across although the thick mud can reach a man's knees. Coming down the river in three boats, collecting by day and camping out at night, they passed countless ponds and marshes, gathering snakes, shells, fish, and birds.[63]

With the spring migration in full swing birds were exceedingly numerous and Stimpson recorded observations of their "habits, migrations & breeding places."[64] Near Waterford, Wisconsin they undoubtedly saw majestic sandhill cranes, still common along this stretch of the river and easily identified by their distinctive rattling calls. While their numbers were declining, passenger pigeons were still abundant and many nested in Wisconsin.

They ended their three-week journey at Ottawa, where they met with members of the new Ottawa Academy of Sciences, which had modeled their constitution on the Chicago Academy's.[65] Using one of his favorite words, Stimpson informed a correspondent that he had the "jolliest" time on the trip.[66] Animated by the spirit of scientific inquiry, Stimpson and Blatchford had forged a close friendship in the field. Indeed, Stimpson made no better friend during his years in Chicago, and Blatchford, in turn, wrote warmly of Stimpson on several occasions. As always, Stimpson felt buoyed by fieldwork, and he returned to Chicago in late May reinvigorated and in the highest of spirits. He would need all the energy and enthusiasm he could muster, as he was about to face his second trial by fire.

FIGURE 1. William Stimpson, circa 1868. From the collections of the Chicago Academy of Sciences/ Peggy Notebaert Nature Museum.

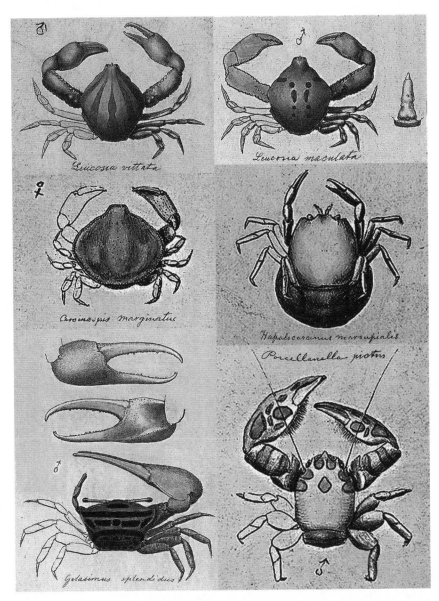

FIGURE 2. Drawings of crustaceans by William Stimpson. Smithsonian Institution Archives. North Pacific Exploring Expedition Collection, 1852–1861 and undated, record unit 7253, box 2, folder 5.

FIGURE 3. Stimpson family, circa 1853. From left to right: Francis (Frank) Stimpson, Mary Elizabeth Stimpson, Sarah Stimpson, Eleanor Hall Stimpson, William Stimpson, Herbert Stimpson, and James Stimpson. Courtesy of Barbara Scudder Wilson, great-great-granddaughter of William Stimpson.

FIGURE 4. William Stimpson, unidentified artist, 1860. Salted paper print, National Portrait Gallery, Smithsonian Institution.

FIGURE 5. Members of the Megatherium Club. Top row: Robert Kennicott, Henry Ulke; bottom row: William Stimpson, Henry Bryant. Smithsonian Institution Archives. Image no. SIA2008-0347.

FIGURE 6. Spencer Fullerton Baird. Smithsonian Institution Archives. Image no. MAH-10737.

FIGURE 7. Robert Kennicott, circa 1860. The Grove National Historic Landmark, Glenview Park District. Courtesy of the Glenview Park District.

DEATH AND DECISION

In the life of every great successful man there are periods when his future hangs by a slender cord.
—Ezra B. McCagg[1]

Few cities endured more fires than nineteenth-century Chicago, whose unprecedented expansion occurred without any sort of regulation. As a result, most structures and the sidewalks were made of wood. As the city grew so did the number of fires, from 243 in 1865 to 660 only six years later. Chicago had more fires than New York City, which had three times as many people.[2]

The list of nineteenth-century American natural history museums that suffered serious losses from fire makes for sobering reading. With three museums suffering heavy damage in a six-week span, 1866 turned out to be a particularly brutal year. On May 21 the collection of the Lyceum of Natural History of New York went up in flames. A Fourth of July celebration gone bad led to a fire that consumed Maine's Portland Society of Natural History, the second time that organization had been burned out.[3] Sandwiched between these two events came a devastating fire at the Chicago Academy's museum.

The rash of fires led America's leading scientific journal to promote the notion of fireproof buildings. "The lesson taught by these disasters should be heeded throughout the land: make all buildings for scientific Museums thoroughly fire-proof."[4] To no avail—in 1869 a blaze largely destroyed the collections of the Academy of Sciences of St. Louis. That same year Ward's Natural Science Establishment of New York burned and even the huckster P. T. Barnum lost heavily in two fires that destroyed many natural history specimens and countless congeries.

The Chicago Academy of Sciences would suffer more from what was colloquially called the "fire fiend" than any other American natural history museum. Around one o'clock in the morning on June 7, 1866, a blaze swept through the Metropolitan Block building where the Academy shared the upper floors with the Eastman Commercial College and a Masonic lodge. The fire originated in the school and crossed over the brick wall separating it from the Academy. As happened with distressing frequency, the firemen were originally sent to the wrong location, and by the time they arrived the building was engulfed in flames.[5]

Stimpson lived on the west side of the city about three miles from the museum and when he arrived around 7 a.m. he experienced the same sick feeling that he had had at the Smithsonian fire. Wreckage and debris littered the site and water stood a foot deep in places. At first holding out little hope that anything had survived, to his relief he found that the intense heat of the fire had caused large zinc sheets from the roof to fall onto some of the display cases, preserving them from the flames.

While it could have been worse, the losses were indeed heavy. Stimpson prepared a circular summarizing the damages that was printed and sent to other scientific organizations around the world. Half of the mammal and bird collections were in ashes, including most of the mounted birds on display. The extensive collection of eggs and nests was destroyed, as were the insects, except the Lepidoptera. The dried crustaceans and echinoderms that he had contributed were gone, and in all over twelve thousand specimens were lost. The herbarium remained intact except for the series of North Pacific Exploring Expedition plants. Also saved were two thousand jars of invertebrates in alcohol and most of the mollusks.[6]

Contemporary accounts placed the blame on an incendiary, as arsonists were commonly referred to. The *Chicago Times* bemoaned the Academy's fate as "peculiarly aggravating.... No fire that has devastated our city for years will be so deeply and bitterly lamented, as will this which has destroyed so much that is rare and beautiful in nature, much of which can never be restored."[7] While the museum was insured, there were some things that money could not make good. Many specimens could be replaced only at great cost and effort, if at all.

The old quarters were temporarily repaired, but Stimpson refused to trust anything of value to them. Bone tired after working for eleven straight days, he finally took time to write a long, disjointed letter to Baird. His mind reeled with a jumble of tasks that needed urgent attention. Baird promised to replace what he could and did his best to cheer his younger compatriot by urging him to look on the fire as the best thing that could have happened to him, occurring as it did at the outset of his museum career before he had accumulated a lifetime of materials. Speaking from experience, Baird pointed out that Stimpson had learned a valuable lesson and would now be forever on his guard against the recurrence of such a calamity. The insurance money could be used to carry on future operations, thus making a gain out of what seemed to be a terrible loss.[8]

The trustees now redoubled their efforts to erect a fireproof structure for the museum.[9] The members had previously rejected the idea of locating on the north or south side of town, but lots in downtown Chicago were prohibitively expensive. Providentially, a man of the cloth came to the rescue. Less than a month after the fire, the Catholic bishop of Chicago, James Duggan, agreed to sell at a discount a lot on Wabash Avenue. Trustee Ezra B. McCagg recalled the circumstances. Bishop Duggan "was most liberal, his main spring of thought and action being embodied in a single sentence, 'The work of the present generation in Chicago is laying foundations.'"[10] The lot was centrally located in the heart of the business district and "in the best possible neighborhood," and Stimpson hoped that they could begin construction immediately. He would have a major say in the construction process, having been named chairman of the building committee.[11]

Settling the insurance claims proved to be agonizingly slow, however. The Academy had thirty thousand dollars' worth of insurance but needed a detailed inventory of what had been lost in order to collect. During that hot cholera summer Stimpson and Blatchford undertook the task of documenting the damages. Stimpson needed help in placing monetary values on some of the collections, as his own estimates had been greeted with skepticism. "My position is a very difficult one, as all rely upon me for a statement of values which they are inclined to consider *fancy*." Placing a dollar amount on dead bats or snakes is an inexact science. In some cases Stimpson used prices from catalogues, in others he calculated the expenses of making the collection, and in still others he used appraisals from other naturalists.[12]

Of all the objects lost in the fire, few were more valuable than the bird eggs and nests. Oology, or the scientific study of the eggs of birds, had become quite popular by the 1860s. Fueled by printed checklists and a lively trade in buying, selling, and exchanging eggs, it had been a major focus of Kennicott's collecting activities and a research interest of Baird's.[13] Stimpson thus asked Baird to provide an estimate. Baird asserted that the Academy's collection had been second in America only to the Smithsonian's in size and importance. Kennicott's Arctic collections had been of great value, "not only on account of the rarity, variety, and number of the species and specimens but also because of their perfect authentication and verification."[14]

Despite this private praise, when it came time to certify a dollar amount Stimpson admitted to being "very much disappointed" at Baird's estimate. After "the d---l's own time" haggling with the insurance companies, in late 1866 the Academy received around twenty thousand dollars, funds that enabled them to begin the rebuilding process.[15] Still, it would take years to recover from the effects of the fire.

For the second time in eighteen months, Stimpson had lost portions of his scientific property to fire, lamenting to Baird that, "My *personal* losses were far greater than at the S. I. fire, in *books & other* effects." Most distressing of all, the manuscript and illustrations for his book on Atlantic coast mollusks, representing so much time and effort, had been "much damaged." With the Smithsonian publication series again up and running Baird had assured him there "would be little or no delay" in publishing. This ostensibly good news had to have been a little disheartening for Stimpson, since he knew he would have no time in the foreseeable future to devote to this goal. He told Baird that he had given all of his time to the interests of the Academy but hoped that after straightening out the "wreck" left after the fire he would again pick up work on his book.[16] For now his research and writing were on hold.

Stimpson's role as the Academy's caretaker had been much more complicated than he could have envisioned when he agreed to come west in 1865. Instead of his usual summer field excursion he spent several frustrating months trying to save the "relics of our library and collection" with precious little to show for his trouble.[17]

A reminder of past associations in the East came in mid-September when he learned of the sudden death of sixty-one-year-old Augustus Gould, struck down by cholera in Boston. Just a few months earlier, Gould had written him asking for information on mollusks for a new edition of his landmark book *Report on the Invertebrata of Massachusetts*. Stimpson mourned Gould, whose encouragement and example had done so much to get him started in science.

Weeks later news of another death in the small fraternity of naturalists reached Chicago, but this time the repercussions would irrevocably alter the course of Stimpson's life. It had been eighteen months since Robert Kennicott left Chicago and during that time Stimpson had barely heard from him. In early October as deaths from cholera peaked in the city, sending a wave of fear and dread throughout Chicago, news came of Kennicott's death. At first they hoped it had all been a mistake, since many explorers had been reported dead and later turned up alive. But there would be no denying the fact that Kennicott's body had been found in May near the banks of the Yukon River at Nulato in Russian America. It had taken almost five months for word to reach his friends and family.

Stimpson, badly shaken by the death of his close friend, wrote Dall, "I do not know when the death of so young a Scientific man has created such a profound sensation of sorrow, not only in his own city, but throughout the country wherever scientific circles exist."[18] Kennicott passed away at the age of thirty, having done and seen more in that time than many granted twice that number of years.

Stimpson realized immediately that he faced a difficult choice that would have profound ramifications for his personal and professional life. The trustees would surely ask him to stay in Chicago permanently. Doing so would severely curtail if not end his career as a marine zoologist. There was a part of him that desperately wanted to return east, but his sense of loyalty and duty would not allow him to abandon the Academy. He also had a wife and child to support, and there were few opportunities for naturalists to make a living in their chosen field.

In late October he traveled east for the first time in almost ten months. He badly needed a few weeks away, and for the next two months he split his time between Washington and Ilchester.[19] He attended to numerous letters and packed more Smithsonian duplicates for shipment to Chicago. As a result he missed a special meeting of the Academy honoring Kennicott. He instead sent a letter that stands as a testament to their friendship. Kennicott had "died at his post in the performance of a dangerous and arduous duty . . . a martyr to his devotion to science, and to the interests of this Academy." Stimpson vowed to carry on Kennicott's work to create a "fit monument to his memory."[20] In November the trustees formally offered and he accepted the position of director of the museum at a salary of two thousand dollars a year.[21] A few months later they also made him a trustee, further cementing his ties to the Academy.

Stimpson returned to Chicago a few days before Christmas in 1866. It had been a year of frustrations and tragedy. The fire, the deaths of two close friends, and the cholera epidemic all conspired to drive him into a severe funk. His recent visit to the Smithsonian only deepened his sense of separation from America's scientific community. "I have been very lonesome . . . and almost entirely without scientific sympathy," he sighed.[22] He had achieved his dream of heading a natural history museum but the fact that it had taken the death of one of his closest friends to achieve it cast a pall over the accomplishment.

With that peculiar brand of optimism generated by the beginning of a new year, he began to fight his way back. Being a husband and a father had inevitably changed him and with maturity came less self-absorption. He resolved to forsake tobacco and coffee. January also brought a pleasant diversion when all Chicago buzzed with anticipation over the Crosby Opera House art lottery. Stimpson bought a number of tickets and boasted that he had a "pretty good gallery."[23]

Reality intruded just a few days later when he helped to bury Kennicott. Stimpson had grown close to the entire Kennicott family, and at various times Robert's brothers, Bruno and Flint, worked at the museum. Stimpson relayed the events of the heart-wrenching burial to Baird. "There were several singular coincidences (?) which made me think that Nature sympathized with us in our sadness. As soon as Bob's remains arrived at the Grove ... the trees became filled with the Black-cap titmouse—they stayed until after the ceremony. At the moment when we lowered the body into the grave, a covey of quails sitting nearby gave forth their sweet song—not the cheerful 'Bob White' of spring time, but the plaintive note by which the scattered covey express their anxiety, or call each other to the evening rest. It did indeed seem as if the Birds that Bob loved so well were mourning his decease, or had come to watch over his grave, to prevent his being lonely."[24]

Having made the decision to remain in Chicago he purchased a home in the country eight miles west of the city in what is now suburban Oak Park. Far removed from the dirt and noise of frenetic Chicago, by mid-April he and Annie had settled into the quiet community. Nestled away in an idyllic setting near a forested strip of land along the Des Plaines River, Oak Park had a mixture of blue- and white-collar workers. Among his immediate neighbors were retired merchants, an English farmer, an Irish laborer, a gardener, and a saloon keeper. Conveniently, the Chicago and North Western Railway had recently added commuter service to the city.[25]

The costs of maintaining the new homestead were heavier than expected. It did not help that Annie brought two domestic servants from Maryland to live with them, a luxury that Stimpson's salary could not easily afford. Later that year Herbert Stimpson paid a visit to his son and helped him out of his pecuniary difficulties.[26] Stimpson certainly did not get rich in Chicago—the 1870 census shows his home valued at three thousand dollars and his personal estate at a very modest three hundred dollars.

The purchase of a home signaled the depth of Stimpson's allegiance to the Academy. It also led to a clash with his father-in-law. James Frisby Gordon had not accepted his daughter's removal to Chicago, and now that it seemed to be permanent he paid a visit to Baird in Washington, asking whether there might be work for his son-in-law closer to home. Chagrined by this unwanted meddling, Stimpson hastily let Baird know that Gordon's visit had not been his idea. With more than a hint of regret he wrote, "Much as I should like to go back to old and pleasant associations & communication with Scientific Men, I think it my duty to remain here, for the present, at any rate." He had made a commitment and would honor it but the "for the present" also showed him hedging his bets.[27]

Baird knew more than anyone just how much Stimpson sacrificed by staying, and he sympathetically but accurately viewed things as Stimpson did. "You are very right in your view as to remaining. It would be fair to you perhaps to come permanently eastward but it would be death to the Ch[icago] Mus [eum]."[28] Staying more out of a sense of obligation rather than personal preference inevitably cloaked Stimpson's years in Chicago with an uncomfortable ambivalence.

Reminders of past associations came in his correspondence with eastern naturalists. In April 1867 Agassiz asked Stimpson to spend a month or two that summer overhauling the crustacea at the Museum of Comparative Zoology. Meek had received a similar offer regarding the fossils. Both men turned Agassiz down, and it is evident that Stimpson enjoyed rejecting this overture from his onetime teacher, telling Meek, "Good for you on Agassiz! I also had a pressing invitation from him a few days ago to come and arrange his large and splendid collection of crabs. Didn't see it."[29]

The opening of the Essex Institute in Salem in 1867 enabled several of Agassiz's former students to gain full-time employment. Stimpson exulted over their good fortune in a letter to Putnam. "Ain't you fellows licking your scientific chops over prospective good times. It serves you right. Wonder how they feel about it in Cambridge."[30] Unspoken was the thought that had Stimpson remained in the East one of those positions might have been his. One of the ways that Packard, Morse, Putnam, and Hyatt celebrated their freedom was by creating a new monthly journal of natural history. The *American Naturalist* debuted in March 1867, and its success in appealing to both naturalists and laymen indicates the growing popularity of natural history in America.[31] Stimpson immediately subscribed and would publish several articles in the coming years.

Stimpson faced three major tasks in 1867—curating the collections, overseeing the Academy's publications, and supervising the construction of the new building. Each presented different challenges and required special skills. None of these undertakings caused him more grief than the publications series. Every detail from proofreading to editing text to dealing with the printers and lithographers fell to him. As noted earlier, any natural history society that wanted to be taken seriously needed to publish and disseminate new contributions to scientific knowledge. Publications were also essential to secure a worldwide system of library exchanges with other similar institutions. Two separate titles were planned: an annual volume of *Proceedings* as a vehicle for short scientific notices and lists of donations, and a more lavish *Transactions* for lengthier illustrated articles.

Stimpson's first attempt fell far short of expectations, although largely through circumstances out of his control. The inaugural issue of the *Proceedings of the Chicago Academy of Sciences* appeared in the spring of 1866. The four scientific contributions were all descriptive natural history papers with a Smithsonian connection. Meek gave a short, technical treatise on the affinities of a family of extinct mollusks. The longest paper, accounting for nearly a quarter of the issue, contained Meek and Worthen's descriptions of new species of invertebrate fossils from Illinois and other western states. Gill contributed a short description of the northern elephant seal, based on a single skull in the Academy's collection.[32] Stimpson penned the final article, comprising descriptions of new species and genera of mud and ghost shrimps. Based on specimens housed at the Smithsonian, these had been written at least six years earlier but had been delayed for various reasons. After less than three pages the paper ends abruptly, cut off in the middle of the word *posterior*, of all things. The fire in 1866 damaged the remaining text, and although Stimpson continued to accumulate new items the Academy never published another installment of its *Proceedings*.[33] Thanks to the Smithsonian, hundreds of European scientific societies were sent copies of the abbreviated *Proceedings* and more were distributed in the United States.

Stimpson had also been laboring over the Academy's *Transactions*. Getting it in print proved no easy task.[34] Given the Academy's geographic location, it is not surprising that most of the volume dealt with the fauna and climate of the western United States. Leading off was an article on midwestern fossils by Joseph McChesney, formerly Illinois's assistant state geologist. In 1862 McChesney left to serve as US Consul to Newcastle-upon-Tyne in England and, with the author's blessing, Meek agreed to revise the paper's outdated nomenclature. Throughout 1867 he and Stimpson amassed an extensive correspondence relating to McChesney's descriptions, and Stimpson announced his pleasure when "this troublesome work" had been concluded.[35]

The second article contained descriptions and figures of rare or noteworthy birds in the Academy's museum, including one collected by Kennicott in Nicaragua. Heeding Baird's counsel, Stimpson agreed to have John Cassin take charge of the illustrations, and in late March of 1865 Stimpson implored Baird to have Cassin get started as soon as possible.[36] When Stimpson turned up in Philadelphia later that year to check on their progress, he found that virtually nothing had been done. Even worse, a miscommunication had resulted in a higher cost for the plates than Baird had originally quoted to him.[37] Eight months later, Stimpson was still waiting on Cassin who, in turn, expressed dissatisfaction with the timeliness of his payments. Stimpson retorted that Cassin had routinely ignored his editorial comments and refused to answer his queries. The choice of Cassin was perhaps problematic since he was part owner of the lithographers Bowen and Company.[38]

Ann Shelby Blum has noted that publishing natural history engravings in nineteenth-century America was both expensive and fraught with complications, and Stimpson's experiences certainly bear this out.[39] The trustees agreed to pay for the plates but these captains of industry were by nature competitive, even over who would sponsor particular illustrations. They "rather stickle for new species," explained Stimpson.[40]

The bird paper proved vexing in other ways. Stimpson assumed that Baird would write the notes for each species, but getting no immediate response Stimpson jotted down a few lines relating to where and when they were collected. Baird added comments but refused to accept authorship, leading Stimpson to protest that it "would be ridiculous for a 'crabologist' to write a Bird paper." Stimpson argued that Baird's name would lend more prestige to the publication, but Baird demurred, and this stands as the only one of Stimpson's scientific papers that deals exclusively with vertebrates.[41]

Three other contributions rounded out the first part of the *Transactions*. Increase Lapham's short piece on the climate of the country bordering on the Great Lakes reflected his lifelong interest in meteorology. He was among the first to document the ways in which the Great Lakes altered weather patterns in the Midwest. Lapham would later play a leading role in establishing the National Weather Bureau.[42]

Meek's article on the Mackenzie River region of Canada stands as the first comprehensive attempt to map the geology of this vast area. Kennicott had been one of the few scientific observers to travel this far north, and he and several Hudson's Bay Company employees amassed a respectable assortment of fossil invertebrates for Meek to describe.

The final paper dealt with fossils found by members of the Iowa Geological Survey. Stimpson clashed with the lead author, the geologist Charles A. White. The nature of the dispute is unclear but Stimpson wrote to Meek concerning White. "He will find that it won't 'break' us to print his paper, nor will it ruin us to have him cease friendly intercourse and refuse to come here to get his recent shells named."[43]

By the end of 1867, it had been nearly three years since Stimpson had begun working on the first volume of the *Transactions*. The text and plates, nearly ready for the press, had been thoroughly soaked in the 1866 fire and represented a near total loss. Once this had been redone Stimpson learned that the printer had mixed two different kinds of type and the last sixty pages had to be reset. Just as this had nearly been completed a fire destroyed the printing company, leading to another "vexatious delay."[44]

When the first part of volume one finally saw the light of day in February 1868, Stimpson felt compelled to insert an apology to the contributors for the long delay. The trustees considered the publication of the *Transactions* as a sound investment as it permanently established the exchange of publications from kindred societies all over the world. The scientific community took note, with the *American Naturalist* praising it as a "splendid volume." "Science is carefully fostered in the West," congratulated the editors. Stimpson probably did not mind much that his name was misspelled in the review (Simpson), an error that occurred frequently throughout his life.[45]

More than any of his other responsibilities, Stimpson enjoyed his role as a museum curator. He had been collecting since childhood and enjoyed organizing specimens. The most significant single addition to the museum since the fire came in October 1866, thanks to the generosity of trustee George Walker. At Stimpson's urging, Walker purchased the late William Cooper's cabinet of West Coast marine shells. Philip Carpenter proclaimed it "*peculiarly valuable*," comprising over 2,600 marine and 400 terrestrial species. The *American Journal of Conchology* extolled it as an "authoritively named and scientifically arranged cabinet" and applauded the Chicago Academy as a "very enterprising Society."[46] Walker's largesse signaled that Chicagoans could successfully compete for private collections with more established eastern institutions.

The growth of the collections in 1867 continued to owe a great deal to the Smithsonian. Rare Arctic birds and eggs came in steadily courtesy of Kennicott's many friends in the Hudson's Bay Company.[47] Perhaps the flashiest additions were a group of artifacts illustrating the ethnology of the South Seas, the American Arctic, and the West Coast of North America. The new discipline of anthropology was just taking its baby steps in the scientific investigation of mankind, and Stimpson especially welcomed these eminently exhibit-worthy items as a supplement to the more common pickled and stuffed animals. Thanking Baird for the "jolly lot" he noted that Dr. Edmund Andrews and others were "ethnologically inclined."[48] With the donation, Baird claimed that the Academy now had the second-best collection (next to the Smithsonian) of Eskimo and Polynesian objects in the country. In this case Agassiz had "very warmly" argued that the new Peabody Archaeology Museum be the recipients of these objects, but Joseph Henry allowed them to go to Chicago.[49]

Local collectors, most of whom have been forgotten, were by far the largest source of specimens, and their contributions were an essential part of the museum's growth. Among the most prolific was Joel Reeves, the "king of the Calumet marshes" south of Chicago and an inveterate bird and egg collector.[50] Following in his brother's footsteps, Flint Kennicott found birds, mammals, and reptiles in abundance near the Kennicott homestead north of Chicago. Civil War veteran James W. Milner gathered birds and eggs in Waukegan, Illinois. Ellis Chesbrough, Chicago's celebrated city engineer, recovered insects from the Lake Michigan water intake crib.[51]

Others helped in organizing the various collections. One of the most ardent volunteers was Kate N. Doggett, who took charge of the herbarium. As the wife of trustee William E. Doggett and a relative of Jared P. Kirtland, she had a lifelong interest in natural history. Kennicott felt that she had the perfect credentials: "a taste for insects, a large fortune, a good deal of ambition, and no children!" Attractively unconventional, she eschewed gloss and pretension while hosting parties for the city's literary elite. Stimpson considered the Doggetts to be his friends and probably attended some of their soirees.[52]

The Academy sponsored or contributed to a number of expeditions that enriched the museum. Charles Sonne and Andrew Bolter ventured to Mammoth Cave in Kentucky and brought back bats, bind fish, and crustaceans. Flint Kennicott made a large haul of birds and eggs from Savannah, Missouri, and valuable items also came in from the ill-fated Russian American Telegraph Expedition.

Not everyone who promised aid paid such handsome dividends. In 1868 a German named Rudolph Borcherdt began working at the Academy as a taxidermist. After donating a few specimens he convinced the trustees to supply him with a scientific outfit, railroad passes, and money in order to collect and mount large mammals from Colorado. Stimpson later admitted to Baird that they had misjudged the man. "He has treated us very shabbily. . . . It turned out that this was only a dodge to get a free passage to the west, and we have received nothing in return for our outlay."[53]

Despite this embarrassment Stimpson was overwhelmed by a deluge of new material, declaring, "Things are coming in more rapidly than I can attend to them properly alone." Finding workers to preserve and catalog the growing backlog proved difficult. Stimpson had relied heavily on Flint Kennicott for assistance, but he had gone off to Missouri. Another assistant had headed west, and two others decided to devote themselves to commercial endeavors. The lack of scientifically inclined citizens left Stimpson to attend to many time-consuming chores that other museum directors such as Baird, Agassiz, and Leidy easily delegated to others.[54]

As a sign of its increasing stature, the Academy began to share its bounty with others. A suite of sixty Illinois game birds was sent to the Exposition Universelle in Paris in 1867, where it received good reviews. Stimpson took special care in packing the shipment and was proud that the collection survived the journey with nary a feather out of place.[55]

Chicago served as a jumping-off point for numerous naturalists exploring the West. Eastern museums were eager to obtain their own specimens, and Stimpson gladly provided aid. Joel A. Allen, another in the seemingly endless line of students trained by Agassiz, arrived in early 1867 while bound for Iowa. Stimpson regretted that the Academy could not afford to help defray Allen's expenses given that they had already expended over $1,500 for explorations for the coming season, but he did obtain a free railroad pass as an inducement to get Allen to share his bounty with the Academy. Allen arrived in late May 1867 to collect for the Essex Institute and spent several days with Academy members blasting away at spring warblers.[56]

While a great deal has been written about John Wesley Powell's pioneering western explorations, none of these accounts mentions Stimpson's small but important role. After losing part of an arm at Shiloh, Powell became a professor at Illinois State Normal University. An active member of the Springfield-based Illinois Natural History Society, Powell in 1867 successfully lobbied the state legislature for an appropriation to fund a museum with himself as a curator. Since he had previously donated fossils and shells to the Chicago Academy, it made sense for him to visit Chicago to see how an established natural history museum was run. In April 1867, he met with Stimpson and explained that his real ambition was to explore the West. Stimpson wrote Powell a letter of introduction to Baird, who as usual gave advice and assistance and helped Powell plan an expedition to the Rocky Mountains. The Natural History Society and the Illinois Industrial University (later the University of Illinois) each gave five hundred dollars and the Academy one hundred dollars for a share of whatever Powell's party collected.[57]

Stimpson went to great lengths to supply Powell with the appropriate equipment. He succeeded in assembling a "perfect" zoological outfit replete with copper cans, seines, bottles, egg drills, and more. Baird sent alcohol for preserving specimens. Fred J. Huse, who had worked at the Academy museum, accompanied Powell. Unfortunately, the returns from Powell's initial western exploration proved to be quite skimpy: a few mammal skulls and bird skins.[58]

In September 1867, Stimpson managed to embark on a brief foray of his own into the field, although it was far removed from his usual line of investigation. During that unparalleled drought year a farmer discovered some large bones while digging a ditch near Fort Wayne, Indiana. A local physician bought the bones and the right to continue excavating and contacted Stimpson for help. He spent several days supervising on site, eventually recovering nearly complete skeletons of an adult and juvenile mastodon in an excellent state of preservation. The partisan Chicago papers crowed over the discovery of "one of the finest collections of mastodon bones ever found."[59] Stimpson laconically reported to Meek that they had accumulated a "big pile of bones," which were donated to the Academy. The mounted skeletons became one of the main attractions in the museum.[60]

By their very nature these contributions from disparate sources constituted something of a hodgepodge. For zoologists to accurately determine the geographic distribution of species, a more systematic approach to collecting was required. To that end the Smithsonian and the Academy partnered to continue their virtual monopoly on collections from the American subarctic. Through Kennicott and his Hudson's Bay Company colleagues, the Academy and the Smithsonian had a history of cooperation in this little-known and relatively unexplored region. With the end of the telegraph expedition in early 1867, Baird looked for a way to keep someone in the field collecting.[61]

He believed he had found just the man in Ferdinand Bischoff, who had accompanied Kennicott on the telegraph expedition and had come to Washington to testify regarding the natural resources of the region. While stationed for a year at Sitka, Bischoff made numerous collections "of great extent and value," and Baird now wanted him to return to the North under the joint auspices of the Academy and the Smithsonian. Fortuitously for science, Bischoff "was made the object of matrimonial attack by a German woman who persecuted him continually," so when given the opportunity the confirmed bachelor fairly leapt at the chance to leave Washington.[62]

Little is known about Bischoff.[63] He may have been one of the many Germans who fled to America after the failed uprisings in 1848. Kennicott referred to him as an insect collector and taxidermist from Peoria, Illinois, and by 1865 Bischoff was working at the Academy, where Kennicott described him as a "glorious old fellow." Shortly thereafter Bischoff joined the telegraph expedition.[64]

Baird confidently predicted that the Bischoff undertaking would be "of transcendent importance" in mapping the geographic distribution of species. "The whole question of the final determination of our Northern Vertebrate zoology depends in large part on the comparison of Kamschatkan species with ours" he wrote Stimpson. Baird hoped that Bischoff's collecting could settle all doubtful questions.[65]

Baird and the Academy trustees proposed to jointly underwrite Bischoff. "I would like to keep the matter in [the] hands of Chicago Acad. and Smithsonian and without funds of anybody else" explained Baird. He was aware that the Boston Society of Natural History had emerged as a suitor, offering five hundred dollars. The Academy trustees agreed to pay $750 in gold over three years and Walker made clear his expectations. "We ... feel that, having thus far controlled the Arctic explorations, we want still to keep them in our hands and not allow other parties to participate in the expense or have any share of the spoils except by gifts. We therefore are very anxious that no one should join in this matter except the S. I. and ourselves."[66]

Given Baird's praise of Bischoff's abilities, Stimpson's expectations ran high as he awaited eight large crates crammed with starfish and crustacea that

Bischoff had previously collected on the telegraph expedition. To his dismay he found an immense collection that "would be very valuable were it in good condition." Potential new species had not been prepared and packed properly, and as a result hundreds of dried crabs arrived with most of the legs broken off. It was nearly impossible to tell which ones went with which body and with all his other duties Stimpson did not have the time to figure out this perplexing three-dimensional puzzle.[67]

One of the birds that Bischoff had collected on the telegraph expedition provoked a storm of controversy that helped sour Stimpson and the trustees on the Bischoff venture. A single small owl proved to be a new species, and Stimpson wanted it described in the Academy's *Transactions*. Baird instead turned the bird over to Daniel Giraud Elliot, who wrote a description and commissioned a drawing. Elliot had a book coming out and wanted to include the new owl, but Baird told him that the Chicago Academy had the privilege of presenting the bird to the world. Stimpson had made this crystal clear to Baird—"Elliot must of course publish *first* in our Trans."—and later reiterated, "We can publish the new owl immediately—don't let Elliot publish it first, which would be contrary to our agreement and have a bad effect here."[68]

Anxious to establish priority, Elliot did publish the description and figure of what is now known as the western screech owl in the *Proceedings of the Academy of Sciences of Philadelphia*. An absolutely furious Stimpson dashed off a note to Baird saying that the Academy did not want to publish it second-hand. The trustees were especially miffed, "as what induced them to subscribe to Bischoff's expedition was the desire to keep the whole North Pacific matter in our hands, from the origin in Kennicott's expedition down, but really the returns have been exceedingly small compared with the outlay, and I find it difficult to explain matters."[69]

Baird rather lamely claimed that Elliot had misunderstood his instructions and attempted to soothe ruffled feathers by sending more specimens to Chicago. The Academy did later republish the description and figure, but Stimpson continued to lambaste Elliot, warning Baird, "we'll know how far to trust him next time."[70]

Bischoff specimens from the new undertaking began to trickle into the Academy in 1868, but Walker and Stimpson expressed disappointment at the results. Walker agreed to furnish the final installment of $250, and Stimpson implored Baird to make sure Bischoff had enough alcohol, but when the next shipment arrived in June it again proved to be a missed opportunity. The invertebrates were dried specimens and "almost utterly worthless" because they had been dried with the salt in them. Stimpson could not understand how Bischoff lacked alcohol.[71] Bischoff's relative fail-ure illustrates the hazards of a museum relying on individual collectors. The large outlay for Bischoff prevented the Academy from funding other worthy explorations, and Stimpson had long since realized that their money would have been much better spent on Dall, who was just beginning a long and brilliant career in natural history.[72]

CHAPTER 11

AMERICAN SCIENCE VISITS THE WEST

What a contrast there is between P. T. Barnum, with his great museum of curiosities, and a patient, unwearied student and naturalist who spends a long life, impoverishing himself and family, in his zeal for a science in no degree lucrative—like the late Dr. Gould of Boston. Whose labors are worth the most to the world, those of Louis Agassiz or the great American showman? We think we would rather take stock in the scholars after all.[1]
—George M. Kellogg (1869)

Known today for its world-class museums, in 1867 Chicago had nothing remotely worthy of the term. When people heard the word *museum* they probably thought of Colonel Wood's Museum. Very much in the style of P. T. Barnum's museums, it contained natural history "curiosities" but mainly featured theatrical performances, concerts, and magicians.[2] In 1868 the Chicago Historical Society erected a "fireproof" building of its own, with a collection of books, historical documents (including an original copy of the Emancipation Proclamation), artwork, and Civil War relics.[3]

Throughout 1867 Stimpson supervised construction of the new Academy museum building. He had very specific ideas about how things should be organized, and he relished the opportunity to help design a museum that combined attractive displays with adequate storage for scientific collections and ample working space for naturalists. He knew that few museum directors ever get a chance to leave their imprint on a brand-new facility. As one contemporary noted, "it requires a naturalist to plan a natural history building."[4]

The Chicago Academy had a prime location, on Wabash Avenue just north of Van Buren St., two blocks from Lake Michigan and about a mile and a half from city hall. The building on the front of the lot was renovated and turned into a boarding house to generate income, while the back lot would house the "strictly fire-proof" museum building. Construction began in June, exactly one year after the 1866 fire. One of Chicago's leading architects, William W. Boyington, received the commission. No stranger to the Academy or to science, Boyington had subscribed five hundred dollars to become a life member in 1864 and had also designed the Dearborn Observatory. He had just begun work on the Chicago Water Tower, today his most famous and enduring building. Incredibly prolific, Boyington had at least seventeen other buildings in the works in 1867, a punishing schedule that prevented him from giving much time to any single project.[5]

As with many construction projects there were countless delays and Stimpson laid the blame squarely on Boyington's overworked shoulders. Fuming, he told Baird that they had "employed an architect to *spoil* the building. *I* could have done better without him." In December, Stimpson finally began the complex task of transferring the collections to the new building. Moving tens of thousands of natural history specimens requires careful planning and great patience, especially in the days before bubble wrap. Four hundred million-year-old fossils, huge bottles filled with rattlesnakes, trays filled with delicate insects—each presented special challenges. As the carpenters and plasterers put the finishing touches on the interior, Stimpson spent two months carefully filling each room. He did not relax until the last objects were taken out of the Metropolitan Block in early January 1868. Two weeks before the scheduled opening date, Stimpson admitted to Meek, "I have been all in confusion" as he prepared the building for its public debut.[6]

Plain but substantial, without exterior frills or ornamentation, the museum was quite unlike Boyington's other more elaborate designs. In fact no known photograph or illustration of the building exists. The dimensions were hardly imposing: fifty-five feet long, fifty feet wide, and fifty feet tall. Forgoing an ornate façade, every possible step was taken to make the building resistant to fire, with the *Chicago Tribune* noting some of the safeguards: "The foundations were carefully laid and the walls made very thick and strong. Heavy iron shutters protect every window, and iron doors have been placed at both points of entrance. The stairways are of iron, thus making each story fire-proof by itself. The floors are all laid in concrete, and supported by iron girders, with brick arches between them." A large bank vault in the basement protected important manuscripts.

The museum consisted of a large hall with a thirty-four-foot high ceiling. Two galleries, one above the other, contained nearly two dozen table cases of black walnut mounted in three tiers and with a total of 780 drawers, all of them dust- and insect-proof. On the south and west ends were cases for the display of specimens in alcohol. The room would be lit by six large windows on the east and west sides and a glass skylight. At night three large chandeliers provided gas lighting.[7]

The opening night gala in late January 1868 drew a large crowd despite frigid temperatures. There were several nods to the Academy's past in the evening's festivities. A portrait of Kennicott, painted by Henry Ulke, looked down upon the proceedings. Stimpson also unveiled a new Academy membership certificate that he had designed which included portraits of Kennicott, Jared Kirtland (the dean of Midwestern naturalists), and Franklin Scammon, an amateur botanist and the older brother of trustee J. Young Scammon. Framing them were illustrations of plants, fossils, and animals from the Midwest, including mastodons, bison, and cranes. Beneath was a view of Chicago and Lake Michigan and the motto "Lacus Nostri Maria," Latin for "Our Lakes are Seas." The certificate symbolized that Kennicott's dream of a "Smithsonian of the West" had begun to be realized.[8]

Stimpson recounted for the members and guests the growth of the museum during the past year: three thousand insects, two thousand crustaceans, and fifteen hundred birds for a total of twelve thousand new specimens. The Academy had shared its good fortune by sending duplicate sets to a dozen museums in the United States, Canada, Europe, and South America, as well as to schools and colleges. The library had likewise increased at a stunning rate, with Stimpson boasting that it had become "the most important collection of its kind in the West." He personally showed his commitment by donating the remainder of his personal library. According to the trustees, the library made the Academy "a more social institution" than it had been before, a place where those with an interest in science could browse the latest journals from around the world. Thanks to the *Transactions*, they now exchanged publications with over sixty kindred organizations, more than half of them in Europe, and planned to open ties with even more. Stimpson again publicly acknowledged a debt to the Smithsonian's "fostering care of our young association, and through it and by its help many of the most important operations of the Academy are carried into effect."[9]

Trustee George Walker followed Stimpson and spoke of the numerous fires that had plagued the city. Now presumably impervious to flames, the Academy could rest easy. "While formerly every sound of the fire-alarm startled the friends of the institution, now you can all rest in security, feeling that the results of years of labor will not be consumed in an hour." Stimpson echoed this confidence, assuring potential donors that everything given to the museum would now be "perfectly safe from fire." For the past year it had been difficult to hold regular meetings and attendance had been sparse. Walker now practically pleaded with members to attend meetings. "Around this building should centre the scientific interest of our city," he stated. Stimpson figured that they would outgrow their snug little structure in five or six years, as it had always been envisioned as an adjunct to a much larger museum complex to be erected on the front part of the lot on Wabash Avenue. The current building served as "the entering wedge which shall open the hearts of the good citizens of Chicago to furnish the necessary means to erect, upon the front of this property, a building which shall always be sufficient for the uses of the Academy."[10]

Given their population and respective histories, it is remarkable that Chicago had a natural history museum building before New York, a fact noted at the time.[11] There were only a handful of natural history museums in America that had their own building and were not associated with a college or university. Located at least seven hundred miles away from the centers of American science, the Chicago Academy was the only institution west of the Alleghenies to fund the building of a large natural history museum between 1850 and 1870. Like other museums of the period, the Academy attempted to balance scientific research on the collections (mainly taxonomy) with public education through exhibits that were instructive and morally uplifting. Joel J. Orosz argues that this template for the modern American museum was in place by 1870.[12]

The other major natural history museums in America had much better sources of funding than the Chicago Academy. The Smithsonian, of course, had the Smithson bequest and some federal money; the MCZ, thanks to Agassiz's lobbying, received backing from the State of Massachusetts and wealthy Boston Brahmins. The Boston Museum of Natural History and the Academy of Sciences of Philadelphia were based in cities with a long tradition of supporting intellectual causes. The Chicago Academy made do with funds provided by a handful of trustees.

The only cautionary note amidst the celebration came in a discussion of the institution's finances. Walker had personally loaned ten thousand dollars to complete the building but there was still much more to be done, including the construction of several display cases. Barely a week after the much-ballyhooed opening and the issuance of the *Transactions*, Stimpson told friends that they lacked money to pay basic expenses and that unless membership increased the Academy might have to "suspend" operations. An appeal to the members for more money went nowhere, and the trustees began to express frustration over paying the lion's share of the Academy's expenses. In its eleventh year of existence the Academy had only fifty resident members in a city now approaching three hundred thousand people.[13]

As is often the case, the Academy's achievements resonated farther from home. A reflection of that increased stature came with the announcement that the American Association for the Advancement of Science (hereafter AAAS) would hold its 1868 meeting in Chicago. Barely thirty years a city, Chicago would now play host to the country's leading scientific organization.

The AAAS was struggling to reemerge as a unifying force for the American scientific community after having suspended operations during the Civil War. The two postwar meetings had each attracted fewer than eighty members, less than half the total for the prewar meetings. The *Chicago Tribune* blustered, "It is reserved for Chicago to infuse new life into this association, to restore to it something like its former extent and importance, and to impress upon it the stamp of Western progress." Here were the goals of postwar Chicago writ large—to transform the nation through Chicago's indomitable energy and spirit of enterprise.[14]

The choice of Chicago reflected a desire by the AAAS to expand the organization's reach. This would be the first meeting held west of Ohio, and while some western scientists resented the eastern dominance of the AAAS, they realized that the meeting would give them a chance to show that they were more than mere data gatherers whose chief role was to provide raw materials for easterners to work up. Realizing the potential for furthering the cause of science in his adopted city, Stimpson looked forward to showing off the museum. While his close friends knew of his work at the Academy, other more casual acquaintances were probably curious as to what he had been doing for the past few years, given that he had not published any scientific papers of late.

Stimpson had not been an especially active member of the Association but fondly recalled the 1849 meeting at the very beginning of his scientific career. In 1867 he had been appointed (along with heavyweights such as James D. Dana, Baird, Jeffries Wyman, and Joseph Leidy) to an AAAS committee charged with updating the rules of zoological nomenclature.[15]

For at least six months prior to the meeting, Stimpson, as secretary of the local planning committee, saw to all the mundane but critical details that go into making a successful national conference. Hoping to attract a cross section of American scientists, he distributed nearly three thousand circulars. In a spirit of reconciliation after the war he particularly wanted southern naturalists to come, and to that end he invited the members of the moribund Elliott Society. When he did not receive a reply to his first circular he implored Lewis R. Gibbes, "We are very anxious to see the South well represented ... and I myself should be very glad to see you once more. Do try to come!" Trustee William Doggett even promised to pay Gibbes's travel expenses and put him up, but it soon became apparent that the southern naturalists were not coming. The war had left many of them destitute and no doubt bitter.[16]

Joseph Henry and Baird knew just how far the Chicago Academy had come in creating an important repository for natural history collections, and Stimpson fully expected a large contingent from the Smithsonian. He began cajoling Henry months before the meeting, writing, "It will be a great benefit

to the Academy to have you come, and you will of course have no difficulty
about passes. Do not disappoint us." Stimpson sent railroad passes for Baird
and Joseph Henry and his daughters, but none of them came to Chicago, leaving him extremely disappointed.[17]

Stimpson became increasingly harried and preoccupied in the days leading
up to the opening session. Adding to his stress, his father, stepmother, and
stepbrother came to stay with him and Annie at Oak Park during the meeting,
while Annie delivered the news that she was pregnant and due early in 1869. It
did not help that July 1868 was one of the hottest on record in Chicago, made
even more miserable by high humidity.

The seventeenth meeting of the AAAS opened on Wednesday, August 5
with the *Chicago Tribune* proclaiming that the course of science, like empire,
had begun moving west. Trustee J. Young Scammon's welcoming address
boasted of the city's cosmopolitan makeup, which he declared a harbinger of
hope for the future. Scammon felt that in time Chicago would become "one of
the fullest developments of humanity." He apologized for the city's few literary
and scientific accomplishments but added, "we rejoice that we can show you a
beginning not unworthy of a city of thirty years."[18]

Benjamin A. Gould, the new president of the AAAS and the superintendent
of the Cambridge Observatory, spoke next and paid homage to the Chicago
Academy with words that must have warmed Stimpson's heart. Gould called
Chicago a "magnificent city—sprung from the prairie shore like a creation of
Aladdin's lamp. . . . Yet here we find an active and earnest scientific spirit, a
scientific academy full of vigorous life, doing its part toward the increase of
human knowledge and toward disseminating the spirit of scientific inquiry."[19]

Never comfortable in the spotlight, Stimpson apparently made no formal speeches during the meeting, concentrating instead on making sure
that everything ran smoothly. He met every afternoon with other committee members at the plush Sherman House hotel, and they were quickly faced
with every convention planner's worst nightmare—the meeting rooms were
too small. The sessions were hurriedly moved to the recently completed First
Baptist Church, which had a spacious lecture room that seated six hundred
comfortably.

Stimpson no doubt attended the sessions dealing with natural history,
where he would have mingled with old friends such as Theodore Gill, John S.
Newberry, William P. Blake, and Alpheus S. Packard. Other notables in attendance included James Hall, Benjamin D. Walsh, B. Waterhouse Perkins, Albert
Bickmore, and Increase Lapham, but Stimpson could not help but notice those
who had not come, including most of the giants of American natural science:
Dana, Leidy, Agassiz, Baird, and Asa Gray.

Many of the presentations had a western flavor. John Wells Foster, soon to play a prominent role in both the AAAS and the Chicago Academy, prepared six papers, including one on a huge fossil beaver skull (*Casteroides ohioensis*) which had been discovered twenty miles west of Chicago. Amos Worthen exhibited invertebrate fossils from the famed Mazon Creek site in Grundy County, Illinois and he and Meek named a new fossil shrimp in Stimpson's honor, *Acanthotelson stimpsoni*. Edward Drinker Cope, on his first visit to the West and not yet feuding with another attendee, Othniel Marsh, gave an overview of the Dinosauria in which he recognized their affinity with birds and noted that only one hundred species of dinosaur were known from North America.[20]

Accounts of the meeting mention the extraordinary hospitality lavished upon the participants. On Saturday afternoon Stimpson led a sightseeing tour of Lake Michigan for 250 guests on the ship *Orion*.[21] Many Chicagoans embraced the proceedings and 135 of them joined the association, including lawyer Robert Todd Lincoln, son of the former president, as well as nine women including Elizabeth Atwater and Kate N. Doggett.[22] The *American Naturalist* enthusiastically applauded the "unusual interest and vigor" of the meeting and called it a "brilliant success." Chicagoans were delighted, "not because we have any very great amount of scientific enthusiasm among our people," apologized a local reporter, "but because everyone seemed to have a good time.[23]

It had been a grueling yet exhilarating week for Stimpson. Relieved when it was finally over, he modestly announced to Baird, "The meeting is considered a success generally I believe." His fellow trustees gave him most of the credit. "Dr. Stimpson devoted nearly his entire time to this matter until the meeting closed and the perfect system in all the arrangements, and the great success of the 'local Committee' was almost entirely due to his continued exertions." The leadership of the AAAS also praised Stimpson by voting thanks to the "indefatigable Secretary" of the Academy for the "unusual personal enjoyments which have characterized this meeting." The only complaint voiced was that the many parties had left participants utterly exhausted.[24]

The scientific convention clearly heightened interest in science in Chicago, and donations to the museum increased significantly in the months afterwards. The Academy took an important step designed to keep the momentum going when it voted in November 1868 to open the museum free to the public every Saturday.[25] In 1869 the Academy received tax-exempt status from the Illinois legislature, which recognized its public education component.

In the years that followed the Academy became a magnet for all manner of scientific inquiry in Chicago, welcoming new constituents under the Academy's umbrella. The Audubon Club of Chicago agreed to house their collection of birds and their library at the Academy; the State Microscopical Society of Illinois held their inaugural and subsequent meetings there; the new Chicago Botanical Society became an affiliate in 1869, the same year that the first women members were accepted; and an Academy report helped provide the impetus for the creation of Chicago's public parks system.[26] The AAAS meeting and the events that followed marked a new awareness of the Academy's role in promoting science in Chicago.

A few weeks after the AAAS meeting the National Academy of Sciences elected Stimpson as one of its fifty members. At the time Stimpson was the youngest (thirty-six) American-born zoologist. Admittedly, the National Academy had only been around for five years and had yet to blossom into an effective scientific body, but Stimpson received the honor long before many of his peers. He had been recognized by an organization that represented not just the natural history community but American science in general. Reflecting the slow westward expansion of science, he was one of just a few members that lived west of Ohio.

Joseph Henry, who had reluctantly assumed the presidency earlier in 1868, had pushed for more rigorous standards for membership based on original research. Henry had been most impressed with Stimpson's Hydrobiinae monograph a few years earlier, and while he agreed with Louis Agassiz that physiology was a "higher order of scientific investigation than the description of species" he argued that "the *amount* of labor as well as the *kind*" needed to be taken into account.[27] While his publications were the main factor, Stimpson was no doubt recognized in part for his efforts towards promoting science in Chicago. Others have noted that Henry tried to bring younger blood into the still struggling organization.[28]

Stimpson was no stranger to the fifty men that constituted the inaugural class of the National Academy in 1863. He had known fifteen of them fairly well, including his skipper on the North Pacific Expedition John Rodgers, as well as Dana, Agassiz, Leidy, and Henry. What Stimpson may not have known is that he was nearly selected two years earlier. In 1866 the choice came down to him or Alexander Agassiz. On hearing the news that Agassiz had been chosen, Henry wrote Baird, "on the whole I would rather the choice had fallen on Stimpson."[29] Henry worried that the vote would reinforce the common belief that the National Academy was nothing more than an extension of the "Scientific Lazzaroni," a small group of influential scientists (including Louis Agassiz and Benjamin Peirce) who met informally and wielded enormous influence on American science.[30]

The AAAS meeting and his election to the NAS highlighted the significant changes that had taken place in Stimpson's scientific life. From 1849 until

1865 he had been known as a field naturalist and taxonomist. Since coming to Chicago he had instead tackled administrative and management responsibilities relating to the Academy. In some ways this was a similar progression to that experienced by the other great American natural history museum builders of the era—Baird, Agassiz, and Joseph Leidy.

In the days immediately after the AAAS meeting he was simply exhausted, but as summer turned to fall, often an abrupt occurrence in Chicago, he began to feel sick and lethargic. He developed a persistent cough and a low-grade fever followed by the most dreaded symptom of all, hemorrhages from the lungs. He had fallen victim to pulmonary tuberculosis, the most feared of all nineteenth-century diseases and the one that had killed his mother. At what should have been Stimpson's moment of triumph he suffered a physical breakdown that would plague him for the rest of his life.

It is unclear exactly when Stimpson contracted consumption, as the disease was colloquially known for its habit of literally consuming its victims. His letters refer to respiratory problems several times, including before, during, and immediately after the NPEE and again in the spring of 1863. Tuberculosis can be highly contagious. Fielding B. Meek, perhaps his closest scientific friend, had contracted the disease, and the two men had spent many a day together in a small workroom at the Smithsonian. The only thing we can say with certainty is that Stimpson battled tuberculosis from September 1868 until his death.

The mental aspects of such a long-term, usually fatal malady are perhaps even more terrifying than the physical manifestations. One twenty-seven-year-old victim lamented his fate, "to have so long a death, and feel its slow, constant, and sure approach." The "Red Death," Edgar A. Poe's metaphor for the disease, accounted for at least 20 percent of all deaths in nineteenth-century America. The unpredictability of the affliction added to the psychological toll, since tuberculosis sufferers might have an attack and go into remission for years or even decades, or they could die quite suddenly from a major relapse.[31]

Stimpson reported coughing up nearly half a pint of blood at a time, leaving him limp and almost totally debilitated.[32] He was confined to his home in Oak Park, and Academy president Dr. Edmund Andrews monitored his condition daily. The etiology of the disease was still unknown, and doctors could only recommend fresh air in a warm climate, rest, and a healthy diet. Chicago's harsh winters and the change of seasons tended to aggravate the condition, with death rates soaring between October and December.

There are few references in Stimpson's letters to his disease, but his friends were more forthcoming. George Walker cautioned Baird, "The Dr. has been very sick indeed. He has had bad *hemorrhages* but I am thankful to say they have been stopped and he is beginning to improve. . . . I feel quite worried about him." Walker insisted that Stimpson needed to stay with friends during an upcoming visit to Washington. "He is very sensitive about accepting private hospitality but I think he ought to be *forced* to do so at this time. He ought not to sleep or live very far from his friends for some time for it would go hard with him if he should have another attack and be at the time with those who would not take proper care of him." Fearing for his friend's life, Blatchford pleaded with Baird. "Do not let our valued friend Stimpson over work himself."[33]

With Annie five months pregnant, Stimpson found himself in the unenviable position of being dependent on his wife when he should have been caring for her. He despised being an invalid and had always prided himself on being a hard worker, a trait that now might prove fatal. The trustees granted him a paid leave of absence, and Stimpson escorted Annie and Willie to Ilchester before going on to the Smithsonian.

Lodging once again in Smithsonian towers he soon learned of the hijinks of a new generation of young Smithsonian naturalists, led by Illinois ornithologist Robert Ridgway.[34] His health somewhat improved, he took up work on his oft-delayed book on Atlantic Coast mollusks with a new sense of urgency. He had a large number of new genera and species that had yet to be published, and at some point he had expanded the geographic scope of the book to include Florida, a daunting but necessary addition to his mind. He shared proofs of several of the illustrations with Lewis R. Gibbes while continuing to express concern over Edmund Ravenel's invaluable shell collection.[35]

Early in 1869, the cold, damp climate of Washington irritated Stimpson's lungs, and doctors advised him to travel without delay to the more hospitable climate of Florida. The Sunshine State was where Ponce de Leon sought the fabled fountain of youth; Stimpson's hopes were more modest but he knew that tuberculosis patients had found relief in the gentle ocean breezes and equitable weather. It had only been seven months since Florida had been readmitted to the Union, and the entire state contained less than two hundred thousand people, significantly fewer than the city of Chicago. Beginning in 1867 Pourtales and Louis Agassiz had begun dredging for marine invertebrates in the Straits of Florida, but little had been done closer to shore.

Reluctantly leaving a very pregnant Annie and three-and-a-half-year-old Willie, he traveled from Washington on January 14, 1869, in the company of two companions who shared his enthusiasm for natural history. Forty-two-year-old Robert E. C. Stearns had, like Stimpson, been born near Boston. He later moved to California, where he became interested in mollusks and joined the California Academy of Sciences. A newspaper publisher by trade, up to this time Stearns could be characterized as a dabbler in natural history. The trip would be transformative for him, as with Stimpson's encouragement he began to publish descriptions of new mollusks. Like Robert Carter, Stearns published an account of his time in the field with Stimpson. The *American Naturalist* published Stearn's four-part "Rambles in Florida," an entertaining series that displays both a keen eye for detail and a dry sense of humor.[36]

The third member of the party, seventy-seven-year-old Ezekiel Jewett, had served in the War of 1812. In the 1840s Jewett made an outstanding collection of fossils in New York and was named curator of the State Museum, working for a year with Meek. At the age of seventy he volunteered to serve in the Union army and was disappointed to be rebuffed. Jewett performed his greatest service to science by encouraging young Othniel Charles Marsh to pursue his passion for paleontology.[37]

The trio took a train to Charleston and from there caught a steamer south to Fernandina Beach on Amelia Island. While the weather was unseasonably cool, Stimpson began to regain some of his strength, allowing him to briefly explore the shell mounds in the vicinity. The next day they took a train across Florida to Way Key, one of the dozen islands comprising the Cedar Keys on the Gulf of Mexico. The accommodations were decidedly primitive, "a practical joke upon the traveler" according to Stearns. The cloudless skies and bright sunshine produced a feeling of "quiet and dreamy beauty," and during a brief stay they again examined shell mounds, sometimes stumbling over the bones of yellow fever victims.

Stimpson had always been a boon companion in the field and Stearns later recalled his impressions of Stimpson. "I shall never forget the delightful season passed in his company, the pleasant toils of each day, and of the rehearsal of each day's triumphs in the evening as we sat in front of the blazing fire of pitch-pine, which lighted up his face with a glow less genial than the smile which played around his lips; or when some joke more pungent than usual was uttered, the explosion of laughter which followed, and which was joined in by none more heartily than himself."[38]

At Tampa, a town of perhaps eight hundred souls "of all sizes and colors," their rented house was dubbed Camp Misery after they were besieged by fleas. Stimpson actively collected bird eggs along the Hillsborough River, where they found thousands of herons, cranes, and egrets among the mangrove swamps, although they did not manage to bag any "pink curlews" as the flamingoes were called.[39] Back at camp they enjoyed feasting on bananas, fresh grapefruit, and oranges, the "golden fruit."

During a three-week stay Stimpson had time to closely examine the numerous shell mounds that dotted the landscape, some covering acres and towering fifty feet high. The only other scientific author who had commented on them, Timothy Conrad, concluded that they were natural formations.

Stimpson felt sure that they were man-made, and the largest ones were different from any shell mounds yet described. He believed that they were not mere refuse heaps (known as *kjoekkenmoeddings*) but seemed to have been built for a purpose. "The shells ... are not such as indicated merely the '*reject-menta*' of aboriginal feasts, being of all sizes from that of *Littorina* to that of *Busycon*, and often showing evidence of having been dead when placed on the mound; some, indeed, showing remains of barnacles attached to their inner surfaces," he remarked. He speculated that the mounds had been built as places of refuge during storms; these platform mounds were indeed used for this purpose.

In a mound near the mouth of the Manatee River he discovered a tool more than a foot in length that resembled an augur and was made from the shell of one of the largest gastropods, the Florida Horse Conch. He also uncovered pottery, manatee and turtle bones, and other shell implements that showed signs of great age. While these forays into archaeology were interesting, Stimpson spent most of his time on the prowl for mollusks, combing the beaches and dredging in shallow Tampa Bay amid beds of waving sea grasses and the playful caperings of manatees. Comparing the shells from the mound with species known to inhabit that section of Florida, he found that some of them were now rarely if ever found alive in the vicinity but were abundant on the barrier islands of the coast. He tried to account for this by theorizing that when the mounds were built the barrier islands were smaller, thus forming a less considerable bar to the waters of the Gulf.[40]

Struck by the fact that the mollusk species from either side of the Florida peninsula were very different, he wrote a brief abstract on the subject for the *American Naturalist*. The large number of species found on only one coast or the other reminded him of the similar, albeit greater, difference in fauna that existed on both sides of the continent. During his brief stay Stimpson found 314 mollusk species, of which less than half were common to both sides of the Florida coasts. The tropical character of Florida's west coast shells he correctly attributed to the influence of the Gulf Stream.[41]

Stimpson credited the climate of Florida with reinvigorating him, and he returned to Ilchester greatly refreshed. At some point Annie must have telegraphed the news that she had given birth to twins, possibly prematurely, two weeks after he had left Washington. Thus Stimpson did not see their babies, a boy and a girl, until they were more than a month old. They christened their daughter Marie Gordon and their son Herbert Baird, the latter after Stimpson's two best friends—his father and Spencer Baird. Flattered, the latter responded in a typically self-deprecating vein, "I hope he may be none the worse for the association."[42]

Returning to Chicago in early April 1869 after a five-month absence, Stimpson had no sooner arrived when news from Baird sent him scurrying back to Washington. "Alex Agassiz and his father will both be here. Come on early to look after your interest," Baird had warned him.[43] The Agassizes had planned to attend the semi-annual meeting of the National Academy of Sciences, and Baird may have feared a raid on the Smithsonian invertebrates. A year earlier, MCZ curator Hermann Hagen had asked to borrow the Smithsonian's crayfish, leading Baird to ask Stimpson whether he had any plans to work on this group. Stimpson replied, "I suppose I shall have to let them go, which I regret very much as I suppose they will go the way of the other things we have sent to Cambridge."[44] Using the power of the sea for an analogy, he delivered the best characterization of Louis Agassiz's reputation for failing to return specimens lent to him when he wrote that certain missing Smithsonian specimens might be "in the great whirlpool at Cambridge."[45]

After attending his first meeting of the National Academy, Stimpson had a long discussion with Baird about the preservation of the Smithsonian's marine invertebrates in alcohol. Ever since coming to Chicago, Stimpson had been anxious about these specimens, as even in the best-sealed jars the alcohol evaporated, meaning the jars needed to be checked and refilled on a regular basis. If these "wet" collections were allowed to dry out, the animals would be useless for anatomical investigations.[46]

The two men hatched a plan to send all the Smithsonian's invertebrates in alcohol to Chicago for Stimpson to curate. Joseph Henry, an ardent proponent of divesting the Smithsonian of its collections, readily agreed. The fact that the Academy had a "fireproof" building certainly made the decision easier.[47] Stimpson would be responsible for preserving and critically examining the collection and for arranging it into sets. The first suite would go to the Smithsonian and the second to the Academy, and remaining sets were to be distributed to museums designated by the two organizations. The Smithsonian paid for packing and shipping the collection to Chicago as well as for alcohol and jars. The Academy agreed to make the specimens available to researchers and to return the collection whenever the Smithsonian requested it.[48]

Stimpson thought this possibility unlikely, telling a fellow trustee, "Now I consider this collection in great measure my own, for I have built it up and spent much time upon it without pay during the past 20 years. It is founded, in fact on my own private collection, presented many years ago, so that there is little probability that the S.I. will demand it of us—and even if they should the moral effect of the existence of so large a collection in our Museum will have been of the greatest value to us, and by that time we will probably have most of the things from other sources."[49]

Stimpson stayed in Washington to personally oversee the packing of the crustacea and later boasted that only 1 percent of the thousands of bottles had arrived broken. Baird's assistant Solomon G. Brown did not take the same care with the mollusks, and a serious loss to science occurred when more than half of the specimens identified by Carpenter arrived "entirely destitute of *alcohol*." Stimpson placed the blame squarely on Brown by telling Henry that the bottles had loose cork stoppers and seemed to have been empty for months. His vexation was apparent when he lamented the "great loss, for the [US] Exploring Expedition things were among them." While many of the animals had disintegrated the shells at least were saved.[50]

It was not until late June that Stimpson made it back to Chicago "with baggage and babies" in tow. With the Academy on summer hiatus, he had time to work on mollusks again and pleaded with Meek to come to join him so they could renew their long-discussed collaboration listing the genera of fossil and recent shells.[51] Meek chose to stay in Washington, so Stimpson spent many hours in the Academy's basement that summer in the essential curatorial task of filling the jars with fresh alcohol. Even healthy individuals can be adversely affected by breathing in these fumes, and Stimpson's tuberculosis left him far from healthy. He began to experiment with alternatives to alcohol, which was extremely expensive and often difficult to procure. In the hopes of finding a cheaper and more reliable substitute he tried carbolic acid, more commonly known as phenol, and published a note on the subject.[52]

Stimpson now found himself with virtually no help in the museum, and his efforts at finding a replacement foundered. He offered forty dollars a month and a good room with fire and lights included but found no takers. Out of desperation he asked Baird if he knew of any "capable and docile young naturalist who would like to take the place of student-assistant to me." It took almost two more years before Stimpson would find a reliable assistant.[53]

The press of Academy business forced Stimpson to decline an invitation to the 1869 meeting of the AAAS held at Salem. He regretted not being able to go, especially when he learned that Louis Agassiz and many of his former students had met and reconciled some of their differences. Stimpson missed out on these "good times," and shortly afterwards Agassiz suffered a cerebral hemorrhage.[54]

By the end of 1869, the Academy museum housed one of the best collections of marine invertebrates in North America, indeed in the world. Trustee George Walker shared Stimpson's enthusiasm for mollusks and gave him the go-ahead to assemble a complete library of works on the subject. Nearly giddy at the prospect, Stimpson wrote Meek that Walker had given him "*carte blanche* to purchase conchological works ... to form a complete conchological library & to go to any *extent* of *expense I choose*! Good, is it not? I think we will soon be up to time, in Chicago, in the shell way."[55]

Perhaps even more impressive were the ten thousand jars of crustaceans, including many types from Stimpson and Dana. Academy president Edmund Andrews noted that the collection was second in size only to that of the Jardin des Plantes in Paris but far superior in value, since most of the animals were preserved in alcohol while the Paris collection consisted mainly of dried specimens.[56] No serious carcinologist or malacologist could now afford to ignore Chicago. While all of this meant more work for him, the presence of these collections promised to be a boon to Stimpson's long-delayed research and writing, as he now had ready access to nearly all of his books and specimens.

The Smithsonian marine invertebrates were by no means the only new collections received in 1869. Baird had sent a suite of South American hummingbirds, hundreds of bird eggs, and bones from the fossil Irish elk (*Megaceros giganteus*), the latter a rare commodity in American museums.[57] Another acquisition was a mummy from Peru that became the subject of a report by Academy member Norman Bridge.[58] One donation came direct from a bloody Western battlefield. A Chicagoan had accompanied troops on a surprise attack against an Apache village that killed a medicine man, whose set of surgical instruments was confiscated.[59] J. Y. Scammon also presented the large herbarium of his late brother Franklin.

Local rock hounds had discovered one of the most extraordinary fossil assemblages ever brought to light in the 300-million-year-old Mazon Creek fossil beds, located sixty miles southwest of Chicago. The fossils contain an astounding variety of rarely preserved animals, including some of the earliest known spiders, cockroaches, and dragonflies. This veritable mother lode naturally attracted Stimpson's attention, and he expressed dismay when a valuable local collection was destroyed in a fire, telling Meek, "Such fine things ... ought to be deposited in a fireproof building, such as our Acad. offers for instance!"[60] Stimpson also wanted to acquire the fossils of the Illinois Geological Survey. Amos Worthen had given him hope that the governor would order the collection moved to the Academy, but the lobbying went for naught and the collection stayed in Springfield, where it suffered major damage in a February 1871 fire.[61]

Pursuit of yet another state collection proved more fruitful. Stimpson had met Illinois' state entomologist Benjamin D. Walsh at the 1868 AAAS meeting. Walsh had moved to the United States from England in 1838, eventually ending up in Rock Island, Illinois. An avid Darwinian, Walsh had come out in favor of evolution and brilliantly skewered Agassiz's arguments against it.[62] Walsh had amassed some thirty thousand insects including many types, which one source called "a standard reference of national significance."[63] As he walked home on the railroad tracks one day in 1869, absorbed in reading a letter, a train struck and killed him.

A battle developed over the collection as Charles Valentine Riley, a protégé of Walsh's, tried to claim the collection, as did Walsh's widow. Riley, affiliated with the Academy of Science of St. Louis, which was trying to rebuild after a devastating fire, may have acted out of genuine fear that the collection would not be properly curated. Mrs. Walsh won custody, however, and sold the specimens to the State of Illinois for $2,500, which, acknowledging that they did not have adequate storage facilities, sent the insects in October 1870 to the Academy for safekeeping. Although only provisionally the Academy's, their presence, like the Smithsonian marine invertebrates, helped establish the Academy as a major repository for natural history collections.[64]

CHAPTER 12

A FISH OUT OF WATER

In fast and friendly Chicago, weeks go by like days, and days like hours, and life is almost too rapid to be chronicled . . . all is astir here. There is no such thing as stagnation or rest. Lake-winds and prairie-winds keep the very air in commotion. You catch the contagion of activity and enterprise, and have wild dreams of beginning life again, and settling—no, circulating, *whirling*—in Chicago.
—Sara Clarke Lippincott (1871)[1]

While all the new additions to the museum barely dented the public consciousness, two events in late 1869 garnered a great deal of publicity for the Academy. At the October meeting John Wesley Powell, recently returned from his epochal voyage down the Colorado River, gave a stirring account of his recent exploration. As on Powell's trip in 1867, the Academy had provided assistance in outfitting Powell, whose exploits captured the imagination of the American public and made him a hero. Stimpson had the difficult job of serving as the opening act that night, and his more scholarly report on Florida shell mounds was overshadowed by Powell's adventure.[2]

Weeks later, Stimpson had his own moment in the spotlight. While on a buggy ride near his home in Oak Park, his discerning eye noticed a series of low mounds near the Des Plaines River. Having seen similar mounds in Florida he recognized them as man-made and returned with his neighbor Charles Kennicott, Robert Kennicott's older brother, for an impromptu dig. (Charles suffered from mental illness and was eventually institutionalized as insane.) Over the course of several days they unearthed the remains of twenty individuals from a grouping of Native American burial mounds, including two complete skeletons. Stimpson noted that the bones were thrown together "promiscuously, as if the rites of sepulture had been hurried-perhaps at the close of a battle." No tools or weapons were found in what came to be called the Stimpson Mound, and subsequent investigations revealed an even more extensive group of mounds about three miles north, dubbed the Kennicott Mounds.[3]

The Chicago papers gave extensive coverage to the finds. The *Chicago Times* proclaimed, "The sneer of European savans that America has no history is rapidly being refuted. . . . A discovery has just been made in the suburbs of Chicago which it is believed, will add much toward the solution of problems connected with the earliest inhabitants of North America." Just a few weeks earlier the New York press had trumpeted the discovery of the "Cardiff Giant," which some believed to be the petrified remains of a ten-foot-tall Indian. The *Chicago Evening Journal* claimed that Chicago would not be outdistanced by New York in the field of archaeology, and people in the Windy City were vastly amused to learn a few weeks later that the so-called giant was a hoax, perpetrated in part by a Chicagoan.[4]

The biggest debate in American archaeology in the 1850s and 1860s concerned Indian mounds, and at the next two Academy meetings John Wells Foster led the discussion.[5] Foster was one of the leading proponents of the so-called Mound Builder myth, which theorized that the mounds had been built by a lost ancient civilization, variously attributed to Vikings, Phoenicians, Hebrews, or people from Atlantis. Native Americans could not have built such imposing structures, the argument went, and they had in fact exterminated the gentle mound builders. Foster characterized Indians as degraded and brutal savages who could no more have built the mounds than they could have built the pyramids of Egypt. Others have noted that the Mound Builder myth reached its peak at the same time that the wars against Native Americans were reaching their tragic conclusion.[6]

Stimpson missed these meetings, having made another trip to the Smithsonian to pack thirteen more boxes of marine invertebrates for transmission to Chicago. He also decided to bring the remainder of his private shell collection, returning to Chicago four days before Christmas.[7] Capping off a busy year, Stimpson set out to complete the publication of the first volume of the Academy's *Transactions*. Historians have noted that having scientists in control of publications was a key sign of the professionalization of American science.[8] This had been a very personal undertaking for him, as fully two-thirds of the two hundred pages were devoted to the legacy of Robert Kennicott and his last expedition. Stimpson spent months preparing Kennicott's biographical sketch, an article that serves as a fitting tribute to "the martyr-naturalist," as Baird dubbed him. The article is doubly important because it includes lengthy excerpts from Kennicott's 1859–1862 journal, a wonderful account of the joys and travails of a young field naturalist in the subarctic. The original was later lost.[9]

Other papers focused on Alaskan natural history. William H. Dall and Henry Bannister's provided the first comprehensive accounting of Alaskan birds, covering 212 species. Highlighting its importance, the *American Naturalist* devoted a separate review to the paper, calling it one of the most important contributions to American ornithology in a decade.[10] Baird followed with additional details on about a dozen species of Alaskan birds, accompanied by eight plates.

Seeing these illustrations through the press proved to be every bit as onerous for Stimpson as the previous plates had. Cassin had begun working on them in 1868 but died in early 1869, a victim of long-time exposure to arsenic and the lingering effects from his time in a Confederate prison camp. Baird's role as an intermediary with the printers again led to sundry delays and misunderstandings. By early September, six plates had been completed except for the backgrounds. When he received the proofs two weeks later Stimpson immediately objected, arguing that the coloring was "*very bad* and cannot be accepted." He demanded new backgrounds, since the "gentleman who pay for the plates are not satisfied with the old coloring." Having had his fill of complaints, Baird claimed that he never received Stimpson's letter (it is, in fact, in the Baird papers at the Smithsonian).[11] The trustees, having subscribed $1,200 for the colored plates, were not especially pleased with the results.[12]

The *Transactions* also featured two illustrated papers by Foster dealing with the antiquity of man in North America and the tools of the Mound Builders, respectively. Foster had become active in Academy affairs in 1868 and served as president of the AAAS in 1869. Two years later he would be elected president of the Chicago Academy. The final article, by Samuel Scudder, described Iowa butterflies collected by Joel A. Allen during his visit west in 1867.

The completion of volume one of the *Transactions* in early 1870 brought forth praise from the *Western Monthly*, Chicago's leading literary magazine. Publications such as this blunted criticism that Chicago was a purely mercenary place, and the writer concluded that, "this volume marks an era in the scientific progress of the West, being the first fully-illustrated work of the kind ever published west of the Alleghenies."[13] The Academy's collections, its professional leadership, the "fireproof" building, and now its scholarly publications were all ample evidence that the study of natural science in Chicago had reached a new level of maturity.

Chicagoans could now peruse the latest scientific literature from around the world at a new reading room in the Academy's library. Copies of the *Transactions* were displayed at the Academy's annual meeting in 1870. The night ended with a bit of macabre "comic relief" when Samuel A. Briggs exhibited the shrunken head of a South American Indian no larger than an apple, "with long, black, glossy hair, and fine hemp drawn through the nostrils." This bizarre and disturbing object was simultaneously both "hideous" and "ludicrous" and had somehow come into Stimpson's possession, no doubt a gift from one of his far-flung friends.[14]

At the beginning of his fifth year in Chicago, in January 1870, Stimpson penned a plaintive letter to Joseph Henry asking to be remembered to old friends. In what was by now a common refrain he spoke of missing the "pleasant halls of the Smithsonian" and wishing he were there "instead of here, so far away from Scientific sympathy." Later that year while Chicago was suffering through a terrible heat wave, Baird asked him for tips on dredging locales in Buzzard's Bay, one of Stimpson's favored spots two decades earlier. Expressing a heartfelt desire to join him, Stimpson replied, "My heart is not in the highlands, but 'on the Buzzard's a dredging the clams!'"[15]

Despite the Academy's many successes Stimpson was still stranded hundreds of miles from the ocean that had formed the basis of his career. He loved running a museum and had begun to put his stamp on the Academy, but he had not expected that he would still have to do virtually everything himself. He described his predicament to Alpheus Packard of the Essex Institute. "My time day and evening is about entirely taken up by Academy duties for I have to do everything that requires any knowledge beyond that of an ordinary clerk—I envy you Salem fellows and often wish I could get together such a coterie of savants here."[16] A scientific spirit was slowly but surely being kindled in Chicago, but Stimpson had not been able to gather a group of young naturalists to assist at the Academy. There would be no western reprise of the Megatherium Club.

More than ever he clung to correspondence to stay connected to the rest of the natural history community. Between 1865 and 1871 he averaged some thirty letters a year to Baird alone. They bear witness to his great affection for his friend. Stimpson also posted numerous letters to Meek, usually involving discussions of invertebrate taxonomy and morphology. A disagreement on crinoid anatomy between Meek and Elkanah Billings (founder of the *Canadian Naturalist*) led to another Stimpson pun. Siding with his friend, whose views ultimately prevailed, Stimpson wrote, "I wonder how Billings feels? Rather Bilious I imagine."[17]

Having long since made the transition from zoologist to museum administrator, Stimpson now at times seemed to live vicariously through the exploits of younger men. He took a keen interest in the dredging and malacological work of Dall, lauding his focus on anatomy as "the kind of work we need in mollusca now a days." Stimpson sent a collection of brachiopods for Dall to describe and Dall, in turn, asked Stimpson to review several of his unpublished manuscripts. He also named a new cetacean parasite in Stimpson's honor, a gesture that only a true naturalist would appreciate.[18]

On the last day of March 1870, during a typically cold and damp Chicago spring, Stimpson suffered a recurrence of hemorrhages in his lungs. He had been symptom-free for fifteen months, so this latest attack must have been disheartening. For a week the attacks came steadily, with even the slightest exertion sending him into a terrifying bout of coughing and blood-spitting.[19] At the height of his intellectual powers, Stimpson's life now began to slide into a depressingly familiar pattern marked by periods of intense weakness and suffering followed by a guarded and uncertain recovery. From his letters it seems clear that he succumbed to bouts of melancholy if not depression, common symptoms among tuberculosis victims.

Fortunately this relapse was of short duration and he had rebounded sufficiently to attend the Academy's regular meeting in mid-April. Once his health stabilized Stimpson began what would turn out to be groundbreaking research on the Great Lakes. The scientific investigation of American lakes was still in its infancy, and as Stimpson put it, "A knowledge of the animals and plants living at the bottom of the great North American lakes, the largest bodies of fresh-water in the world, has long been a desideratum." While these lakes were a relatively poor substitute for the Atlantic Ocean, Stimpson would have a chance to renew his lifelong love of dredging.[20]

Once again the Academy trustees provided the means to make it happen. On a fine May morning, George Walker's steam tug *Constitution* chugged out eighteen miles into the lake. Surprisingly shallow at that point (eighty-four feet), the lake's sand and gravel bottom yielded little save a new genus of leech, but this was enough to justify further study. Stimpson vowed to try again at Grand Traverse Bay in Michigan, some 650 feet deep. "I should like to get into that hole, though I should scarce expect to find marine things as they did in Lakes Wenner and Wetter" he confided to Baird. Stimpson had read Sven Lovén's landmark paper identifying marine-like shrimp from two Swedish freshwater lakes, but he thought the chances of finding them in Lake Michigan were remote.[21]

Ninety miles to the north, the physician/naturalist Philo Romayne Hoy had been attempting to ascertain the diet of the whitefish. While examining the stomachs of fish taken off Racine, Wisconsin, he found what he thought were the partial remains of a crustacean.[22] Stimpson confirmed the fact and traveled to Racine with Edmund Andrews and Blatchford, the latter agreeing to underwrite the costs of the venture. They set out in late June, accompanied by Hoy and Increase Lapham. A dredge haul from forty-five fathoms brought up living examples of the crustacean genus *Mysis*, which until Lovén's discovery had been found only in the colder parts of the North Atlantic and Arctic Oceans. An elated Stimpson knew this was the first time that a marine invertebrate genus had been noted in the American Great Lakes.[23] Hoy wrote that when they found the shrimp alive, "Stimpson was in extacies [sic]" and his bunkmate Lapham did not get much sleep that night.[24]

While he initially talked of presenting his findings to the National Academy of Sciences, Stimpson opted for a short note in the American Naturalist. He erred in believing the Mysis a new species as it was identical to those found by Lovén, but it was still an important find. His paper is the earliest on the Great Lakes deepwater fauna.

Stimpson helped reconcile himself to living in Chicago in part through his investigations of the natural history of the region. The bones of extinct ice age mammals were found regularly, and in 1870 he made a quick trip to Indiana in search of mastodon bones, returning with a large limb bone from an animal he estimated would have stood twelve feet tall. When the remains of a large mastodon were uncovered in Wheaton, Illinois, an Academy member observed that, "there was nothing extraordinary in the mastodon as they were now becoming so common that almost every college had one in its museum." The bones were later acquired by the Academy.[25]

Another Academy member found the remains of a far more rare Pleistocene mammal in 1870, but like so many of the Academy's accomplishments in this period it has been little noted. James Milner had moved to Waukegan (about fifteen miles north of Chicago) and discovered the skeleton in a peat swamp a mere six feet from the shores of Lake Michigan. Stimpson aided in reconstructing the enormous animal, which resembled a cross between a moose and an elk. He asked Baird for a moose skeleton for comparison and Baird complied, providing helpful suggestions regarding the proper articulation and restoration of the skeleton. Andrews believed the find to be a new species, but no formal description was written and the skeleton was later destroyed. It was not until 1885, some fifteen years later, that more bones turned up and the stag moose was finally described.[26]

As he had the previous summer, Stimpson devoted time to the care of the Smithsonian marine invertebrates, arranging, labeling, and organizing them to make up sets for distribution to other institutions. After two long years he could finally inform Baird, "You would not know the collection now it is renovated. I have saved some things which I at first thought were gone up."[27]

Stimpson fulfilled other professional commitments in 1870 by serving on two committees for the National Academy of Sciences. Baird, Stimpson, and John L. LeConte were asked to investigate the prospects for silk culture in the United States. Stimpson made a few inquiries and the committee filed a report that was never published.[28] A second committee was charged with providing instructions for Charles Hall's expedition to the Arctic seas. Stimpson prepared a list of instructions for the collection of marine invertebrates, but the Hall expedition would become mired in controversy with the mysterious death of its leader in November 1871.[29]

The Hall expedition might have reminded Stimpson again just how much he missed marine zoology. These were especially heady days for marine scientists as advances in equipment and techniques had rapidly begun to open up the ocean's depths for exploration. In 1866 the father-and-son team of Michael and George Sars had made headlines by dredging animals in 300–450 fathoms in quiet Norwegian fjords. Such ideal conditions were rarely achievable, and in the rougher waters off the coast of England John Gwyn Jeffreys did well to reach 170 fathoms in 1867.

That same year Agassiz's colleague Louis Francois de Pourtales of the US Coast Survey found animal life abundant at 270 fathoms in the Straits of Florida. Pourtales continued his operations in 1868 and 1869 (accompanied by Agassiz the latter year), reaching depths of 800 fathoms. But the real breakthrough came aboard the British ship *Porcupine*, which successfully dredged animals at an astounding 2,345 fathoms in 1869. These discoveries ushered in a new era of deep-sea dredging.[30] Nearly all Stimpson's dredging had been at depths of less than one hundred fathoms, usually less than fifty. Those numbers now represented the minor leagues of marine exploration, and it must have been a bitter pill for Stimpson to swallow, knowing that he was no longer on the cutting edge of his favorite type of fieldwork.

But Stimpson would still have a role to play in this emerging story thanks to Alexander Agassiz. Three years younger than Stimpson, Alex had taken a more active role in the management of the MCZ as his father's health faltered. In late 1868 he asked Stimpson to describe the new Pourtales crustacea and mollusks, with the results to be published as part of a planned illustrated catalog along with papers by both of the Agassizes, Theodore Lyman, and Pourtales.[31]

As we have seen, with the exception of a few brief notices Stimpson had done very little taxonomic writing since taking over the Academy. Grateful for the chance to do the sort of science at which he excelled, he could undertake such work in Chicago now that he had the requisite books and the Smithsonian invertebrates at hand. Pourtales had found an astonishing number of species of *Brachyura*, or true crabs, leading Stimpson to devote the entire paper to this large group. Among the new species he described were examples of spider, purse, mud, elbow, and swimming crabs.

Pourtales had discovered essentially an entirely new deepwater fauna off Key West and the Dry Tortugas, with thirteen species taken at depths exceeding 100 fathoms and some reaching to 150 fathoms, which is usually the demarcation point for the beginning of the deep sea. It took Stimpson months

of concentrated effort to produce meticulous descriptions of eighty-one species and nineteen new genera. Nearly two-thirds (fifty-two) of the species
were new to science and Stimpson acknowledged Pourtales and Agassiz by
naming species in their honor. Baird hailed Stimpson's "elaborate memoir" in
the pages of *Harper's Weekly*, as Stimpson had the honor of being the first to
describe deep-sea crabs.[32]

Louis Agassiz placed great importance on Pourtales's work, recognizing
that it had opened "a new era in zoological and geological research." As a result
he pushed hard for the completion of the various reports, leading to another
bit of unpleasantness between Stimpson and Agassiz. Stimpson wanted illustrations to accompany his descriptions, as he felt uneasy about publishing so
many new genera without them. Agassiz suggested that the work be done in
Chicago and asked for an estimate. We do not have Stimpson's reply but he
seems to have floated the idea of doing less elaborate drawings, judging by
Agassiz's tone in his next letter. "It seems to me useless to publish indifferent
figures. They add nothing to the descriptions when these are good and they
are a disgrace to the institution and the country from which they are issued. I
would not mind so much the price of the work, as its character."[33] The paper
appeared without figures.

Shortly after completing the Pourtales paper in December 1870 Stimpson's
respiratory distress again flared up. During the worst of these bouts, eating
and drinking became next to impossible and breathing painful. Unable to lie
down, he dozed fitfully all night in a chair in an attempt to keep his lungs clear.
He awoke bathed in sweat from the night fevers that accompanied the attacks.
One of the factors that can activate tuberculosis is confinement to a vitiating
atmosphere, which describes almost exactly the conditions in the storerooms
of natural history museums. While hunched over the Pourtales crabs for days
on end Stimpson was exposed to alcohol fumes. His illness forced him to miss
the Academy's December board meeting, and he told George Walker, "I have
suffered a good deal in my lungs during the past week, with some little hemorrhage, and consulting Dr. [Edmund] Andrews last Saturday he advised me
to go south as soon as possible."[34]

For some time Stimpson and Eliphalet Blatchford had been making preparations to mount a major scientific expedition to the Florida Keys. In one
of the many twists and turns that characterized their complex relationship,
Agassiz had again played a part in shaping Stimpson's future. While dining at
Blatchford's home, Agassiz spoke of Pourtales's success dredging in the deeper
waters of Florida but lamented that the shallower waters were still largely
unexplored. When Agassiz suggested that the Chicago Academy might underwrite such investigations, Blatchford agreed to finance them. In recalling the

incident years later Blatchford said, "I well remember his [Agassiz's] enthusiasm as he rose from his chair, came around the table, took my hand and called his wife and mine to witness the promise made."[35]

Stimpson's health crisis threatened to derail his carefully laid plans with Blatchford. Herbert Stimpson arrived in Chicago to take his eldest son for rest and recuperation at his brother James's plantation in Alabama, with the hope that he would recover in time to join Blatchford later that winter. There would be no Christmas together with Annie and the children this year, as they returned to Ilchester; Stimpson would not see them again for five months. Tuberculosis forced him to lead what one commentator called "a decently restrained vagabond life," cut off for long periods from family and the comforts of home in the hopes of staving off another attack.[36]

For the first two weeks of his stay Stimpson rested before gradually taking short trips to collect invertebrate fossils from the rich Eocene tertiary beds nearby. He still planned on going to Florida with Blatchford if he could and continued gathering information on their itinerary. To that end he had written Dr. Joseph B. Holder, whom he had known growing up in Massachusetts. Holder had served for years as surgeon at the military prison on Dry Tortugas and sent a detailed letter and hand-drawn map of the region. He warned Stimpson to keep their purposes veiled as their appearance at Key West as men of "periwinkle and pickled snake" would elevate prodigiously the cost of hiring a sailing craft.[37]

Stimpson remained in Putnam, Alabama, for nearly two months, but he was still not completely healthy when he left for Florida.[38] In early February 1871, Stimpson met Blatchford and his eleven-year-old son Paul at New Orleans, and by the time they reached Key West Stimpson's health had improved.[39] Blatchford hired the fast-sailing schooner *Liberty* with a crew of nine native-born Bahamian "conchs" who knew every rocky inlet and shoal along the coast. From mid-February to the end of March they dredged at various spots along two hundred miles of coastline from Cape Florida just south of Miami to the Dry Tortugas. In the process they sampled the three major ecosystems along the Keys: mangrove swamps, sea grass beds, and coral reefs.

Historians until recently have written little about "the craft and social history of scientific collecting." These expeditions were very complex socially and provided a way for different types of people to experience nature, mixing both scientific and recreational cultures. Robert E. Kohler's book *All Creatures* does a marvelous job of explaining what he calls the "natural history survey" mode of collecting that began in the late nineteenth century, which differed from the earlier exploratory type of expedition in that it was more "methodical and guided by scientific agendas" as well as being more

organized and systematic.[40] The Stimpson-Blatchford expedition of 1871 can
be seen as an early example of natural history survey mode collecting. The
degree of planning and organization were far greater than for most museum
expeditions of the period, as was the duration of the trip, six weeks. The zoological objectives were clearly defined. Stimpson systematically dredged the
shallower waters of the Florida coast, complementing the work of Pourtales,
who had dredged the deeper water further from shore. Another goal was to
collect exhibit-worthy animals for the Academy's museum. For Stimpson, the
collections would help fulfill his ambition to document all of the East Coast
marine invertebrates, especially mollusks and crustacea.[41] According to two
modern-day zoologists, Stimpson was the first to attempt a formal scientific
inventory of Florida mollusks.[42]

Blatchford deserves credit as a scientific patron willing to spend large sums
while also availing himself of the recreational aspects of the field experience,
especially fishing. He purchased much of their equipment, including a three-
hundred-foot seine and a twelve-by-thirty-foot trawl. They also had a fyke
net, a type of fish trap that Stimpson believed had never before been used
along the American coast. They fished for both sport and science, at night
using reflectors to lure fish to the boat. Every effort was made to get at least
one example of each species of shark that inhabited these waters, as mounted
specimens were sure to attract visitors to the museum. Blatchford struggled to
land a ten-foot-long "man eater" and after being harpooned the shark attacked
the boat, leaving two of its teeth embedded in the stern. He later purchased
the stern and displayed it in the museum next to the mounted shark.[43]

Dredging in both the reef channel and in the Gulf Stream, the sailors soon
came to enjoy the "novel and steady work of lowering and hauling in the
dredges." Stimpson carefully numbered each dredging and entered its location
on the charts. He often toiled long into the night writing up the results of the
day's work in a massive logbook. The delightfully equable weather suited him
and he noted that the air and water temperatures were virtually identical.

They spent two days in Miami, getting their first look at the Everglades
and comparing notes with the ornithologist Charles J. Maynard. They added
to their collection of fish and captured several large tarantulas, a nonnative
species probably brought by ships arriving from Mexico. At Marquesas Key,
Paul Blatchford climbed a tree to secure a living young eaglet while the others
manned guns to protect him from the outraged parents. Paul's presence might
have led Stimpson to think of the day when his own children might accompany him in the field. From Key West he wrote a playful letter to his son Willie,
who had just turned six, relating the capture of the eagle and giving an account
of a successful shark hunt.

Perhaps the most striking vistas they encountered were at Dry Tortugas. Seventy miles from Key West, these seven islands comprise only one hundred acres of dry land. Named for the abundant sea turtles common in the area, there are few places on earth that can rival it for remoteness, stark beauty, and the variety of marine life. The Tortugas provided a fascinating window into the world of shallow water coral reefs teeming with brightly colored fish, mollusks, and crustaceans. Stimpson had seen many a coral reef on the North Pacific Expedition, but here in the Keys he had leisure to investigate the reef fauna more fully and they secured several exceptionally large brain corals.

He and Blatchford received a cordial welcome from the officers and crew stationed at massive Fort Jefferson. Known for its isolation, it also served as a military prison and had until recently been home to Dr. Samuel Mudd, one of the conspirators in the Lincoln assassination. At a dinner party held at the fort, Blatchford marveled at the ingenuity of the commander, who had captured several sharks and transferred them to the moat surrounding the fort to discourage escape attempts.

At Key West the peculiar habits of naturalists attracted unwanted attention from those suspicious of the methods and motives of men who spent all of their time looking for "useless critters." In a lively account of the six-week excursion, Blatchford related the following anecdote: "We had an amusing experience with the United States Officer of Customs, on putting in to Key West from our eastward trip. Smuggling was frequent on these uninhabited and dangerous coral coasts, and our commodious little craft with its unique and bulky-looking cargo could offer hiding for a rich cargo of choice Cuban tobacco and cigars. With better grace than our Captain and crew, Dr. Stimpson and I submitted to the officer's examination and gave replies to his 'bluff game.' . . . He evidently failed to realize any value in our Expedition for gathering such a 'lot of stuff.'"[44] Elated at their success, Stimpson wrote Baird from Key West. "We have obtained some new fishes . . . a great number of new invertebrates, and some fifty species of shells not hitherto reported from the Florida Coast. . . . Our collections now standing ready for transportation fill 62 barrels & boxes & weigh about three tons," approximately fifteen thousand specimens in all. Many of the new shells were dredged in deeper waters and some belonged to genera that had been thought to be extinct.[45] Baird deemed the trip worthy of an extended notice in *Harper's Weekly* "in consequence of its magnitude and the thorough nature of the examination made of the marine fauna of the Southern coast. . . . The collections . . . will add greatly to the already rich cabinet of the Chicago Academy of Sciences—an institution which has assumed a high rank among sister establishments."[46]

In early April 1871, Stimpson landed at New Orleans intending to stay for a week before returning to Chicago. But he ran into Academy trustee Daniel Thompson, who had invested in the sugar business with Charles H. Walker, a former president of the Chicago Board of Trade and brother of Academy trustee George Walker. Thompson convinced Stimpson to extend his southern sojourn by accepting an invitation to stay at Walker's sugar cane plantation on the Bayou Teche at Pattersonville (now Patterson), about one hundred miles west of New Orleans. A narrow, serpentine waterway over one hundred miles long, the Bayou cuts through the heart of Acadia country. Stunned by the lush beauty of the immense forests of cypress, magnolia, and oaks that lined the banks of the waterway, Stimpson praised the region as "by far the most fertile and beautiful country I had seen in the South."

They soon embarked on a ninety-mile cruise in the Gulf of Mexico on Walker's custom-built yacht. Between Vermilion Bay and Isle Derniere they captured numerous sea turtles. Stimpson was amazed when Walker caught a thirteen-foot-long, nine-foot-wide sawfish weighing nearly one thousand pounds. This monster was soon on display at the Academy's museum.[47]

Arriving back in Chicago in May he pronounced himself in "excellent health" and ready to again take up the challenges of running the Academy. After an absence of four months he had much to catch up on. At the January annual meeting he had been appointed the Academy's librarian, formalizing a position that he had long held anyway. As he again settled into a routine he probably spent more time than he would have liked giving tours of the museum for organizations such as the American Home Baptist Mission Society and the General Assembly of the Presbyterian Church. These interruptions were part of the job but kept him from more scientific pursuits. Despite the limited public hours, over two thousand people visited the Academy museum in a five-month period in 1870.[48]

Stimpson also had correspondence to catch up on. He wrote Louis Agassiz about the Florida trip and asked him to identify the new echinoderms. In the past Stimpson would have described them himself, but he knew that he had more work than he could handle with the crustaceans and mollusks. This was a way to reach out to Agassiz, who responded warmly to the bid. "It has given me the greatest pleasure to see your handwriting again & to hear such good news of your health and success," he wrote. Agassiz's own health had improved since his stroke but he admitted that he still felt weak.[49] The anger and frustration that both men had felt in the past towards each other had been replaced by the recognition that both of them were in failing health.

Stimpson learned that Baird, in his new role as head of the US Fish Commission, had appointed James Milner as deputy commissioner for the

Great Lakes. Baird had mapped out an ambitious program for deepwater dredging in the lakes, and Stimpson pledged to provide Milner with all possible aid. Milner would do most of the dredging while Stimpson worked up the results. Stimpson characterized Milner as "a very intelligent naturalist and a thorough worker" and furnished him with dredges, alcohol, and advice.[50] Stimpson still held out hopes that he might do a little dredging himself, telling Baird that he and Blatchford wanted to plumb the depths of the one-hundred-fathom-hole off Traverse Bay (Michigan) "without fail," as Stimpson was "determined to have a dip in it." A press of other work prevented him from carrying out that plan.

One project that kept him busy was pushing forward with the second volume of the *Transactions*. Two papers had already been printed and distributed as separates. One by Edmund Andrews described the geology of the Great Lakes while the second, by Alpheus Packard, described insects collected by Dall in Alaska. Philip R. Uhler and Hermann Hagen added a section on the Neuroptera.[51]

Stimpson spent a large portion of the spring and summer of 1871 cataloging and curating his recent Florida materials. He had a great deal of work to do before he could begin making up sets to exchange with other museums.[52] The sawfish, sharks, birds, and other large animals had to be mounted and put on display, and Stimpson personally supervised the work. By this point the Academy building, only three years old, was already crammed from cellar to garret with specimens, forcing the trustees to again discuss the topic of financing a new and larger building.

With his health uncertain Stimpson's main priority was to finally see through the press several long-delayed manuscripts. However, his attempts at writing were done "amid constant interruptions to which I must submit in view of the pecuniary interests of the Academy." Although he would have loved nothing better, he declined an invitation from Baird to join him on a summer cruise to Woods Hole, explaining that he had "spent so much time already in such work that I must bore down to closet work for the rest of the year." He desperately wanted to carve out some time to put the finishing touches on the books that he felt would secure for him a permanent place in the history of science.[53]

CHAPTER 13

LOSING A FAUNA

The study of natural history in the leisure of my life, since I was four-teen years of age, has been to me a constant source of happiness, and my experience of it such that, independently of its higher merits, I warmly recommend it as a pastime, which, I believe, no other can excel it.
—Joseph Leidy (1879)[1]

There were a variety of reasons that Stimpson had so many important unpublished manuscripts. The Civil War and its aftermath had delayed publication of the report on the Zoology of the North Pacific Exploring Expedition, as well as other government reports. Stimpson's removal to Chicago had forced him to put his own research and writing on hold. The fires at the Smithsonian and the Academy, in 1865 and 1866, respectively, had destroyed or damaged some of his manuscripts. With Kennicott's death he took on more administrative duties for the Academy that left little time for quiet contemplation. Finally, tuberculosis had forced him to cut back on work and to spend time traveling for his health.

Earlier in his career Stimpson had made other choices that hindered his ability to finish what he started. For a number of years his drinking may have kept him from working to his full potential. He had also placed a priority on scientific sodality with his friends in the Megatherium Club when he might instead have isolated himself in order to complete projects. His failure to stick with one specialty for any length of time, while characteristic of many of his peers, led him to disperse his energies across many different phyla. In an attempt to be as comprehensive and thorough as possible, his pursuit of every "American species" might be seen as an obsessive and never-ending quest for perfection. A modern paleontologist has written that a key to success is "a will for finishing things and an ability to snarl at intruders. The ability to discourage unwanted distractions is vital to the completion of the Great Work."[2]

Perhaps the undertaking closest to his heart was the monograph on Atlantic Coast mollusks, in preparation since 1849. For over two decades he had labored to master the subject, undertaking extensive fieldwork, diligently researching the scientific literature, and visiting leading museums to examine specimens. He had expanded the book's geographic range, eventually settling on the Atlantic Coast from Greenland to Texas. By 1871 he had completed over two hundred drawings on wood. "Nearly every species was illustrated by specimens from every locality in which it occurs, not only on our own shores, but on those of Europe and the Arctic Sea, and in the Tertiary and Quaternary formations, shewing the effect of climatic influences, geological age, etc."[3]

His recent discoveries in Florida gave him a host of new species to add but he had not worked actively on mollusks since about 1865. He yearned for Meek's counsel and told him, "I would give a good deal to be able to work them up at the other end of a table at which you worked." He hoped to meet with Meek later that year in Washington for a "pow-wow over the recent progress in Conchology."[4] Linked to the mollusk book was a series of dredging papers documenting the contents and locations of countless dredge hauls from Nova Scotia to Florida and the Gulf of Mexico, invaluable data for re-creating the geographical and bathymetrical distribution of marine animals. Their publication would quantify the results of his fieldwork in a permanent manner.

Even more epic in size and scope was Stimpson's report on the Invertebrate Zoology of the North Pacific Exploring Expedition. Totaling almost two thousand pages and illustrated by nearly three thousand drawings, many in color, it included a section on shells written by the late Augustus Gould as well as Stimpson's contributions on the crustaceans, annelids, and other invertebrates. The expedition itself represented three years of his life, and the time spent organizing the collections and writing his Prodromus probably totaled another three years. The year 1871 marked fifteen years since the expedition returned home, yet Stimpson still held out hope that Congress would appropriate funds to finally publish the report.[5]

Other unpublished manuscripts that Stimpson had in hand included a work on the crustacea of North America, the second part of the paper on the Pourtales crustacea, a forty-page synopsis of the mollusk genus Murex, and a fuller accounting on the marine invertebrates of the North-West Boundary Survey.[6] Every one of these contributions was based on the collections housed at the Academy. Given that his attacks were coming more and more frequently, he felt a growing sense that he needed to push things forward more rapidly. He reasoned that he might not be able to leave his children money but he could leave them these writings as a lasting legacy of their father's scientific achievements.

Stimpson had made some progress towards his goal in the summer of 1871 when intellectual endeavors were again interrupted by Academy business.

Member Ellis Chesbrough raised almost $1,200 to buy a valuable cabinet of minerals from the estate of George W. Hughes of Maryland, and in early August Stimpson traveled east to handle the arrangements. He visited many of his old haunts and gathered up his remaining books, papers, and specimens from his father's home in Cambridge, the Smithsonian, and Ilchester. He also spent considerable time going through the crustacea at the Museum of Comparative Zoology and arranged to have one-third of the collection shipped to him for further study at Chicago in early October. But at some point his travels brought on more hemorrhages, and he arrived in Chicago in early September in a severely weakened condition.[7]

One of the responsibilities of a curator of a world-class collection was to make it available to researchers. In September 1871, John Gwyn Jeffreys of England visited North America "to see the major collections in Montreal, Boston, and Washington." Chicago was hastily added to the itinerary when Jeffreys learned that the Academy contained many of the Smithsonian's marine invertebrates. Jeffreys's multivolume series on British conchology was already a standard reference, and he particularly wanted to see the specimens that Pourtales had dredged, which would serve as a means of comparison with European species that Jeffreys had dredged at similar depths.[8]

Jeffreys stayed in Chicago nearly a week and found plenty to occupy him among the Academy's thousands of lots of shells. He and Stimpson were similar in that both were field naturalists acclaimed as among the foremost dredgers of their day, but Jeffreys was basically a conchologist while Stimpson favored malacology. They engaged in a lively discussion comparing the mollusks of Europe with those from eastern North America. Louis Agassiz had earlier met with Jeffreys and gave him permission to borrow any of the Pourtales specimens he wanted. Stimpson complied, telling Baird, "In accordance with Agassiz's instructions and your own suggestions I allowed him to take whatever he wanted both from the Pourtales and my own collections, including many uniques." Jeffreys wrote Stimpson from Montreal on October 6 to say, "My visit to North America although short has been most agreeable & profitable and I especially enjoyed & value my conference with you."[9]

Playing host to Jeffreys further taxed Stimpson and an abrupt change of seasons intensified his respiratory distress. Plagued by hemorrhages and bereft of energy, in early October he wearily admitted to Baird, "I am suffering a great deal from lung disease." His physicians urged him to leave the city as soon as possible. He began searching for options as to where he would spend the winter. One possibility came from Benjamin Peirce, the superintendent of the US Coast Survey, who wanted him to join a dredging expedition to the Sargasso Sea.[10]

Stimpson may have had forebodings that his affliction had taken a serious turn for the worse. In late September he sent the third part of a paper entitled "Notes on North American Crustacea" to the *Annals of the Lyceum of Natural History of New York*. The first two parts had been published in 1860, and the third installment had been written at that time. In a brief introduction, Stimpson noted that many of the descriptions were more than ten years old and had not been revised, but he was now determined to see them in print (most of the forty-one new species and six new genera described in the paper are still valid today). The New York Lyceum duly noted the paper as having been received and read on October 2, 1871.[11]

The summer and fall of 1871 had been extraordinarily dry in Chicago and the month of October began with a rash of fires. On Saturday night, October 7, a fire broke out just west of the south branch of the Chicago River. It quickly consumed four city blocks about one mile west of the Academy museum. In extinguishing the blaze several fire trucks were destroyed or damaged.[12] The next day the editors of the *Chicago Tribune* commented prophetically on the fire. "It was one of the most disastrous which had ever visited a city which had already enrolled in her annals numbers of such visitations, many of them so terrible that they could serve as eras in her history. For days past alarm has followed alarm, but the comparatively trifling losses have familiarized us to the pealing of the Court House bell, and we had forgotten that the absence of rain for three weeks had left everything in so dry and inflammable a condition that a spark might start a fire which would sweep from end to end of the city."[13]

Temperatures that Sunday exceeded 80 degrees, with strong winds from the southwest. The fire that began about 8:30 that night in Mrs. O'Leary's barn soon spiraled out of control and within twenty-four hours most of the city's business district had been incinerated. At least three hundred lives were lost and one hundred thousand were rendered homeless. Largely overlooked today is the wildfire that occurred that same day in and around Peshtigo, Wisconsin, which killed perhaps five times as many people.[14]

Stimpson would probably have been awakened the night of the fire by the eerie glow emanating from the east. His home in Oak Park was in no danger, but he must have watched anxiously as the blaze continued throughout the day on Monday. Perhaps he had faith in the Academy's "fireproof" building, but as the fire intensified and increased in size he must have realized that a holocaust of immense proportions had swept over his adopted city. The scene that day on Wabash Avenue near the Academy beggared description. The social order broke down as the streets were clogged with desperate, fear-maddened people trying to escape the inferno. Children separated from their parents wandered the streets crying, pushed along by the frenzied mob. To many it seemed that Judgment Day had arrived.[15]

For a time it appeared that the Academy museum would be safe from the flames as the first wave of fire swept to the north. Several Academy members and trustees lived within a few blocks of the museum, including George Walker, J. Y. Scammon, and Edmund Andrews. Instead of attempting to remove the more valuable objects they chose to leave everything in the building. "Those present at the museum closed every avenue of attack by the fire, removed from the walls whatever would readily burn, piled the library and valuable manuscripts on the floor, and departed to a place of safety."[16]

Sometime that day a group of men made a last-ditch effort to save what remained of the southernmost section of the downtown area. A few blocks south of the Academy a cache of dynamite was appropriated and several buildings were blown up to create a firebreak. As the fire drew nearer one last desperate attempt was made to save the Academy. Ten teams of horses were hitched to ox chains and pulled down the large frame boarding house on the front of the lot.[17] The wreckage remained, however, serving as fuel to the fire.

In the end the Academy was partially doomed by its proximity to a carriage paint shop located across the alley at the rear of the building. The store was loaded with a full supply of oils and varnishes for the fall season, and an eyewitness later gave the only known account of the destruction of the Academy. "It [the carriage house] was all on fire and burning fiercely with the strong wind blowing the flames directly against our [the Academy] building. Our tall windows were protected by thick, iron shutters specially made for us in Buffalo. Twice, in severe gusts of the gale, he detected the flames lick our alley front and thought 'Well, you are safe.' He recognized, however, a third blast of fierce flame which carried down our tall shutters and admitted the relentless fire. The expansion of the high iron shutters had torn them from their hinges."[18]

The Academy's vaunted fireproof building was consumed as easily as the wooden shanties where the fire had originated. By 3:00 p.m., the museum and all it contained were utterly destroyed. Thirty-six hours after the fire had begun a light rain started falling, ending the most famous fire in US history. A letter written by Academy member James Milner a few days later relates the terrible devastation: "The world as it is to the people of this vicinity has changed; an age has closed, and a new epoch, obscured in doubt and uncertainty, is about to begin. The fate of Gomorrah has come upon a city ... the busy, peopled streets, the pleasant stores with their wide inviting doors ... are now stumbling pathways, ... heaps of blackened bricks and dusty ashes, with silent people wandering among the ruins. You can scarcely imagine the desolation. If a man wants his mind impressed with what the end of the world will be, let him come here."[19]

Stimpson ventured into the city on Tuesday. He eventually found Blatchford and they slowly made their way through block after block of smoking ruins before reaching the site of the museum, when Stimpson's worst fears were realized. Blatchford related the scene.

> Few words were spoken and those in subdued tone, as if in the presence of the dead. With difficulty we worked our way down the iron staircase into the basement, both having the same thought of one place the fire fiend could not reach–the vault of the Academy. ... We were struck with amazement as we found the iron door of the vault burst open and hanging by its upper hinge. Inside were huge broken blocks of stone, and all else was a blackened mass! The truth flashed upon us. The heavy stone cornice on the northeast corner of the building had fallen and crashed through the vault's roof, a slab of limestone ten inches thick. The fire followed and fed by the inflammable contents, completed the cruel work of destruction. At the right of the door, fallen from its shelf, lay the choicest portion of Stimpson's life work, a blackened pile of carbonized manuscript! I carefully lifted one filmy sheet between my fingers. It fell to pieces. No word was spoken; no hope was in our hearts; its last trace had been ruthlessly taken from us.

Blatchford continued, "I laid my hand on the shoulder of my dear stricken friend, as we threaded our way out to the desolate avenue. The only word he spoke, as I remember, was 'That was for my family.' I longed to give comfort; but words would not come. ... Here was my first and last vision of a 'broken heart.'"[20] Stimpson's despair was that much greater when he saw that a mere block and a half to the south, buildings stood untouched by the fire.

The first person he notified was Joseph Henry, in a grim letter that begins, "It becomes my painful duty to inform you of the total destruction of the Smithsonian Collection of Marine invertebrates, in the great fire of yesterday. All is lost, together with the rest of the scientific materials gathered in the Academy's building, in value about $200,000. Our building, like all of the other so-called 'fire proofs' in the city, collapsed like a bubble in the intense heat. All of my collections, books, manuscripts and drawings, the results of twenty years of labor are destroyed. ... Everything of much value that I had in the world was deposited in the building for *safety*, and I am now left destitute." Stimpson went on. "It seems however selfish in me to allude to such losses when I see so much physical suffering around me. . . . We who have roofs and clothing and food still left, are doing what we can to help the sufferers."[21]

Henry shared the letter with Baird, who tried his best to sound encouraging. Imploring Stimpson not to lose heart, he wrote, "I think the greatest loss is that of your manuscripts & drawings which have occupied you so many years, & which I hoped would have seen the light before long. Still, perhaps these can, to some extent at least, be done over, & possibly better than before." He also promised to send a set of the specimens that he had collected at Wood's Hole that summer.[22]

The Museum of Comparative Zoology had lost heavily as well, most notably the Pourtales crustacea and mollusks, the first tangible results from America's initial foray into deep-sea dredging. Agassiz immediately wrote Stimpson and paid tribute to what his former student had accomplished in Chicago. "The interests of science are not so immediately connected with the welfare of the community, that in the first hours of distress an appeal in behalf of the Academy of Science of Chicago may meet with any success. It is however the duty of men of science to testify to the great success you had achieved in making Chicago one of the prominent centres of activity in unfolding the Natural History of the United States."[23]

The newspaper accounts coming out of Chicago about the fire did not specifically mention the Academy, but word spread quickly in the small natural history community. In the days and weeks after the catastrophe, Stimpson and the Academy received letters from a veritable Who's Who of American naturalists including Alexander Agassiz, James D. Dana, Isaac Lea, Addison Verrill, Theodore Gill, Benjamin Silliman Jr., Fielding Meek, Alpheus Packard Jr., and many others. Dozens of organizations and individuals offered to send specimens and publications.

Verrill, whom Stimpson had adopted as a protégé of sorts, did not mince words when he wrote, "The loss of the Academy is surely a great misfortune to the world at large and to science generally, as well as an almost overwhelming one to you."[24] Meek grieved "to my very hearts core over the terrible news of the awful calamity that has befallen Chicago." He urged Stimpson to return to Washington where he would be surrounded by a large circle of friends.[25]

The freshwater conchologist Isaac Lea promised to send books and articles and in a letter to John W. Foster he wrote, "I cannot go to my bed although now 11 O'clock without saying to you and my friend Dr. Stimpson that you and all the good workers of your excellent Society have my fullest heartfelt sympathy. ... The accumulation of such rare specimens, collected by the hands of such men of Science as Kennicott, Stimpson, Pourtales, is indeed an irreparable loss—a loss which can only be understood by those who have worked like these energetic and devoted men." He added, "I have peculiar sympathy with my friend Stimpson whose course I have watched since he was a lad. To have lost the fruits of so many years of hard labor—*specimens, Manuscripts* and all—all past recovery! I beg you to say to him how sincerely I feel for him in his great loss."[26]

In the face of this utter devastation some looked to God for answers. Philip P. Carpenter was a deeply religious man with the soul of a poet. He did his best to console his distraught friend. "Perhaps I understand your loss so far as the E. Coast Mollusks are concerned better than *any one else.* . . . To a non-Christian, it looks like life thrown away; but my faith is that our work *here* is . . . to form the *infernal man* for better work hereafter. The lesson *I* draw from your loss is that scientists ought not to wait till they have perfected & finished things; but ought to publish papers & prodromi, as they are going on. Then the world wouldn't lose a whole fauna, all in a swoop."[27] Carpenter's words "a whole fauna" capture the enormity of the loss.

William H. Dall reached a similar conclusion. "When Stimpson by the Chicago fire lost practically most of the manuscripts and specimens which represented years of collecting and research, and never recovered from the loss, I made up my mind not to undertake a magnum opus of that kind, but to publish promptly researches which seemed to include something new."[28]

Frederic W. Putnam hoped that Stimpson would not feel guilty about what had been lost, telling him that he had "the heartfelt sympathy of every scientific man in the country."[29] Ezekiel Jewett and others recognized that the greatest loss to science came from the annihilation of Stimpson's manuscripts. One could always get more specimens—"there is left in the sea things equal to those lost" he consoled.[30] James D. Dana, one of Stimpson's earliest supporters, spoke of the "vast" losses to science and promised to send books and specimens. The recently departed Jeffreys was "shocked at hearing of the dreadful calamity which has happened in Chicago." Alexander Agassiz also expressed horror. "What a frightful calamity. . . . The reality must be something fiendish."[31]

The publisher of the *American Journal of Science*, Benjamin Silliman Jr., told Foster, "It makes me sick at heart to look over the list of your losses in the destruction of the Academy for the second time by fire." He later inserted a notice in the journal asking naturalists the world over to aid Stimpson. "Dr. Stimpson is one of the ablest and most energetic workers in zoology in the country; and he deserves something more than ordinary consideration. Should a scientific library be restored to him by gifts from others over the world, and from owners of duplicate copies of zoological works, it would not be more than a just return for all his unwearied labor in the cause of science."[32] Matching the word to the deed, Silliman sent a full set of back issues of the journal.

On November 9, *The Nation* published a piece by Frederick Law Olmsted giving his observations of postfire Chicago. While Olmsted did not mention Stimpson he paid tribute to the "Kinnicut brothers," a garbled reference to Robert Kennicott and his family. In response Kennicott's brother-in-law, Frank W. Reilly, wrote movingly of Stimpson. "There seems absolutely no recompense for such a loss as befalls him who, steadily putting aside every temptation—where temptation, as in Chicago, was most rife—to commercial gain and profit; and, devoting himself solely to the service of science, sees in an hour the fruits of years of study, ripe for, but yet ungiven to, the world, vanish like the baseless fabric of a dream." Reilly lauded Stimpson's devotion to fulfilling Kennicott's aspirations for the Academy. "Even the proverbially warm sympathy for each other of the larger men of science, of which Stimpson himself is an example, in his chivalrous devotion to the memory and reputation of his predecessor, Kennicott, is weak and inadequate for such a case as this."[33]

This outpouring of support did little to cheer Stimpson. He simply could not escape the fact that his life's work had been reduced to ashes. His characteristic energy and enthusiasm were gone, replaced by a numbing sense of despair. He informed Baird that the Academy would go on "notwithstanding this terrible blow."[34] On October 21 Stimpson attended a meeting of the Academy held at the Chicago Theological Seminary. The mood was somber and only five others showed up. They approved a petition to the governor of Illinois asking for a set of duplicates from the State Geological Survey, and Stimpson was appointed to a committee empowered to get specimens from "scientific collectors everywhere."[35]

Two days later Stimpson met with three other trustees to discuss the future. In a spectacularly unwise move, no insurance had been carried on the museum or its contents due to the misguided faith in their "fireproof" structure. Fortunately the boarding house on the front of the lot was insured for ten thousand dollars. It was a foregone conclusion that the affairs of the Academy were secondary to the monumental job of rebuilding the city, so the main order of business involved Stimpson. The trustees would continue paying his salary and hoped that he would return in the spring, when perhaps there would be an opportunity to start anew.[36] The trustees managed to scrape up five hundred dollars and presented it to a surprised and grateful Stimpson. He thanked them less for the money than "on account of the expressions of goodwill and confidence which accompany the gift, which afford me the highest happiness."[37] He also paid tribute to those he had worked with in Chicago when he wrote, "had I lost twice as much I shall never regret coming to Chicago for I have found there noble and generous friends not only to myself but friends of Science such as no city in America can boast. Of more value to me than worldly possessions will be the memory of the friendly experiences I have had with yourself and the other Trustees and the friends of the Academy, while we together built up a monument which though now levelled with the dust will long live in scientific history. May our past be an earnest of our future."[38]

With winter fast approaching he knew that he would have to leave the city soon, but he first had to complete the inexpressibly melancholy task of cataloging what had been lost. Only he knew the full scope of the collection and it was essential to inform the scientific community of what had been destroyed, if for no other reason than to save time looking for specimens known to have been in existence. One thousand copies of the circular were printed and sent to institutions all over the world.[39]

The following partial list gives an idea of the immensity of what had been consumed. Kennicott's Arctic collections made during 1859–1861; the invertebrates from the North Pacific Exploring Expedition; Stimpson's collection of marine shells from the Atlantic coast, undoubtedly the most complete assemblage of this fauna in the world; all of the other Smithsonian marine invertebrates in alcohol; the Walsh collection of insects; Pourtales's crustacea and mollusks; the Chicago Audubon Club's collection of mounted birds; the cabinet of shells collected by William Cooper; and Stimpson's Florida collections made between 1869 and 1871. In addition to these special collections were the more general zoological collections of the Academy. In all at least two hundred thousand specimens were lost, among them many type specimens. The archaeology and ethnology collections had numbered several thousand artifacts. Gone too was the Academy's fine scientific library, including the expensive carcinological and malacological publications that Stimpson had so painstakingly assembled. The Audubon Club of Chicago also lost its library, including a copy of Audubon's great work. In all over two thousand books and five thousand pamphlets were destroyed.

Given this inventory, the destruction of the Academy museum in the Great Chicago Fire must at the very least be counted as an enormous setback for the study of natural science in Chicago. An article in *Appleton's Journal* proclaimed that, "since the world began, there has probably never been suffered a loss greater to the cause of science, or one representing so large an expenditure of money, time, and scientific knowledge ... than has been suffered by the Chicago Academy of Natural Sciences." The American natural history community had never before suffered so serious a blow from a fire.[40] Stimpson's losses were beyond computation. The *Prairie Farmer*, a popular agricultural journal, captured one of the essential truths of the Great Fire. "Of all the lives that were wrecked by the hideous calamity at Chicago, there was, perhaps, none more valuable than his."[41]

CHAPTER 14

A CLEAN SWEEP

With other ministrations thou O Nature
Healest thy wandering and distempered child.
Thou pourest on him thy soft influences,
Thy sunny hues, fair forms and breathing sweets,
Thy melody of woods, and winds, and waters,
Till he relent, and can no more endure
To be a jarring and a dissonant thing,
Amid the general dance and minstrelsy.[1]
—Samuel T. Coleridge

With a welter of conflicting emotions, Stimpson left Chicago in late October. While he had felt great pride in what the Academy had achieved, those accomplishments seemed more like a fleeting mirage than reality. On reaching Ilchester, he did not have the strength to continue on to the Smithsonian and instead unburdened himself in a letter to Baird. The shock brought on by the fire was beginning to wear off, to be followed by a crippling depression. He fatalistically began to refer to his scientific career in the past tense. "I am feeling a good deal depressed, although I know that I ought not repine at unavoidable misfortunes but to prepare for fresh efforts as you suggest." He told Baird that it was "terribly difficult to undertake for a third time the restoration of my work. The works I had in hand would have been published long ago had I remained in Washington; I could not do this in Chicago at first for want of books and other facilities. These facilities I had at last gathered around me . . . and was rapidly doing the final work on my materials and getting them ready for publication. It is agonising to me to reflect on how little my published work amounts to, and how inferior they are to what I had ready for the press, which I had looked forward to to retrieve my reputation. I am indeed followed by an adverse fate which it seems useless to fight against."

There had been a "*clean sweep*" of not only his manuscripts but also the collections that were the basis for those works. He realized with stark clarity that if the East Coast mollusk book and the North Pacific report had been published, his fame as a naturalist would have been assured. He feared his life would be judged a failure, and all the words of sympathy and encouragement from his fellow naturalists failed to dispel that thought. In the midst of this gloomy reverie, Stimpson found one small measure of solace in that the specimens he had loaned to Jeffreys had been spared. As he put it, "what I gave I have; what I kept I lost."[2]

In the weeks and months after the fire he tried to come to terms with his loss. Nothing wounds as much as regret unless it is guilt, and Stimpson had plenty of both. Despite telling his fellow trustees that he would never regret coming to Chicago, the truth was that he had never really wanted to leave Washington or the East Coast. Guilt nearly overwhelmed him. Friends and colleagues had given or loaned the Academy tens of thousands of specimens with the promise that they would be safe and well cared for. That trust had been broken and he felt responsible, even as he was virtually powerless to make amends.

His frail condition meant that he again had no choice but to return south for the winter. Before it had become clear that the Academy would continue paying him he had received several propositions. The first came from Louis Agassiz. For all the arguments and ill will that had marked their relationship, both men retained a strong affection for the other. While the ashes of the Academy were still cooling Agassiz offered him a job at the MCZ organizing the crustacea, promising two hundred dollars per month for as long as he could afford it. There was one catch in that he would have to make up his mind quickly, as Agassiz was preparing to depart for a year's cruise on the US Coast Survey ship *Hassler* and needed to discuss these and other matters with Stimpson before his departure. To Stimpson working *for* Agassiz, even in absentia, was the same as working *under* him, and accepting this position would have felt like a gigantic step backwards, so he declined it.[3]

Not one to take no for an answer, eight days later Agassiz made another proposal. "It is not enough that you should be able to begin anew; there is a vast amount of work done, which is now only recorded in your mind and nowhere else. That must first be saved and made more tangible & permanent." Agassiz suggested that Stimpson go to Florida to collect, with the MCZ paying expenses. He could then return to Cambridge to work up the results before his presence would be needed again in Chicago. The Academy would receive a set of specimens while the MCZ would retain the rest. As with the previous offer Stimpson had to decide quickly, for Agassiz planned to leave Cambridge around November 1. After deliberating with the trustees, he declined this offer as well.[4]

Another option soon presented itself. Benjamin Peirce of the US Coast Survey offered Stimpson the chance to dredge on the steamer *Bibb*, scheduled to make soundings that winter off the Florida coast. After the return of the *Bibb* another vessel would be similarly engaged further north in the Atlantic, work enough to keep him busy for the better part of the coming year.[5] Agassiz probably learned of this through his friendship with Peirce and enlisted Pourtales to convince Stimpson to accept the posting on the *Bibb*. "The importance of this work as complementary of our Florida work is very great, and you *must* arrange to do it; it would be worse than a sin to miss that chance, which will not come again.... Prof. Agassiz tells me to impress the great importance of this occasion upon you with all my powers." Whether intentionally or not, Pourtales played on Stimpson's feelings of guilt, but he and Agassiz may have been unaware of just how physically debilitated Stimpson was at this point.

At any other time in his life Stimpson would have jumped at the chance to embark on a deep-sea dredging expedition but now he hesitated, perhaps unsure whether he could withstand the rigors associated with such work. He preferred to return to Key West where he could collect nearer to shore and with less pressure on him. But driven to replace what had been lost, he agreed to go with the *Bibb*. Baird supplied alcohol and dredges and the MCZ provided the rest of the outfit.[6]

In late November a dejected Stimpson traveled to the Smithsonian to meet with Baird. An unnamed observer recounted that, "his health is very bad, he could scarcely look worse and be alive; it is feared he will never regain his health—his severe affliction and great loss by the Chicago fire has too much changed the once energetic Stimpson for him to ever recover, I fear."[7]

More misfortune awaited him on his return to Ilchester. On a cold day in early December, Stimpson took a buggy ride with his father-in-law. The buggy's shaft bolt came loose, spooking the horse and causing him to bolt. Both men were thrown to the ground, and James Frisby Gordon sustained a fractured skull and was bedridden and delirious for months. Stimpson escaped with severe bruises and a sprained knee and ankle. On learning of the mishap Baird felt that in the end the fall might do "more good than harm," as he hoped that Stimpson would give up his plans to go to Florida and instead spend the winter recuperating at Ilchester.[8] Stimpson stubbornly refused to alter his plans. He was annoyed that he was now unable to make another visit to Washington to meet with Baird one more time, so Carlile P. Patterson of the Coast Survey came to Ilchester to finalize arrangements for the voyage.

Although still very lame, Stimpson hurried to Baltimore and arrived on December 15 in time to make the steamship *Maryland* bound for Key West.

The Academy trustees made sure that he would not be alone. Jacob W. Velie had begun working as Stimpson's assistant just two months before the fire. Unlike the mostly young men who had assisted him at the Academy, Velie was two years older than Stimpson. A dentist and druggist by trade, he had been recruited to science by the entomologist Benjamin D. Walsh in the early 1860s. Velie's greatest claim to fame had been the five months he spent exploring the Rocky Mountains in 1864 with Dr. Charles C. Parry, who named a peak in his honor. Velie was an accomplished taxidermist and an enthusiastic collector but did not make much of a mark as a naturalist. While Stimpson dredged on the *Bibb*, Velie would stay ashore to collect vertebrates. The two men planned to team up and work together in the spring.[9]

With six days to recover from his injuries on the trip down to Key West, Stimpson found time to reflect on the latest developments in the study of oceanic life. When he began his career around 1850 he could go out dredging alone and fairly close to shore and still make significant discoveries. At that time the Englishman Edward Forbes had theorized that life could not exist below three hundred fathoms, and most naturalists accepted this idea until the late 1860s. By 1870, the advent of deep-sea dredging required an entire team of people, larger ships, and crew, as well as serious funding, either governmental or institutional. In England plans were afoot for the now-famous *Challenger* expedition, which sailed late in 1872. At the same time, Anton Dohrn was laying the foundations for a marine biology lab in Naples, and Baird was increasingly focused on his duties as the head of the US Commission of Fish and Fisheries at Wood's Hole. These changes in the way both fieldwork and lab work were conducted is another illustration that the manner in which natural history was done by Stimpson's generation was changing. A new era in the study of the seas was at hand, just when Stimpson's physical and emotional reserves were at their lowest ebb.

Then and now the ocean depths were thought to be one of the keys to understanding life on Earth. On the eve of his own voyage on the *Hassler*, Agassiz had boldly predicted that deep sea dredging would uncover coelacanths and other animals hitherto known only as fossils, an idea that supported his argument that species did not change over time. While wrong on evolution, sixty-seven years later Agassiz's prediction regarding the coelacanth came true.[10]

Despite everything he had been through, Stimpson must have been looking forward to his first real opportunity to dredge in the deep sea. He had seen a glimpse of what he might find through his study of Pourtales's collections, and now he would have a chance at making his own discoveries. Landing at Key West on December 21, Stimpson expected to begin work immediately, but the *Bibb* had not yet arrived. A day passed, and then several, and he and Velie waited with increasing impatience. Since the ship was expected at any time they could not undertake excursions along the Keys. It could not have been a very happy Christmas and New Year's for Stimpson, and as the days turned to weeks he became exasperated. A smallpox outbreak at Key West only added to his despondent mood.

The *Bibb* finally chugged into sight in mid-January, a full month after Stimpson had expected it. Acting Master Robert Platt, a thirty-six-year-old Civil War veteran, commanded the vessel, and he and Stimpson learned that they had several things in common. Like Stimpson, Platt had sailed along the Maine coast, and during the Civil War he had served under John Rodgers, Stimpson's commander on the North Pacific Expedition. After the war Platt had skippered the *Bibb* when Pourtales was making his groundbreaking discoveries, so he had more experience in deep-sea dredging than any other American naval officer.

Platt's orders were to make soundings and charts for the deployment of a telegraph cable connecting Cuba and Mexico's Yucatan peninsula. The survey would take them from Cape San Antonio off Cuba to Cape Catoche on the Yucatan coast, a distance of about 135 miles. The International Ocean Telegraph Company paid for the ship's coal.[11] When the *Bibb* finally left Key West on January 22, it must have been with a great sense of relief and anticipation that Stimpson again found himself at sea. Unfortunately the ensuing voyage quickly devolved into a series of disappointments and failures. On arriving in Havana the ship was promptly placed in quarantine for three days due to the smallpox outbreak at Key West. More delays followed as Platt spent nearly a week negotiating with the Spanish authorities over the details of the cruise. Stimpson occupied himself by visiting the local fish markets in search of rarities.[12]

They departed Havana on January 31 but while passing ancient Morro Castle at the entrance to Havana Harbor, Stimpson was stricken with a violent hemorrhage. Platt put in at Mariel for a day to fix a faulty boiler and to tend to Stimpson, but as soon as they put to sea a fierce gale began to blow and the hemorrhaging recurred. Stimpson vowed to carry on, but as the storm continued Platt decided to wait it out at Port Cabanas, where Stimpson's condition improved slightly.[13]

The *Bibb* next sounded Guadiana Bay, but stormy weather again forced them into port. When the seas subsided Platt cast off once more, finally ready to sound the Yucatan Channel, which connects the Caribbean and the Gulf of Mexico. Stimpson had recovered sufficiently to oversee the dredging, but their first few casts revealed an ocean floor nearly devoid of life.[14] Between the weather that kept them idle and the lack of success in dredging, it was almost as if Nature herself was conspiring against Stimpson.

During a brief stop in Cozumel they examined several mounds to collect copper tools, statues, and pottery. The weather remained turbulent, however, and fully three-quarters of the trip passed in laying at anchor waiting for better weather. Tired, sick, and dispirited by his lack of success, Stimpson found little reason to celebrate his fortieth birthday on February 14. Platt sent a report to his superiors from Mujeres (Island of Women) not far from Cancun. "Dr. Stimpson has been very ill but is now better," he wrote, and he noted that the dredge had brought up "several new shells." After making the soundings the *Bibb* began the return passage to Havana.[15]

In all, Stimpson managed to dredge in deepwater for perhaps half a dozen good hauls. They reached one thousand fathoms, by far the deepest dredging of his career. At a depth of 424 fathoms (almost half a mile) they recovered what turned out to be a new echinoderm, but the most significant find occurred in 250 fathoms near Havana when they brought up a splendid foot-long crinoid. Shortly thereafter a storm began raging, and they "had a very rough time" between Havana and Key West before reaching the latter on March 22.

It had been an extremely difficult two months, with Stimpson enduring at least four episodes of hemorrhage, attacks that left him exhausted. As he tried to rest, his thoughts turned to family, and in a touching letter to his "dear son Willie" he admonished him to "learn to read as fast as you can, and then all kinds of learning will be open to you." Stimpson's illness had made it impossible to share these sentiments in person.[16]

In late March and early April he had recovered sufficiently to dredge anew on both the *Bibb* and the steamer *Bache*. Milder weather prevailed and one observer remarked of Stimpson that the "energies of that able man of science, though greatly wearied by disease, were then in full exercise." In one hundred to two hundred fathoms off Sombrero and Sand Key he found a few new echinoderms. These collections were made in the same area that Pourtales had dredged in 1868, so these and other specimens effectively served as types of the collection that had been lost. But after a few days Stimpson again began coughing up blood and had to be taken ashore.[17]

At Key West he stayed at the Russell House hotel, just a block from the seashore. On April 19, he wrote Baird in a firm, clear hand, but his words were those of a defeated man. "I have now scarce breathing room enough left to climb a short stairway without much distress," he declared, admitting that he felt sicker now than when he had left Maryland. He had been "very unfortunate in the way of collecting," and while Velie had managed to get a number of birds, the overall results were pitifully small compared to what they had hoped for. Stimpson did succeed in getting many of the species of Gulf Stream mollusks and crustaceans that had been burned in the fire.[18]

The one unqualified success was the discovery of the rare crinoid *Pentacrinus caput-medusae*. Stimpson believed it was "the only one ever obtained so near our coast. I don't think there is one in any of our museums, as Pourtales did not get it." Very few living crinoids were known, and they were among the most sought-after prizes in zoology. Crinoids were much more abundant in the fossil record so getting one of these "living fossils" was indeed noteworthy.[19]

Florida had not healed him this time, and while bedridden Stimpson jotted a few lines to Stearns. "My health is very poor—lungs badly filled up with tubercles, etc., and have frequent hemorrhages—cannot do anything requiring any physical exertion without great distress." He had continued to push on until his body simply refused to go any farther. He asked to be brought home and the *Bibb* sailed for Norfolk around May 9. Five days later Velie accompanied him to Ilchester, but Stimpson insisted that Velie return to Key West to resume collecting.[20]

Stimpson's appearance must have been frightening to Annie and the children. Patients in the terminal stage of pulmonary tuberculosis have been described as hollow-cheeked and emaciated, with sunken eyes. While happy to be reunited with his family he grew weaker by the day. Seeing his three young children (William had just turned seven while the twins were three years old) must have been both a comfort and a torment, as he faced the reality that if he died now his children would remember virtually nothing of their father as they grew up.

By this point he probably lacked the energy to do much more than read his mail and perhaps dictate a few letters. His thoughts turned to the Academy. After the fire Stimpson had been adamant that the new museum building be situated far from the city's center. The trustees hoped to acquire ten acres south of the business district where Stimpson felt they would be safe "from the fell destroyer which has so persistently followed us." But when Walker informed him that the Academy had decided to rebuild on the old foundations, Stimpson admitted that his first sensation was a thrill of joy.[21]

Eliphalet Blatchford, his closest Chicago friend, sent several touching letters. "You know how warmly goes out to you our sympathy as we learn of the days of suffering you have experienced in this prolonged and stormy cruise: and of the weak state in which you must now be from these experiences." Blatchford passed along greetings from a mutual friend, Charles Jackson Paine, general superintendent of the Lake Shore and Michigan Southern Railroad, who wished to convey the following sentiments to Stimpson: "I wish I might send my word of sincere admiration which could gratify him for a moment. If virtue is its own reward, it is nevertheless desirable that it should be rewarded with recognition; and all honest men who have ever met the good doctor will agree with me that he is one of those noblest works of God. Standing upon the shore of science, I could look out and see how illimitable (to my view) was the great sea of his knowledge, and picked up many a precious pearl which he little suspected he had cast upon my strand."[22]

This message may or may not have arrived in time for Stimpson to read it. His disease had entered its final inexorable stages. He grew more feeble and had constant pain in his joints. Confined to bed and plagued by a recurrent cough, his slightest movement caused great discomfort. At night his fever spiked and he sweated profusely. Breathing was increasingly labored and difficult, and the hemorrhages were often accompanied by vomiting and diarrhea.[23]

The end came twelve days after his return to Ilchester on Sunday, May 26, 1872, at 11:30 p.m. He died a few months after turning forty, just seven-and-a-half months after the fire that had eradicated his major contributions to science. In maudlin fashion, many subsequently attributed his death to the fire and a broken heart, but he had already been very ill in the days and weeks leading up to the fire. It was his spirit and will to live that began to fade away on that hot October day.

It is bitterly ironic that for a man who loved social life as much as Stimpson, none of his scientific colleagues appear to have attended his funeral. Baird claimed to be too ill to travel and word apparently did not reach others in time. They instead paid tribute to his memory in print, invariably recalling not his scientific attainments but his personal qualities, including his droll sense of humor, his humility, and his great love for animals.

Stearns published a brief account of his friend, lamenting, "As we approach his grave we think not less of his high intellectual attainments and scientific ability, we think more highly of the friendship we enjoyed, of the friend that is gone . . . his life was a continued effort to increase the sum of human knowledge."[24] Stearns was also probably responsible for an article in the *Overland Monthly*, followed by a poem whose author admired Stimpson's "modest demeanor and generous nature, quite free from the petty jealousy which is frequently met with in scientific men."[25]

A stirring eulogy came from the Reverend William W. Everts of the First Baptist Church in Chicago. A week after Stimpson's death Everts delivered a sermon entitled "The Church and Science" in which he argued that men of science such as Stimpson had been overlooked or taken for granted by Chicagoans. "They are great men, ministering wisely," Everts wrote. "We hardly notice them; we ought to revere them. They will be noted in history as having wrought more grandly than the men who were engaged in mere outward commercial enterprises." Everts went further. "I cannot see that any benefactor could outrank such a man. . . . He was one of the leading spirits in this noble Academy. . . . Professor Stimson [*sic*] will long be remembered, and his work deserves our highest commendation. I affirm that such as he are building more grandly for Chicago and the Northwest than the *material* builder. The man who advances the civilization of the age is the great benefactor."[26]

Just as they had done six years earlier for Kennicott, the Academy held a special meeting to honor a fallen leader. Attendance since the fire had been spotty, but over two dozen men came to pay their respects. William W. Calkins praised Stimson's readiness to help young naturalists while Blatchford focused on his "qualities of unselfish deference, geniality, moral uprightness, sincerity and gentlemanly demeanor," maintaining that Stimson had lived in a manner that inspired in others the same ardor that animated his own life. The physician Hosmer Johnson added, "Science had many worshippers but few priests, but the Academy had so soon been called upon to bewail two; first a Kennicott, and now a Stimson."[27]

Others in the scientific community weighed in. Dall also used a religious analogy to describe Stimson's dedication to science in Chicago, calling him "a scientific missionary, a biological bishop, in partibus infidelium, in the land where the almighty dollar reigned supreme."[28]

Of all the homages, none is more perceptive or meaningful than Joseph Henry's, who wrote that Stimson "was among the few who had the gift of enlarging the bounds of human thought; of adding by persevering investigation and sagacious suggestions new facts and principles to the domain of science. Although his life was short it was one of important results; although not permitted to complete the great work he had in hand at the time of his death he lived to accomplish sufficient to connect his name conspicuously with the history of science in this country. His character was such as to command respect and excite love and admiration." Henry went on. "Of all the young men who have been connected with the Smithsonian Institution there is none to whom I gave my confidence more unreservedly or on whose judgment and integrity I relied more fully . . . in case of the separation of Professor Baird from the Institution there is no one I would prefer to Dr. Stimson to take his place."[29]

Stimson's life in science had many facets. It is impossible to quantify his leadership of the Chicago Academy, or to calculate the effect of his reviews in the *American Journal of Science* in helping to promote nationalism and professionalism, or to trace the positive example his taxonomic writings or his dedication to fieldwork had on younger men such as Verrill, Dall, Morse, Gill, and Kennicott. His impact on American natural history was acknowledged at the time but later forgotten.

Much of what he did publish is in a sense preliminary, a prelude to larger works that never saw the light of day. As a result his bibliography lacks a major defining work and his published output is undoubtedly inferior to that of many of his better-known colleagues. But judging Stimpson solely on his publications diminishes his other contributions. Nathaniel Shaler, a friend and fellow naturalist, recalled that Stimpson's "keen interest in animals of all kinds, his real love for them, made him something much better than his printed work. It was his affection for creatures as well as his general wit that quickly brought us together."[30]

Stimpson is perhaps best characterized as one of a relatively small group of nineteenth-century naturalists who tried to discover and accurately classify the myriad forms of life on Earth. His genius, if it can be called such, lay in creating the levels of classification that built a solid foundation for future investigators. His uncanny ability to discriminate between closely allied species was the result of patient museum study as well as his extensive field experience. He excelled at identifying key anatomical features and had few peers in working at the higher levels of classification.

As a curator and museum director, he labored tirelessly to create a great natural history museum while also supervising the Academy's scientific publications and fostering enthusiasm for science in an intellectually barren part of the country. Unlike the collections of his peers Baird, Agassiz, and Leidy, those that he assembled have not survived and this is one of the primary reasons his life is not remembered.

William Stimpson is buried in an unmarked grave in St. John's Cemetery of Howard County in Ellicott City, Maryland. His wife and daughter are with him, also in unmarked graves. His body rests barely forty miles away from the Smithsonian Castle where he achieved much of his fame and where he felt most at home. In death he is separated from his birth family including his parents and siblings, all interred at Mount Auburn Cemetery in Cambridge along with Louis Agassiz and other famous Bostonians.

Herbert Stimpson outlived his oldest son by fifteen years, dying at the age of eighty-four. William's daughter Marie (known as Daisy) died in 1883 at the age of fourteen. Annie Stimpson died at the age of sixty-eight in 1904, thirty-two years after becoming a widow.[31]

In a testament to their friendship for Stimpson, Baird, Blatchford, and George Walker helped pay for the education of both of Stimpson's sons, and Baird hired them to work at the Smithsonian. William Gordon Stimpson served as an assistant in the mammalogy department and at the age of sixteen was elected a member of the newly founded Biological Society of Washington, where he was welcomed by many of his father's old friends. He seriously considered becoming a naturalist before leaving the Smithsonian in 1887 and going on to a distinguished career as a surgeon with the US Health Service. He wrote several books on medical topics and died in 1940 at the age of seventy-four, survived by three daughters.[32]

Beginning at age thirteen, Herbert Baird Stimpson worked at the Smithsonian, staying for ten years. He became a lawyer and published two novels while also editing the *Conservative Review*. Herbert also helped edit the National Academy of Sciences Biographical Memoir of his father. He died in Baltimore in 1937 at the age of sixty-eight. He married late in life and apparently had no children.[33]

William's brother James stayed in Alabama until his death in 1915, and his branch of the Stimpsons has Mobile, Alabama as their base. Family patriarch Billy Stimpson was named for William Stimpson and coauthored and published a history of the Stimpson family.

At one time both Baird and Gill planned to write a biography of Stimpson.[34] Baird never found the time but Gill did; however, its publication was scuttled by Annie Stimpson, who objected to the account of Stimpson's early struggles, probably an allusion to his problems with his father and Agassiz. Referring to the situation many years later, Dall called Annie "a good mother but an intolerable and silly snob." Gill's biography has since been lost.[35] In 1888 Dall produced a laudatory if sometimes inaccurate sketch of Stimpson's life. The National Academy of Sciences Biographical Memoir of Stimpson was almost an afterthought, coming forty-six years after his death.[36]

It is perhaps fitting that Stimpson and Louis Agassiz share one final link from beyond the grave. Agassiz died in 1873, a little over a year after Stimpson. Subsequently Theodore Lyman published an article on the echinoderms of the *Hassler* Expedition, Agassiz's final fieldwork. Lyman also included specimens that Stimpson had collected in Florida in 1872, so the paper contains descriptions of new animals dredged on what turned out to be each man's final voyage. Lyman credited Stimpson with the discovery of six new species.[37]

One of Stimpson's most significant published works, and by far his longest, appeared in 1907, thirty-five years after his death. Smithsonian curator Mary Jane Rathbun discovered a portion of his manuscript on the crustacea from the North Pacific Exploring Expedition encompassing 358 species in 235 pages, with 26 plates. Covering the species and genera described in parts 3 through 7 of the Prodromus but this time written in English, it contains a great deal of additional information on these genera and species not included in the Prodromus. This book provides a taste of what the final NPEE report might have looked like.

And what of the Chicago Academy that Stimpson did so much for? John W. Foster succeeded Stimpson as secretary but died in June 1873, just a little over a year after Stimpson. In the span of seven years the Academy had endured two fires, lost all its collections, and suffered the deaths of its three scientific leaders. Jacob Velie preserved the collections that had been donated after the fire and made several collecting trips to Florida, but the Academy remained essentially homeless until 1894. That year they moved into a new building on the north side of Chicago. The opening of the Field Museum that same year overshadowed this new beginning and relegated the Academy to the status of the "other" natural history museum in Chicago. Yet it carved out a niche focusing on Midwestern natural history, amassing important collections and creating a publication series and innovative dioramas. Many excellent scientists led the Academy in the twentieth century, including the herpetologist Howard K. Gloyd, the ornithologist Alfred M. Bailey, and the malacologist Frank C. Baker. In 1999 the Academy moved into a new building which today is known as the Peggy Notebaert Museum of the Chicago Academy of Sciences.

WILLIAM STIMPSON AND AMERICAN NATURAL HISTORY

By and large, as the mass of knowledge grows, men devote little attention to the dead. Yet it is the dead who are frequently our pathfinders, and we walk all unconsciously along the roads they have chosen for us.[1]
—Loren Eiseley

There have been a number of recent books that have covered the early years of American natural history from the colonial period until about 1840.[2] Likewise there are numerous studies of American biology covering the late nineteenth century. But we have fewer works that detail American natural history and zoology between 1850 and 1870. Several excellent studies cover specific disciplines during this period, including paleontology, entomology, and ornithology.[3] In addition, Lester D. Stephens and Richard G. Beidelman have written important regional studies of southern and California naturalists, respectively.[4] There are a number of biographies of naturalists of the era, including Spencer F. Baird, Asa Gray, Louis Agassiz, Joseph Leidy, and Ferdinand Hayden, but many others are deserving of biographies, including Theodore Gill, William H. Dall, Addison E. Verrill, Fielding Meek, and Robert Kennicott.[5] There are still large gaps in our knowledge of American natural history in these years, and biographies of these men and others would go a long way towards illuminating that story.

What did the American scientific community look like during Stimpson's lifetime? Robert V. Bruce's *The Launching of Modern American Science* (1987) gives us a clearer picture of American science between 1846 and 1876, years that roughly parallel Stimpson's career.[6] The 1840s saw several events that would shape both American science and Stimpson's life. The founding of the Smithsonian Institution in 1846, the arrival of Agassiz that same year, and the establishment of the American Association for the Advancement of Science in 1848 were all developments crucial to the professionalization of American natural history, and Stimpson had a significant role in all of them. By 1876 science in the United States had begun to shed its second-class status compared to European science, thus marking this era as a maturational and transitional phase in American science.

Bruce argues that American science became a collective enterprise during this period, one that required institutions, government support, and money. The major natural history museums in the United States in the 1860s included the Smithsonian, Agassiz's Museum of Comparative Zoology, the Boston Society of Natural History, and the Academy of Natural Sciences of Philadelphia. Stimpson worked or volunteered at the first three and did research at the latter. When he embarked on his own museum career, in the span of just a few years he established the museum of the Chicago Academy of Sciences as the fifth most significant collection in the United States, and he did it with much less public support than those other institutions. Stimpson's articles appeared in all of the leading journals of the day, including the *American Journal of Science*, the proceedings of both the Academy of Natural Sciences of Philadelphia and the Boston Society of Natural History, the *Smithsonian Contributions to Knowledge*, the *Annals of the Lyceum of Natural History of New York*, and both the *American Naturalist* and the *Canadian Naturalist*.

Many nineteenth-century taxonomists, like Stimpson, have been forgotten. An esteemed historian of the field has recently summarized progress but admits that, "our knowledge remains scattered and fragmentary." Taxonomy, or systematics as it is often referred to today, has been relegated to "the low end of the totem pole of prestige in science."[7] Robert Kohler's book *All Creatures: Naturalists, Collectors, and Biodiversity, 1850–1950*, provides a framework for understanding the significance of Stimpson's taxonomic work. Kohler challenges the tendency to view taxonomy as less than first-rate science. "In fact the work of identifying, describing, and categorizing natural kinds is highly exacting and creative. It is not just facts that are thus produced ... but the very categories by means of which we perceive and understand our world. Those who regard categorizing as a mere preliminary implicitly assume that describers simply sort items into preexisting stable categories—but they do not. Categories are created, recreated, and even destroyed as classifiers apply them to material things."[8]

Taxonomy has had its detractors, especially from those associated with the Agassiz school of natural history. A contemporary of Agassiz's, Nathaniel Southgate Shaler, recorded one of the main criticisms of Stimpson and others focused on taxonomy, noting that "William Stimpson was a naturalist of no mean capacity. If he had not been turned to species-describing, a task akin to 'gerund grinding,' he would have come to largeness."[9] But there is much more to taxonomy than "species-describing." Stimpson worked at the higher levels of classification by instituting over 160 families, subfamilies, and genera that are still valid today. These critiques of taxonomy became common beginning around the 1890s. Taxonomy came to be seen as less prestigious than theoretical and laboratory-based biology, but this is flawed thinking in Kohler's view. "But to think therefore that the one is a lesser preliminary to the other immodestly presumes that change in science is progressive, and that what comes after is by definition superior to what came before. It is not, or not always: just different," he concludes.[10]

Others have made the same point. Writing in 1941, two careful observers of Pacific Coast marine invertebrates noted that the most useful references on the fauna of the Gulf of California were the oldest and the newest ones. Citing Stimpson's work along with others from the period 1841–1871, the novelist John Steinbeck and his biologist friend Ed Ricketts claimed that during this time these invertebrates "were known as well as, or possibly better than they are today, and by a larger proportion of zoologists."[11]

Another major topic of discussion among historians is the fate of natural history. Lynn Nyhart, Keith Benson, and Peter Bowler have all successfully challenged the view that natural history died out around 1890, to be replaced by laboratory-based biology.[12] Natural history never truly went extinct (some of the most popular museums in the world today still incorporate the phrase into their name), but it has certainly fallen out of favor in academia, where the focus shifted to genetics or embryology. Some zoologists rarely if ever work with living animals in their natural habitats. Natural history is deemed old-fashioned and the term *naturalist* evokes images of a ridiculous, perhaps effeminate Victorian prig chasing butterflies with a net so that he can pin them and add them to his "collection." Edward O. Wilson is one of the few prominent zoologists today to proudly consider himself a Naturalist, the title of his delightful 1994 autobiography.

Lynn Nyhart contends that beginning in the 1840s natural history began to be transformed by its more advanced practitioners into a new "scientific zoology," characterized by comprehensive life histories and new methods, including an increasing use of technical terminology, more utilization of the microscope, and embryological studies. While the majority of Stimpson's published writings are devoted to taxonomy, he published several articles that bear the hallmarks of this new approach.

Stimpson used the terms *naturalist* and *zoologist* but he usually referred to himself as a zoologist and his published articles and reviews often mention zoologists or zoology. This makes sense given his tutelage under Agassiz, who enshrined the phrase in his own Museum of Comparative Zoology. In

1860 Stimpson himself used the term *scientific zoologist*, and it appears that he sometimes used the word *zoologist* as a way to distinguish between what he considered to be "professional" versus nonprofessional writers. He also wrote of "biological writings" in 1860 and preferred the terms *malacology* and *carcinology* to distinguish these subfields within zoology. In 1853 he used the term *practical naturalist* to distinguish field naturalists from closet naturalists.

Robert Kohler has noted that, "little has been written about the craft and social history of scientific collecting." One fine example is Anne Larsen's work which examined English scientific zoologists between 1800 and 1840, in which she documented their field practices and exchange networks.[13] The arguments about field and closet naturalists have generated considerable debate. One historian has argued that by 1850 the field collector had been largely superseded by academic researchers in American science. In the first half of the century American field naturalists had been tainted in the eyes of Europeans by the excesses of Rafinesque, who had a positive mania for naming new species and was largely ignorant of the labors of those who had gone before him. Audubon was derided during his life by "sedentary" naturalists that questioned his scientific acumen.[14]

Many historians of natural history have linked fieldwork with amateurs and viewed it as a less sophisticated type of science than laboratory work. Historians of English natural history have generally made a clear distinction between those doing fieldwork as opposed to museum-based taxonomists, as if the two were mutually exclusive. One writes of a "total divorce between field and closet research" and maintains that "very few naturalists" both found new species in the field and described them.[15] The split widened over the course of the nineteenth century but to a much lesser degree in the United States. The American experience was different in part because there were still so many new species to be discovered.

This dichotomy between field naturalists and closet workers was noted at the time. In 1886 the American entomologist Augustus Grote criticized Europeans for following books instead of living nature.[16] If there is such a thing as a distinctive form of American natural history, it might be that Stimpson and many Americans active in zoology in the period 1850–1870 excelled in both fieldwork and the closet work of taxonomy in museums.[17] Others that fit this mold included Addison E. Verrill, William H. Dall, Joel A. Allen, Edward D. Cope, and Robert Kennicott. This generation can be seen as the transitional link between the older style of natural history of the 1840s based on fieldwork, and the new lab-based biology of the 1890s.[18] Of course all such generalizations must be viewed with caution, as each specific discipline developed differently. In Stimpson's specialty of marine invertebrate zoology the field tradition continued to be crucial well into the twentieth century since the depths of the sea were still largely unknown.[19]

In an otherwise excellent overview of the history of American marine zoology one historian wrote that by 1870, "Only the Coast Survey's short cruises with Louis Agassiz and his pupils and the navy's occasional multipurposed expeditions had produced any marine zoological collections, and these were at best haphazard."[20] This overlooks Stimpson's pioneering dredging program along the Atlantic Coast in the 1850s and 1860s, which produced a collection that helped form the nucleus of the Smithsonian's department of marine invertebrate zoology. Unfortunately it too was destroyed in the Chicago Fire.

Stimpson's career has many parallels with that of the Englishman Edward Forbes (1815–1854). Each was renowned as the best scientific dredger in his country and both published on many different phyla of marine invertebrates. Each reveled in the society of their fellow naturalists and brought them together in formal and informal gatherings. Both suffered from poor health and died young, Forbes at thirty-nine, Stimpson at forty. Finally, the majority of each man's personal papers have not survived.

Stimpson is best known among zoologists today for his work in carcinology. A historian of the field laments the perception that Stimpson and others are regarded as "eternal understudies. ... We tend to forget, however, that success often results merely from good luck, working in the right place at the right time, which does not necessarily reflect innate talents. After all, what could a 'lesser figure' like William Stimpson have achieved, if he had lived longer, or if he had not lost almost everything in the Great Chicago Fire?"[21]

Historians of carcinology have been particularly active in assessing the taxonomic work of their predecessors. Austin B. Williams looked at crab research in North America since 1758. Before 1850 there were only eighty-three species known and most (80%) had been described by Europeans. Noting the "great exploratory thrust" achieved during the second half of the nineteenth century, Williams recorded 273 new species described between 1850 and 1899, with Stimpson accounting for 92 (33.6%). Alphonse Milne Edwards was next with fifty-four (19.7%). Between 1900 and 1949 only seventy-two new species were described, and the field of brachyuran studies changed after 1900 as "descriptive efforts, always a mixture of several disciplines, became more analytical," with biogeographic studies and ecological and experimental studies emphasized instead of purely descriptive taxonomy.[22]

Stimpson's Taxonomic Work

The ultimate proof of professionalism in taxonomy is whether your taxa withstand the test of time. The advent of sophisticated online computer databases gives historians a tool to measure this quantitatively. The World Register of Marine Species allows us to quantitatively analyze and compare how the judgments made by some nineteenth-century naturalists hold up today. Although he did not live nearly as long as most of his peers, Stimpson was one of the most prolific taxonomists of his generation. While he did significant work on freshwater mollusks, most of Stimpson's career revolved around marine invertebrates. According to the World Register of Marine Species, Stimpson named 827 currently accepted taxa in the following categories:

Superfamily—2
Family—18
Subfamily—11
Genus—129
Subgenus—1
Species—655
Subspecies—11 (Stimpson did not designate subspecies; these are species he named that have been reclassified as subspecies.)

Stimpson named taxa from eleven different phyla:

Arthropoda—559
Platyhelminthes—65
Mollusca—54
Nemertea—35
Echinodermata—35
Annelida—33
Cnidaria—27
Chordata—12
Bryozoa—4
Brachiopoda—1
Sipuncula—1

Significantly, 161 of Stimpson's taxa, or nearly one in five (19.4%), are at the genus level or higher. Taxonomists working at these higher levels of classification were presumably more "professional" than those who focused on describing new species. Stimpson's judgments have survived for a century and a half because his descriptions were thorough and detailed.

By way of comparison, James D. Dana, whose taxonomic work in zoology preceded Stimpson's by just a few years and who was also mainly devoted to carcinology, is credited with 774 accepted taxa, 189 of which, or 24.4 percent, are genus or higher. Accuracy is also something that can be measured. Overall, Stimpson proposed 1,626 taxa and 827 are accepted, for a success rate of 50.8 percent. Dana's numbers are 774 out of 1,781, for a success rate of 43.4 percent.[1]

Another way to quantify Stimpson's scholarship is through Google Scholar citations. Stimpson's works have been cited 2,455 times, including 635 times since 2011. Of the top five most cited papers, three relate to specimens from the North Pacific Exploring Expedition while his 1853 Grand Manan article ranks fifth with 143 citations.[2]

A closer look at the seemingly mundane subject of scientific nomenclature reveals significant information relating to the social aspects of science. Robert Kohler noted that "no other community of scientists preserves such a deep sense of its collective identity and past" as taxonomists, thus keeping "the past forever present."[3] Patronyms, or names of species that honor an individual, are a way for taxonomists to distinguish patrons or to reward collectors that provide them with specimens. Stimpson named over sixty taxa (not all of them valid today) honoring forty-nine individuals. (James D. Dana and John Xantus were favorites, with nine species combined). A significant number were military officers while others were physicians assigned to exploring expeditions, especially the Pacific Railroad Surveys. Others were amateur collectors.

The WORMS website contains seventy-three valid marine taxa named in Stimpson's honor, the earliest from 1852 and the latest in 2013. Addison E. Verrill leads the way with seven, while William H. Dall has six and Alphonse Milne Edwards five. Twenty-nine are crustacea and nineteen are mollusks. Thirty of the names were bestowed during Stimpson's lifetime by his peers, with the remainder by the taxonomists that came after him.

Notes

Introduction

1. Lynn K. Nyhart, "Natural history and the 'new' biology," in *Cultures of Natural History*, ed. N. Jardine, J. A. Secord, and E. C. Spary (Cambridge: Cambridge University Press, 1996), 427.

2. Robert Kohler, *All Creatures: Naturalists, Collectors, and Biodiversity, 1850–1950* (Princeton, NJ: Princeton University Press, 2006).

Chapter 1

1. Samuel Eliot Morison, *The Maritime History of Massachusetts, 1783–1860* (Boston: Northeastern University Press, 1979), 7.

2. Charles Francis Adams Jr., "Boston," *North American Review* 106, no. 218 (January 1868): 4, Cornell University Making of America website.

3. William H. Stimpson and Richard W. Price, *A Stimpson Family in America* (North Salt Lake, UT: DMT Publishing, 2004).

4. Stimpson and Price, *A Stimpson Family in America*, 26–29.

5. "Herbert H. Stimpson," *Annals of the Massachusetts Charitable Mechanic Association, 1795–1892* (Boston: Rockwell and Churchill, 1892), 474–75, http://archive.org/stream/annalsmassachus01asso-goog#page/n10/mode/2up. On Emerson buying a Stimpson stove, see William H. Gilman and J. E. Parsons, *The Journals and Miscellaneous Notebooks of Ralph Waldo Emerson*, vol. 8 (Cambridge, MA: Belknap Press of Harvard University Press, 1970), 576.

6. For Mary Ann D. Stimpson, see Stimpson and Price, *A Stimpson Family in America*, 114. Her death is noted in the *Boston Daily Atlas*, "Died," April 21, 1842.

7. "Memoirs and Writings of William Stimpson M. D.," Stimpson papers, Shay Collection, CAS Archives.

8. See Alfred G. Mayer, "William Stimpson, 1832–1872," *National Academy of Sciences Biographical Memoirs* 8 (1918), 419–20; "Herbert H. Stimpson" (1892), 474–75.

9. "English High School," *Boston Evening Transcript*, July 29, 1847; "Award of the Lawrence Prizes," July 10, 1848; "Exhibition of the Latin and English High Schools," *Boston Daily Atlas*, July 27, 1848, America's Historical Newspapers, http://infoweb.newsbank.com. The Franklin Medal is in the Stimpson Papers at the Chicago Academy of Sciences.

10. Henry F. Jenks, *Catalogue of the Boston Public Latin School, established in 1635, with an Historical Sketch* (Boston: Boston Public Latin School Association, 1886), 198. Entering the same year as Stimpson was a student with a distinguished pedigree—Charles Francis Adams Jr., great-grandson of John Adams and grandson of John Quincy Adams.

11. For background on Cambridge's social life at this time, see Edward Lurie, *Louis Agassiz, A Life in Science* (Baltimore: Johns Hopkins University Press, 1988 [1960]); A. Hunter Dupree, *Asa Gray, American Botanist, Friend of Darwin* (Baltimore: Johns Hopkins University Press, 1988 [1959]).

12. "In Memoriam: The Late Dr. William Stimpson, of Chicago," *Chicago Tribune*, June 12, 1872.

13. For the Boston Society of Natural History, see Sally Gregory Kohlstedt, "From Learned Society to Public Museum: The Boston Society of Natural History," in *The Organization of Knowledge in Modern America, 1860–1920*, ed. Alexandra Oleson and John Voss (Baltimore: Johns Hopkins University Press, 1979), 386–406.

14. Sally Gregory Kohlstedt, *The Formation of the American Scientific Community* (Urbana and Chicago: University of Illinois Press, 1976), 121.

15. Lurie, *Louis Agassiz*, 162.

16. William Stimpson, "Description of a new species of *Helix*," *Proc. BSNH* 3 (1848–1851): 175. This is not considered a valid species today.

17. For a discussion of Stimpson's role in the development of the aquarium, see ch. 4.

18. "In Memoriam: The Late Dr. William Stimpson, of Chicago," *Chicago Tribune*, June 12, 1872.

19. Stimpson, "Descriptions of two new species of *Philine* obtained in Boston Harbor." *Proc. BSNH* 3 (1848–1851): 333–34. Stimpson was curator from December 4, 1850, until May 18, 1853. See Thomas T. Bouve, "Historical Sketch of the Boston Society of Natural History; with a Notice of the Linnaean Society which Preceded it," in *Anniversary Memoirs of the Boston Society of Natural History published in celebration of the Fiftieth Anniversary of the Society's Foundation, 1830–1880* (Boston: Published by the Society, 1880), 246; "Curator Reports," *BSNH* 4 (1851–1854): 58. See also Stimpson's "Animals presented to the BSNH," *Proc. BSNH* 4 (1851–1854): 120.

20. For Agassiz, see Lurie, *Louis Agassiz*, and Christoph Irmscher, *Louis Agassiz: Creator of American Science* (Boston: Houghton Mifflin Harcourt, 2013).

21. Robert Kennicott to his family, February 15, 1858, GNHL.

22. Elizabeth Agassiz, *Louis Agassiz, His Life and Correspondence*, vol. 2 (Boston: Houghton and Mifflin, 1886), 434–35.

23. Lurie, *Louis Agassiz*, 177.

24. Irmscher, *Louis Agassiz*, 195.

25. Nathaniel Shaler, *The Autobiography of Nathaniel Southgate Shaler* (Boston: Houghton and Mifflin, 1909), 101, 188.

26. Lane Cooper, *Louis Agassiz as a Teacher* (Ithaca, NY: Comstock Publishing, 1945), 41.

27. Mary P. Winsor, *Reading the Shape of Nature: Comparative Zoology at the Agassiz Museum* (Chicago: University of Chicago Press, 1991), 16.

28. Ignas K. Skrupskelis and Elizabeth M. Berkeley, *The Correspondence of William James*, vol. 4, *1856–1877* (Charlottesville: University Press of Virginia, 1996), 63.

29. James R. Jackson and William C. Kimler, "Taxonomy and the Personal Equation: The Historical Fates of Charles Girard and Louis Agassiz," *Journal of the History of Biology* 32 (1999): 509–55.

30. One example is a beautiful drawing of the gastropod *Cardium mortoni*, preserved in the Stimpson Papers, SIA RU 7093, October 29, 1850.

31. Lurie, *Louis Agassiz* 285–286.

32. "Notes on *Vermetus lumbricalis*," Stimpson Papers, SIA RU 7093.

33. See uncredited reviews of *The Aquarium, an Unveiling of the Wonders of the Deep Sea*, by Philip Henry Gosse, and *The Book of the Aquarium and Water Cabinet*, by Shirley Hibbard, in the *North American Review* 87, no. 180 (July 1858): 152, Cornell University Making of America website. Stimpson's name is misspelled throughout as Stimson.

34. For Agassiz's influence, see Winsor, *Reading the Shape of Nature*, 44.

35. Stimpson, "Observations on the identity of *Nucula navicularis* and *N. thraciaeformis*," *Proc. BSNH* 4 (1851–1854): 26–27.

36. Stimpson, "On two new species of shells and a holothurian from Massachusetts Bay," *Proc. BSNH* 4 (1851–1854): 7–9. Totten served as a regent of the Smithsonian and was one of the founders of the National Academy of Sciences.

37. Stimpson, "Monograph of the genus *Caecum* in the United States," *Proc. BSNH* 4 (1851–1854): 112–13.

38. Stimpson, "List of fossils found in the post-Pliocene deposit, in Chelsea Mass., at Point Shirley," *Proc. BSNH* 4 (1851–1854): 9–10.

39. Stimpson, "Observations on the Stations of the Echinodermata," *Proc. BSNH* 4 (1851–1854): 55.

40. Stimpson, "Observations on the fauna of the islands at the mouth of the Bay of Fundy and of the extreme northeast coast of Maine," *Proc. BSNH* 4 (1851–1854): 95–100; Stimpson, "A New species of *Pentacta*," *Proc. BSNH* 4 (1851–1854): 66–67.

41. Stimpson, "Notices of several new species of Testaceous Mollusca new to Massachusetts Bay," *Proc. BSNH* 4 (1851–1854): 12.

42. Stimpson, *Revision of the Synonymy of the Testaceous Mollusks of New England, with notes on their structure, and their geographical and bathymetrical distribution* (Boston: Phillips and Sampson, 1851).

43. Stimpson, *Revision of the Synonymy of the Testaceous Mollusks*, 21.

44. Stimpson, *Revision of the Synonymy of the Testaceous Mollusks*, 11. For Burnett, see Jeffries Wyman, "Notice of the Life and Writings of the late Dr. Waldo Irving Burnett," *Proc. BSNH* 5 (1854–1856): 64–74. For microscopy, see Leonard Warren, *Joseph Leidy, The Last Man Who Knew Everything* (New Haven: Yale University Press, 1998), 69–72, 261–62.

45. Stimpson, *Revision of the Synonymy of the Testaceous Mollusks*, iv. For more on Forbes, see Eric L. Mills, "A view of Edward Forbes, naturalist," *Archives of Natural History* 11 (1984): 365–93; Mills, "Edward Forbes, John Gwyn Jeffreys, and British dredging before the *Challenger* expedition," *J. Soc. Bibliography of Natural History* 8 (1978): 507–36; and Philip F. Rehbock, "The Early Dredgers: 'Naturalizing' in British Seas, 1830–1850," *Journal of the History of Biology* 12 (1979): 293–368.

46. Agassiz to Dana, January 26, 1852, quoted in Daniel C. Gilman, *The Life of James Dwight Dana* (New York and London: Harper and Brothers, 1899), 319; Dana's review appeared in the *American Journal of Science* 13, 2nd series (1852): 150.

47. Stimpson, *Revision of the Synonymy of the Testaceous Mollusks*, v.

48. Winsor, *Reading the Shape of Nature*, 90–91.

49. "Natural History," *Norton's Literary Gazette* 2, no. 1 (January 1852): 6, American Antiquarian Society Historical Periodicals Collection: Series 3, online database.

50. Agassiz to James D. Dana, January 26, 1852, quoted in Elizabeth Agassiz, *Louis Agassiz,* vol. 2 (1886), 493–95.

51. Stimpson to My dear Father, March 13, 1852, Stimpson papers, Shay Collection, CAS Archives.

52. J. D. Kurtz and William Stimpson, "Descriptions of several new species of shells from the Southern coast," *Proc. BSNH* 4 (1851), 114–15; for Kurtz, see *American Malacologists: A national register of professional and amateur malacologists and private shell collectors and biographies of early American mollusk workers born between 1618 and 1900* (Falls Church, VA: American Malacologists [1973], and Philadelphia: Consolidated/Drake Press), 124; John D. Kurtz, *Appleton's Encyclopedia,* http://www. famousamericans.net/johndkurtz/. Kurtz would later serve as the assistant to the chief engineer for the Union during the Civil War.

53. Stimpson, "On some remarkable marine invertebrata inhabiting the shores of South Carolina," *Proc. BSNH* 5 (1854–1856): 110–17.

54. Mayer, "William Stimpson," 422.

55. Stimpson, "Descriptions of two new species of Ophiolepis, from the Southern coast of the United States," *Proc. BSNH* 4 (1854): 224–26.

56. Mayer, "William Stimpson" (1918), 422. Stimpson never enrolled in the Lawrence Scientific School although many of Agassiz's other students did.

57. Addison E. Verrill journal, March 1, 1860, Harvard University Archives; for the moral dangers of describing new species, see Lynn Barber, *The Heyday of Natural History* (Garden City, NY: Doubleday, 1980), 58–59.

58. Lurie, *Louis Agassiz,* 161.

59. Winsor, *Reading the Shape of Nature,* 4.

60. *Museum of Comparative Zoology, Annual Report for 1863* (1864), 8.

61. Shaler, *Autobiography,* 103–4, 128.

62. Quoted in *Audubon's America,* ed. Donald Culross Peattie (Boston: Houghton Mifflin, 1940), 226.

63. Stimpson, "Observations on the fauna of the islands at the mouth of the Bay of Fundy and of the extreme northeast coast of Maine," *Proc. BSNH* 4 (1851–1854): 99.

64. Stimpson's observations were later cited by Joseph Henry in "Report of the Operations of the Light-House Board Relative to Fog-Signals," *Annual Report of the Light-House Board of the United*

States to the Secretary of the Treasury, for the fiscal year ending June 30, 1874 (Washington, DC: Government Printing Office, 1874), 84.

65. *Smithsonian Annual Report for 1852*, 55–56.

66. Stimpson to Joseph Leidy, October 18, 1852, Academy of Natural Sciences of Philadelphia, no. 1.

67. *Life, letters, and journals of Sir Charles Lyell*, vol. 2, ed. Mrs. Katherine Lyell (London: John Murray, 1881), 184–85.

68. See Robert H. Silliman, "The Hamlet Affair: Charles Lyell and the North Americans," *Isis* 86 (1995): 541–61.

69. K. M. Lyell, ed., *Life, Letters, and Journals of Sir Charles Lyell*, vol. 2, 184–85.

70. Lyell's notebooks record that he paid Stimpson $16, perhaps for expenses incurred while dredging or for specimens. I am indebted to Leonard G. Wilson for the information concerning Lyell's notebooks.

71. For Leidy, see Leonard Warren, *Joseph Leidy: The Last Man Who Knew Everything* (New Haven, CT: Yale University Press, 1998).

72. Stimpson to Leidy, April 7, October 18, December 17, 1852, Academy of Natural Sciences of Philadelphia, no. 1. A few months later the Essex Institute noted a Stimpson donation of Florida shells and sea stars. This is the only confirmation that Stimpson made a visit to Florida in either late 1852 or early 1853. *Proceedings of the Essex Institute*, vol. 1 (1856): 52.

73. Gould to Baird, October 1852, SIA RU 52.

74. Louis Agassiz to John P. Kennedy, November 1, 1852, Enoch Pratt Library, vol. 1, Kennedy incoming. The praise is genuine but for Agassiz it was a win-win situation—he could get rid of a troublesome student who had shown dangerous signs of independence while at the same time enhancing his own cachet as a scientific mentor.

75. Dana to Ringgold, November 1, 1852, NA RG 45.

76. Baird suggested that the scientific corps be given some title other than *naturalist* to avoid "unnecessary criticism on the part of illiberally disposed individuals." The word *naturalist* was tainted in the eyes of many navy men, in part due to the fact that they were upset over the use of civilians instead of navy men to collect natural history specimens. Baird to Ringgold, November 12, 1852, SIA RU 7253, box 1.

Chapter 2

1. Sir Charles Francis Darwin, ed., *Charles Darwin's Autobiography* (New York: Henry Schuman, 1950), 38–39.

2. Janet Browne, *The Secular Ark: Studies in the History of Biogeography* (New Haven, CT: Yale University Press, 1983), 29; Philip F. Rehbock, "The Early Dredgers: 'Naturalizing' in British Seas, 1830–1850," *Journal of the History of Biology* 12, no. 2 (1979): 293–368.

3. Stimpson to Leidy, December 17, 1852, Academy of Natural Sciences of Philadelphia, no. 1.

4. Stimpson, "Synopsis of the Marine Invertebrata of Grand Manan," 13. Robert E. Kohler has noted that the differences between lumpers and splitters is generally based on the "size and depth of the collections they have to work with." Robert Kohler, *All Creatures: Naturalists, Collectors, and Biodiversity, 1850–1950* (Princeton, NJ: Princeton University Press, 2006), 242.

5. John Hersey, *Blues* (New York: Vintage Books, 1988), 19

6. Stimpson, "Synopsis of the Marine Invertebrata of Grand Manan," 5–6.

7. Elizabeth Agassiz to Joseph Henry, February 3, 1853, American Philosophical Society, Joseph Henry papers. Mrs. Agassiz was serving as her husband's amanuensis while Agassiz was ill. Consciously or not, the quote reflects a major cultural phenomenon of the 1840s and 1850s, the Young America movement, which was among other things intensely nationalistic and antiaristocratic.

8. Philip P. Carpenter, "A Supplementary report of the Present State of Our Knowledge with regard to the Mollusca of the West Coast of North America," *Report of the British Association for the Advancement of Science for 1863*, 580.

9. William H. Goetzmann, "Paradigm Lost," in *The Sciences in the American Context: New Perspectives*, ed. Nathan Reingold (Washington, DC: Smithsonian Institution Press, 1979), 25. For nationalism and exploration, see Goetzmann, *New Lands, New Men: America and the Second Great Age of Discovery* (New York: Viking, 1986); and Helen M. Rozwadowski, *Fathoming the Ocean: The Discovery and Exploration of the Deep Sea* (Belknap Press of Harvard University Press, 2005), 63–65.

10. Tony Rice, *Voyages of Discovery: Three Centuries of Natural History Exploration* (New York: Clarkson N. Potter, 1999). For the NPEE, see John D. Kazar Jr., "The United States Navy and Scientific Exploration, 1837–1860" (PhD diss., University of Massachusetts, 1973), University Microfilms edition, 186–232. Also see Vincent Ponko Jr., *Ships, Seas, and Scientists: U.S. Naval Exploration and Discovery in the Nineteenth Century* (Annapolis, MD: Naval Institute Press, 1974), 206–30; Rozwadowski, *Fathoming the Ocean*, esp. 50—57. For a bibliography, see Ronald S. Vasile, Raymond B. Manning, and Rafael Lemaitre, eds., "William Stimpson's Journal from the North Pacific Exploring Expedition, 1853–1856," *Crustacean Research, Special Number #5, Carcinological Society of Japan* (2005) (hereafter referred to as *Stimpson Journal*); Kuang-Chi Hung, "'Plants that Remind Me of Home': Collecting, Plant Geography, and a Forgotten Expedition in the Darwinian Revolution," *Journal of the History of Biology* 50 (2017): 71–132.

11. Herman Viola and Carolyn Margolis, eds., *Magnificent Voyagers* (Washington, DC: Smithsonian Institution Press, 1985), 9. William Stanton's *The Great United States Exploring Expedition of 1838–1842* (Berkeley: University of California Press, 1975) is the definitive treatment of the voyage, but also entertaining is Nathaniel Philbrick, *Sea of Glory: America's Voyage of Discovery, the U.S. Exploring Expedition 1838–1842* (New York: Penguin Books, 2003).

12. For Perry, see Samuel Eliot Morison, *"Old Bruin" Commodore Matthew C. Perry, 1794–1858* (Boston: Atlantic Monthly Press, 1967).

13. "Annual Report of the Secretary of the Navy, December 4, 1852," *Daily National Intelligencer*, December 9, 1852; "Surveying Expedition," *Brooklyn Daily Eagle*, May 2, 1853.

14. Cole, *Yankee Surveyors*, 5. For the American whaling trade, see Eric Jay Dolin, *Leviathan: The History of Whaling in America* (New York: W. W. Norton, 2007); "Marine Disaster," *Daily National Intelligencer*, January 5, 1853.

15. For Baird, see E. F. Rivinus and E. M. Youssef, *Spencer Baird of the Smithsonian* (Washington, DC: Smithsonian Institution Press, 1992); William H. Dall, *Spencer Fullerton Baird* (Philadelphia: J. B. Lippincott, 1915); for Stimpson's supplies, see Stimpson to Ringgold, April 7, 1853, SIA RU 7253.

16. Stimpson to Ringgold, December 10, 1852, SIA RU 7093; "United States Exploring and Surveying Expedition," *The Times of London*, May 18, 1853, reprinted from the *New York Journal of Commerce*.

17. Cited in Ponko, *Ships, Seas, and Scientists*, 208.

18. Kazar, "The United States Navy and Scientific Exploration," 198.

19. Dall, *Spencer Fullerton Baird*, 297–98.

20. Stimpson to Ringgold, December 15, 1852, SIA RU 7253.

21. Stimpson to "My dear Father," January 15, 1853, CAS, Stimpson papers, Shay Collection, CAS Archives.

22. Stimpson to "My dear Mother," February 13, 1853, CAS, Stimpson papers, Shay Collection, CAS Archives.

23. Stimpson to "My dear father," March 27, 1853, CAS, Stimpson papers, Shay Collection, CAS Archives.

24. Stimpson to "My dear parents," February 7, 1853, CAS, Stimpson papers, Shay Collection, CAS Archives.

25. "Ringgold's Exploring and Surveying Expedition," *Boston Daily Atlas*, May 6, 1853, reprinted from the *New York Journal of Commerce*. For the *Vincennes*, see James L. Mooney, ed., *Dictionary of American Naval Fighting Ships*, vol. 7 (Washington, DC: Naval Historical Center, Dept. of the Navy, 1981), 525–27.

26. Stimpson to "My dear parents," May 17, 1853, Stimpson Papers, Shay Collection, CAS Archives.

27. Stimpson to Baird, May 25, 1853, SIA RU 52.

28. Wright to Asa Gray, May 28, 1853, Gray Herbarium Library, Harvard University; Stimpson to Kane, February 4 and 15, 1853, American Philosophical Society Library, Kane papers, BK 132. Goodfellow proved to be a disaster.

29. For data on the ships, see Ponko, *Ships, Seas, and Scientists,* 270–73.

30. Stuart journal, June 29, 1853, NA MC 88, roll 7; for men in irons, see the Logs of the *Vincennes,* March 21, 1853–March 24, 1854, NA MC 88, roll 8. For Ringgold, see Olive Hoogenboom, *ANB* 18 (1999): 525–26.

31. Hung, "Plants that Remind Me" (2017), 86.

32. Helen M. Rozwadowski, "Small World: Forging a Scientific Maritime Culture for Oceanography," *Isis* 87, no. 3 (1996): 412–13; Wright to Gray, May 28, 1853, Gray Herbarium Library, Harvard University. For ill treatment of the US Ex. Ex. scientists, see William Stanton, *The Great United States Exploring Expedition* (1975), 83, 93–94.

33. Stuart journal, June 13, 1853, NA MC 88, roll 7.

34. Charles Wright to Asa Gray, April 9, 1854, Gray Herbarium, Harvard University.

35. Stimpson to Dall, February 9, 1866, SIA RU 7073.

36. The official version is in the Smithsonian Archives. A portion of the unofficial journal is in the Chicago Academy of Sciences Archives.

37. For the Couthouy episode, see Nathaniel Philbrick, *Sea of Glory,* 139–40.

38. *Stimpson Journal,* July 4, 1853, 17; Robert V. Hine, *Edward Kern and American Expansion* (New Haven, CT: Yale University Press, 1962), 103. This was one of the first exploring expeditions to have a photographer, but none of Kern's photographs have surfaced.

39. Stimpson to "My dear parents," July 16, 20, 1853, Stimpson Papers, Shay Collection, CAS Archives.

40. *Stimpson Journal,* July 9–18, 1853, 18–19; Stimpson to "My dear parents," June 20, 1853, Stimpson Papers, Shay Collection, CAS Archives.

41. For *Micropisa* see Stimpson, *Report on the Crustacea (Brachyura and Anomura) Collected by the North Pacific Exploring Expedition, 1853–1856* (Smithsonian Misc. Collections 49 [1907]), 10–11; *Stimpson Journal,* July 29, 1853, 20; Stimpson to "My dear parents," August 14, 1853, Stimpson Papers, Shay Collection, CAS Archives.

42. *Stimpson Journal,* August 7, 1853, 21.

43. Stuart journal, August 10, 1853, NA MC 88, roll 7.

44. Stuart journal, August 30, 1853, NA MC 88, roll 7; *Stimpson Journal,* August 31, 1853, 23. Helen Rozwadowski has noted that many sailors considered it bad luck to kill an albatross. See *Fathoming the Ocean,* 202–3.

45. *Stimpson Journal,* August 19, 1853, 22.

46. Alexander W. Habersham, *My Last Cruise* (Philadelphia: J. B. Lippincott, 1857), 14.

47. Stimpson to Baird, October 23, 1853, SIA RU 52. See Charles W. Eliot, "Francis Humphreys Storer," *Proceedings of the American Academy of Arts and Sciences* 54 (1919): 415–18.

48. Stimpson to "My dear parents" and "My dear father," March 24, April 1, 1854, Stimpson Papers, Shay collection, CAS Archives. Coolidge is discussed in Nathaniel Shaler, *The Autobiography of Nathaniel Southgate Shaler* (Boston: Houghton and Mifflin, 1909), 121–24.

49. Stimpson to Baird, October 23, 1853, SIA RU 52. Also see Ames to Ringgold, September 25, 1853, NA RG 45.

50. Charles Wright to Asa Gray, November 4, 1853, Gray Herbarium; Kazar, "The United States Navy and Scientific Exploration," 192.

51. Stimpson to Baird, June 10, 1854, SIA RU 52. Stimpson inadvertently gave the year as 1853; Wright to Asa Gray, October 14 and October 23, 1853, Gray Herbarium Library, Harvard University.

52. Robert V. Bruce, *The Launching of Modern American Science 1846–1876* (Ithaca, NY: Cornell University Press, 1987), 66. For Wright's early career, see the chapter "Charles Wright" in Samuel Wood Geiser, *Naturalists of the Frontier* (Dallas: Southern Methodist University, 1937), 215–52. Also see Anna M. M. Reid, "Charles Wright," *ANB* 24 (1999): 6–7.

53. Bruce, *The Launching of Modern American Science*, 213.

54. *Stimpson Journal*, September 1853, 25.

55. *Stimpson Journal*, October 1853, 26–27.

56. Stimpson to Baird, October 23, 1853, SIA RU 52.

57. Rodgers to Ringgold, October 12, 1853, and Ringgold to Dobbin, October 15, 1853, NA MC 88, roll 2.

58. George M. Brooke, "John Mercer Brooke, Naval Scientist" (PhD diss., University of North Carolina, 1955), 338–40.

59. Stimpson to Ringgold, October 3, 1853, NA RG 45; *Stimpson Journal*, October 1853, 27; Stimpson to "My dear father," November 3, 1853, Stimpson Papers, Shay Collection, CAS Archives.

60. *Stimpson Journal*, December 26, 1853, 32.

61. Robert Hughes, *The Fatal Shore: The Epic of Australia's Founding* (New York: Knopf), 1987.

62. For the quinary theory, see Mary P. Winsor, *Starfish, Jellyfish, and the Order of Life: Issue in Nineteenth-Century Science* (New Haven, CT: Yale University Press, 1976), 82–97.

63. *Stimpson Journal*, December 28, 31, 1853, 33–34.

64. *Stimpson Journal*, December 28, 31, 1853, 33–34. For Wall and the Sydney Museum, see Ronald Strahan et al., *Rare and Curious Specimens, An Illustrated History of the Australian Museum, 1827–1979* (Sydney: The Australian Museum, 1979), 18–24.

65. *Stimpson Journal*, December 30, 1853, 33.

66. *Stimpson Journal*, January 6, 1854, 35. For MacGillivray, see Adrian Desmond, *Huxley: From Devil's Disciple to Evolution's High Priest* (Reading, MA: Perseus Books, 1999), 217, 376; Stimpson to Dana, April 1, 1854, Yale University Library.

67. George M. Brooke Jr., *John M. Brooke, Naval Scientist and Educator* (Charlottesville: University Press of Virginia, 1980), 92; Stimpson to "My dear parents," March 24, 1854, Stimpson Papers, Shay Collection, CAS Archives; Stimpson to Ringgold, March 21, 1854, SIA RU 7093.

68. For microscopy and the sea, see Rozwadowski, *Fathoming the Ocean*, 2005.

69. *Stimpson Journal*, January 31, 1854, 37; Stimpson to Ringgold, March 21, 1854, SIA RU 7093; Brooke, diss., 354–55. The specimens were sent to Jacob W. Bailey at West Point, who named a species for Brooke. See Bailey, "On some specimens of deep-sea bottom, from the sea of Kamschatka, collected by Lieut. Brooke, U.S.N.," *AJS* 21 (1855): 284–85.

70. Robert Carter to Louise Carter, January 6, 1854, Shirley Plantation Collection, Colonial Williamsburg Foundation Library.

71. *Stimpson Journal*, February 24, March 3, 1854, 39–40; Wright to Gray, April 6, 1854, Gray Herbarium Library, Harvard University.

72. *Stimpson Journal*, April 23–26, 1855, 80.

73. George M. Brooke, "John M. Brooke," PhD diss., 355; Stimpson to Baird, June 10, 1854, SIA RU 52.

74. Morison, "*Old Bruin*," 350–53; Frederic Trautmann, ed., *With Perry to Japan, A Memoir by William Heine* (Honolulu: University of Hawaii Press, 1990), 92.

75. A. F. Monroe to Ringgold, April 21, 1854, and Ringgold to Dobbin, May 5, 1854, NA MC 88, roll 2.

76. Robert E. Johnson, *Rear Admiral John Rodgers 1812–1882* (Annapolis, MD: United States Naval Institute, 1967), 112.

77. *Stimpson Journal*, March 27, 1854, 43.

78. Wright to Asa Gray, April 9, 1854, Gray Herbarium Library, Harvard University.

Chapter 3

1. Joseph Conrad, *A Personal Record and the Mirror of the Sea* (New York: Penguin Books, 1998 [1906]), quote from *Mirror of the Sea*, 249.

2. Stimpson to home, April 20, 1854, Stimpson Papers, Shay Collection, CAS Archives.

3. Stimpson to Dana, April 1, 1854, Yale University Archives, Dana collection. On Dana, see Michael L. Prendergast, "James Dwight Dana: The Life and Thought of an American Scientist" (PhD diss., University of California, 1978); Gilman, *The Life of James Dwight Dana*, 1899.

4. Viola and Margolis, *Magnificent Voyagers*, 14.

5. Wright to Asa Gray, April 9, 1854, Gray Herbarium; Stimpson, "Descriptions of some new Marine Invertebrata," *Proc. ANSP* 7 (1855): 385–95.

6. Stimpson to Baird, September 5, 1854, SIA RU 52; *Stimpson Journal*, April 29, 1854, 47.

7. *Stimpson Journal*, April 4, 1854, 44; Stimpson to Baird, August 4, 1856, SIA RU 52.

8. *Stimpson Journal*, April 5, 1854, 44.

9. *Stimpson Journal*, March 25, 1854, 42–43.

10. Stimpson to Dana, April 1, 1854, Yale University Archives.

11. *Stimpson Journal*, April 7, 1854, March 1855, 44,78.

12. *Stimpson Journal*, May 3, 1854, 48–49.

13. *Stimpson Journal*, May 4–5, 1854, 50–52.

14. *Stimpson Journal*, May 6, 1854, 52.

15. *Stimpson Journal*, May 24, 1854, 54.

16. Wright to Asa Gray, April 9, 1854, Gray Library, Harvard University.

17. *Stimpson Journal*, April 23, May 7–9, 1854, 46, 53; James R. Troyer, "John Charles Bowring (1821–1893): Contributions of a merchant to natural history," *Archives of Natural History* 10, no. 3 (1982): 515–29; Stimpson, "On the genus Bipaliura," *AJS* 31 (1861): 134–35.

18. Stimpson's Private North Pacific Journal, Stimpson Papers, Shay Collection, CAS Archives.

19. Stephen R. Platt, *Autumn in the Heavenly Kingdom: China, the West, and the Epic Story of the Taiping Civil War* (New York: Knopf, 2012).

20. Wright to Gray, October 14, 1854, Gray Herbarium Library, Harvard University.

21. Wright to Asa Gray, August 12, 1854, Gray Herbarium.

22. Stimpson to "Dear parents," July 16, 1854, Stimpson Papers, Shay Collection, CAS Archives.

23. Morison, "*Old Bruin*," 397–98.

24. *Stimpson Journal*, April 25, 1854, 47.

25. Charles Wright to Asa Gray, April 9, August 12, 1854, Library of the Gray Herbarium; Stimpson unpublished North Pacific Journal, June 13–15, 1854, Stimpson Papers, Shay Collection, CAS Archives.

26. *Stimpson Journal*, June 26–July 3, July 20–22, 1854, 58; Stimpson to Baird, June 10, 1854, SIA RU 52.

27. *Stimpson Journal*, July 23–24, 1854, 58; Wright letter cited in Hung, "Plants that Remind Me," 95.

28. Rodgers to Perry, July 27, 1854, NA MC 88, roll 6.

29. Robert E. Johnson, *Rear Admiral John Rodgers* (1967), 111–13; Ringgold to Dobbin, September 4, 1854, NA MC 88, roll 2. For Rodgers, also see Craig L. Symonds, "John Rodgers," *ANB* 18 (1999): 725–26.

30. Johnson, *Rear Admiral*, 115.

31. *Stimpson Journal*, September 1–11, 1854, 60–61; Hung, "Plants that Remind Me," 98–99; Stimpson to Baird, March 28, 1855, SIA RU 52.

32. Stimpson to "Dear parents, brothers and sister," September 6, 10, 1854, Stimpson Papers, Shay Collection, CAS Archives.

33. "From Washington—the New Steam Frigate—The Ringgold Expedition," *New York Times*, October 24, 1854.

34. John Rodgers and Anton Schönborn, "On the Avoidance of the Violent Portions of Cyclones; with Notices of a Typhoon at the Bonin islands," *AJS* 23 (1857): 205–11; William C. Redfield, "On the Cyclones or Typhoons of the North Pacific Ocean," *AJS* 24 (1857): 21–28.

35. There are different numbers cited as to how many men were lost, but in a note of March 14, 1855, NA MC 88, roll 4, Rodgers lists fifty-two men on the *Porpoise*. The vessel contained seventy men at the outset of the voyage but desertions and disease had lowered the number.

36. Stimpson Journal, October 19–November 5, 1854, 64–66; Morison, *"Old Bruin,"* 311–15.

37. Frederic Trautmann, ed., *With Perry to Japan* (1990), 48–49. In 2011 the islands were named a UNESCO World Heritage site, and they have been called the Galapagos of the East for the many endemic species living there.

38. Stimpson to "My dear parents," November 12, 1854, Stimpson Papers, Shay Collection, CAS Archives.

39. *Stimpson Journal,* October 19–November 5, 1854, 64–66.

40. Morison, *"Old Bruin,"* 300–11.

41. Stimpson to Baird, March 28, 1855, SIA RU 7253.

42. *Stimpson Journal,* November–December 1854, 68.

43. *Stimpson Journal,* November–December 1854, 69–70.

44. *Stimpson Journal,* December 28, 1854, 71.

45. Johnson, *Rear Admiral,* 120–24. For Japanese-American relations, see Foster Rhea Dulles, *Yankees and Samurai: America's Role in the Emergence of Modern Japan, 1791–1900* (New York: Harper & Row), 1965.

46. Kazar, "The United States Navy and Scientific Exploration," 216; *Stimpson Journal,* November–December 1854, 71–72.

47. *Stimpson Journal,* 72; the "passport to observation" quote is cited in Hung, "Plants that Remind Me," 106. Rodgers also wrote, "the Japanese will not receive the schoolmaster unless he wears a revolver." Kazar, "The United States Navy and Scientific Exploration," 212.

48. *Stimpson Journal,* December 1854, 71–72. There had been Western publications on the fauna and flora of Japan thanks to the collecting efforts of Philipp Franz von Siebold, a German physician who lived in Nagasaki in the 1820s while working for the Dutch East India Company.

49. *Stimpson Journal,* January 1854, 73; Helen M. Rozwadowski, *Fathoming the Ocean: The Discovery and Exploration of the Deep Sea* (Cambridge, MA: Belknap Press of Harvard University Press, 2005), 52.

50. Wright to Asa Gray, February 1, 1855, Gray Herbarium Library, Harvard University; Stimpson to "My dear parents," November 12, 1854, Stimpson Papers, Shay Collection, CAS Archives.

51. Stimpson's private journal, February 12, 1854, Stimpson Papers, Shay Collection, CAS Archives.

52. Stimpson to Baird, March 28, 1855, SIA RU 52.

53. Stimpson to Dana, February 21, 1855, Yale University Library. Baird began using such a system in the 1850s but it was not until the 1880s that this practice became more widespread. See Robert Kohler, *All Creatures: Naturalists, Collectors, and Biodiversity, 1850–1950* (Princeton, NJ: Princeton University Press, 2006), 253. Ultimately, none of Stimpson's published articles used subspecies.

54. Stimpson to Dana, February 21, 1855, Yale University Library.

55. *Stimpson Journal,* January–February 1855, 77–78; Stimpson, "Descriptions of some of the new Marine Invertebrata from the Chinese and Japanese Seas," *Proc. ANSP* 7 (1855): 375–84.

56. *Stimpson Journal,* April 29, 1855, 81–82.

57. *Stimpson Journal,* May 2, 1855, 82–83.

58. Wright to Asa Gray, June 8, 1855, Gray Herbarium Library, Harvard University; *Stimpson Journal,* 85.

59. For the genus Tatsnoskia, see Stimpson, "Prodromus descriptionis ... Part 2, Turbellarieorum Nemertineorum" (1857), 165. Tatsunosuke is mentioned in Allan B. Cole, "The Ringgold-Rodgers-Brooke Expedition to Japan and the North Pacific, 1853–1859," *Pacific Historical Review* 16, no. 2 (1952): 158, 152–62.

60. *Stimpson Journal,* May 1855, 85–86.

61. Alexander W. Habersham, *My Last Cruise* (Philadelphia: J. B. Lippincott, 1857), 274–81.

62. *Stimpson Journal,* June 16, 1855, 90.

63. *Stimpson Journal,* June 11, 1855, 88–89.

64. *Stimpson Journal,* June 23, 1855, 90.

65. *Stimpson Journal,* June 23, 1855, 90.

66. George M. Brooke Jr., *John M. Brooke, Naval Scientist and Educator* (Charlottesville: University Press of Virginia, 1980), 120–22.

67. Brooke, *John Mercer Brooke* (1980), 122; *Stimpson Journal*, July 19, 1855, 93.

68. Stimpson to "My dear parents," September 30, 1854, Stimpson Papers, Shay Collection, CAS Archives.

69. Rodgers to Brooke, August 6, 1855, NA, MC 88, roll 5.

70. Stimpson notes the skull in a letter to Rodgers, October 16, 1855, SIA RU 7253.

71. *Stimpson Journal*, August–September 1855, 94–97; George M. Brooke, *John M. Brooke, Naval Scientist and Educator* (1980), 127–28.

72. "Notes on the Voyage of the Vincennes," *Sacramento Daily Union*, December 10, 1855.

73. Stimpson's vocabulary is preserved as manuscript 311 in the Smithsonian's National Anthropological Archives. Kern's painting of a whale hunt is one of the few images known from the expedition. It is part of the US Naval Academy Museum Collection.

74. Stimpson to "Dear parents," September 16, 1855, Stimpson Papers, Shay Collection, CAS Archives.

75. George M. Brooke Jr., *John M. Brooke: Naval Scientist and Educator*, 127–28.

76. For more on the Arctic Cruise, see "Expedition to the Arctic Ocean," *Friend's Intelligencer* 12, no. 36 (November 24, 1855), 571, APS Online.

77. In 1858 Brooke and Kern returned to make this survey. See George M. Brooke, ed., *John M. Brooke's Pacific Cruise and Japanese Adventure, 1858–1860* (Honolulu: University of Hawaii Press, 1986).

78. Stimpson, "On some California Crustacea," *Proceedings of the California Academy of Natural Sciences* 1 (1856): 95–99.

79. Stimpson to Baird, January 17, 1856, SIA RU 7253.

80. Charles Wright to Asa Gray, February 13, 1856, Gray Herbarium, Harvard University; Stimpson to Baird, January 31, 1856, SIA RU 52.

81. Stimpson, [*Hapalocarcinus marsupialis*, a remarkable new form of brachyurous crustacean on the coral reefs at Hawaii]. *Proc. BSNH* 6 (1859): 412–13.

82. Theodore Gill described the species *Sicyopterus stimpsoni* in 1860, and it is commonly called the Nopili rock-climbing goby. See Peter T. Sherman and Perri K. Eason, "Climb Every Waterfall!," *Natural History* 113, no. 8 (October 2004): 33–37.

Chapter 4

1. *Childe Harold's Pilgrimage*, Baron George Gordon (Lord) Byron, Project Gutenberg, http://www.gutenberg.org/files/5131/5131-h/5131-h.htm.

2. Stimpson to Baird, July 26, 1856, SIA RU 52; Baird to Stimpson, July 14, 1856, SIA RU 7002.

3. It did not help that reports from the Wilkes Expedition were still being published, or that Perry's exploits had been the subject of three lavish volumes. See John D. Kazar Jr., "The United States Navy and Scientific Exploration, 1837–1860" (PhD diss., University of Massachusetts, 1973), University Microfilms edition, 221–22.

4. Stimpson to Jeffries Wyman, December 10, 1856, Francis A. Countway Library, Harvard Medical Library.

5. Kazar, "The United States Navy and Scientific Exploration," 222.

6. Rodgers to Isaac Toucey, April 17, 1857, NA, MC 88, roll 26; Stimpson had hoped for $1,800 per year. Stimpson to Rodgers, November 12, 1856, SIA RU 7253. Ringgold and Rodgers later agreed that he was entitled to a set of the expedition shells for his personal cabinet, an acknowledgment that he had used his own funds to make some of the collections. Stimpson to Baird, May 28, 1862, SIA RU 7002.

7. Stimpson, "Notice of the Scientific Results of the Expedition to the North Pacific Ocean, under the command of Com. John Rodgers," *AJS* 23 (1857): 136–38.

8. See Carpenter, "Supplementary Report" (1864), 582–83.

9. Rodgers to Toucey, April 17, 1857, NA MC 88, roll 26.

10. See Stimpson, "Appropriation for the N. P. Exp. Nat. History," SIA RU 7093.

11. For extant NPEE specimens see William A. Deiss and Raymond B. Manning, "The Fate of the Invertebrate Collections of the North Pacific Exploring Expedition, 1853–1856," in *History in the Service of Systematics*, ed. Alwyne Wheeler and James H. Price, *Society for the Bibliography of Natural History, Special Publication* 1 (1981), 79–85.

12. Fellow invertebrate experts Dana, Gould, and W. G. Binney all utilized this format.

13. Stimpson to Leidy, January 14, 1857, Academy of Natural Sciences of Philadelphia, no. 1; Stimpson, "Prodromus descriptionis animalium evertebratorum quae in Expeditione ad Oceanum Pacificum Septentrionalem, a Republica Federata missa, Cadwalader Ringgold et Johanne Rodgers Duce, observavit et descripsit," Part 1, Turbellaria Dendrocoela, Part 2, Turbellarieorum Nemertineorum, *Proc. ANSP* 9 (1857): 19–31, 159–66.

14. Isaac Lea, *A Synopsis of the Family Unionidae*, 4th ed. (Philadelphia: Henry C. Lea, 1870), xv.

15. Daniel Lewis's book *The Feathery Tribe: Robert Ridgway and the Modern Study of Birds* (New Haven, CT: Yale University Press, 2012), looking at the development of American ornithology in the last quarter of the nineteenth century, noted (xii–xiii) that one of the markers that came to distinguish professionals from amateurs was an increasing use of technical terms. In the field of marine invertebrate taxonomy, Stimpson was using highly technical terminology by the 1850s.

16. Stimpson, "On the crustacea and echinodermata of the Pacific shores of North America," *Journal of the Boston Society of Natural History* 6 (1857): 444–532, plates 18–23.

17. Stimpson to John L. LeConte, June 21, 1858, American Philosophical Society Library, LeConte Papers; LeConte to Stimpson, June 23, 1858, NA RG 57.

18. A. Hunter Dupree, *Asa Gray, American Botanist, Friend of Darwin* (Baltimore: Johns Hopkins University Press, 1988 [1959]), 249–56.

19. *Stimpson Journal*, June 1855, 90.

20. Carpenter, "Supplementary Report" (1864), 582–83.

21. Vincent Ponko Jr., *Ships, Seas, and Scientists: U.S. Naval Exploration and Discovery in the Nineteenth Century* (Annapolis, MD: Naval Institute Press, 1974), 228.

22. Helen M. Rozwadowski, *Fathoming the Ocean: The Discovery and Exploration of the Deep Sea* (Cambridge, MA: Belknap Press of Harvard University Press, 2005), 56–57.

23. Stimpson, "On the crustacea and echinodermata of the Pacific shores of North America," *Journal of the Boston Society of Natural History* 6 (1857): 444–532, plates 18–23.

24. John Steinbeck and Edward F. Ricketts, *Sea of Cortez* (Mamaroneck, NY: Paul A. Appel, 1971 [1941]), 418.

25. Charlotte M. Porter, *The Eagle's Nest: Natural History and American Ideas, 1812–1842* (Tuscaloosa: University of Alabama Press, 1986), 7–8.

26. Nathan Reingold, ed., *Science in Nineteenth-Century America: A Documentary History* (Chicago: University of Chicago Press, 1985), 87; Robert V. Bruce, *The Launching of Modern American Science 1846–1876* (Ithaca, NY: Cornell University Press, 1987), 26.

27. Stimpson to Leidy, December 17, 1856, Academy of Natural Sciences of Philadelphia, no. 1; Stimpson to Ravenel, June 1, 1860, Charleston Museum.

28. Amos Binney, *The Terrestrial air breathing mollusks of the U.S.*, ed. Augustus Gould (Boston: Little and Brown, 1851), 61.

29. There is a vast literature on nationalism in American science during this period. Some of the most recent works include Bruce, *The Launching of Modern American Science*; Philip J. Pauly, *Biologists and the Promise of American Life: From Meriwether Lewis to Alfred Kinsey* (Princeton, NJ: Princeton University Press, 2000), 15–43; W. Conner Sorensen, *Brethren of the Net: American Entomology, 1840–1880* (Tuscaloosa: University of Alabama Press, 1995).

30. Frederick Burkhardt and Sydney Smith, eds., *The Correspondence of Charles Darwin*, vol. 6, 1856–1857 (Cambridge: Cambridge University Press, 1990), 380–82, 399–400. Darwin's monograph on the cirripedes was published between 1851 and 1854, and he noted Stimpson's observation on *Balanus eburneus*. See Darwin, *A Monograph of the Sub-class Cirripedia, with figures of all the species. The Balanidae (or Sessile Cirripedes); the Verrucidae, etc.* (London: The Ray Society, 1854), 248.

31. Philip P. Carpenter, "Lectures on Mollusca," *Smithsonian Annual Report, 1860*, 151–283, quote on 283.

32. Philip Carpenter to William H. Dall, March 14, 1870, SIA RU 7073.

33. Bruce, *The Launching of Modern American Science*, 12.

34. Sorensen, *Brethren of the Net*, 12.

35. Stimpson to Putnam, March 1, 1864, Harvard University Archives. See Vernon N. Kisling Jr., "The Naturalists' Directory and the evolution of communication among American naturalists," *Archives of Natural History* 21, no. 3 (1994): 393–406.

36. Stimpson to Sars, October 10, 1857, University of Oslo Library.

37. Stimpson to Sars, January 31, 1862, University of Oslo Library.

38. Stimpson to Steenstrup, April 21, November 1, 1857, Det Kongelige Bibliotek, Copenhagen.

39. Stimpson to Steenstrup, December 16, 1860, Det Kongelige Bibliotek, Copenhagen; Stimpson to Joseph Henry, February 9, 1861, SIA RU 7002.

40. Stimpson to Lütken, October 18, 1859, Det Kongelige Bibliotek, Copenhagen; Stimpson to Lütken, January 8, 1861, Det Kongelige Bibliotek, Copenhagen; Stimpson, "On new genera and species of star-fishes of the family Pycnopodidae (Astercanthion Müll and Trosch.)," *Proc. BSNH* 8 (1861): 261–73; Lütken to Stimpson, July 3, 1863, Det Kongelige Bibliotek, Copenhagen.

41. See Chandos Michael Brown, *Benjamin Silliman: A Life in the Young Republic* (Princeton, NJ: Princeton University Press, 1989).

42. George H. Daniels, *American Science in the Age of Jackson* New York: Columbia University Press, 1968), 18.

43. One of Dana's reviews had been attacked by Louis Agassiz in the pages of the *Journal*. See Edward Lurie, *Louis Agassiz, A Life in Science* (Baltimore: Johns Hopkins University Press, 1988), 271–75.

44. William Stimpson, "On Botanical and Zoological Nomenclature," *AJS* 29 (1860): 292.

45. W.[illiam] S.[timpson], review of *Naturhistoriske Bidrag til en Belkrivelle of Grönland*, by J. Reinhardt, J. C. Schiödte, O. A. L. Mörch, C. F. Lütken, J. Lange, and H. Rink, *AJS* 25 (1858): 124–26.

46. See Allen, *The Naturalist in Britain*, 132–40; Lynn Barber, *The Heyday of Natural History* (Garden City, NY: Doubleday, 1980), 115–24.

47. Anon., review of "The Aquarium, an Unveiling of the Wonders of the Deep Sea," by Philip Henry Gosse; and "The Book of the Aquarium and Water Cabinet," by Shirley Hibberd, *The North American Review* 87, no. 180 (July 1858): 145–46, Cornell University Making of America series. Stimpson's name is misspelled as Stimson; Robert Carter, *A Summer Cruise on the Coast of New England* (Boston: Crosby and Nichols, 1864), 2. Christopher Hamlin, "Robert Warington and the Moral Economy of the Aquarium," *Journal of the History of Biology* 19, no. 1 (1986): 131–53, argues that Warington invented the aquarium between May 1849 and March 1850.

48. "The Smithsonian Aquarium at Washington," *Scientific American* 13, no. 15 (December 19, 1857): 113, Cornell University Making of America online; "An Aquarium at Washington," *The Farmer's Cabinet*, December 9, 1857.

49. W.[illiam] S.[timpson], review of *Life Beneath the Waters, or, The Aquarium in America*, by A. M. Edwards, *AJS* 26 (1858): 284–85.

50. Stimpson, review of "A. M. Edwards," 284–85.

51. W.[illiam] S.[timpson], review of *Memoires pour servir a l' Histoire Naturelle du Mexique, des Antilles et des Etats-Unis*, by Henri de Saussure, *AJS* 27 (1859): 445–47.

52. Stimpson, "On the genus Bipaliura," *AJS* 31 (1861): 134–35.

53. W.[illiam] S.[timpson], review of *Catalogue of Acanthopterygian Fishes in the collection of the British Museum*, by A. Gunther, *AJS* 30 (1860): 140.

54. Robert Merton, *The Sociology of Science: Theoretical and Empirical Investigations* (Chicago: University of Chicago Press, 1973), 291.

55. W.[illiam] S.[timpson], review of *Die Klassen und Ordnungen des Thier-Reichs, wissenschaftlich dur gestellt in Wort und Bild, AJS* 29 (1860): 130; W.[illiam] S.[timpson], review of *Les genres Loriope et Peltogaster*, by H. Rathke and W. Liljeborg, *AJS* 29 (1860): 293; W.[illiam] S.[timpson], review of *Bidrag till Kaînnedomen om Skandinaviens Amphipoda Gammaridea*, by R. M. Bruzelius, *AJS* 28 (1859): 445.

56. Wm. Stimpson, "On the 'genus Diplothyra,'" *AJS* 35 (1863): 455.

57. W.[illiam] S.[timpson], "On the genus Peasia," *AJS* 31 (1861): 134; W.[illiam] S.[timpson], review of "*Leptosiagion* Trask, nov. gen.," *AJS* 25 (1858): 295.

58. George H. Daniels, "The Process of Professionalization in American Science: The Emergent Period, 1820–1860," *Isis* 58, no. 2 (1967): 150–66.

59. Stimpson to John W. Dawson, April 17, 1861, McGill University Archives, Montreal.

60. W.[illiam] S.[timpson], review of *Post-pliocene Fossils of South Carolina*, by Francis S. Holmes, *AJS* 33 (1862): 298–300. For Holmes, see Lester D. Stephens, *Science, Race and Religion in the American South: John Bachman and the Charleston Circle of Naturalists* (Chapel Hill and London: University of North Carolina Press, 2000).

61. W.[illiam] S.[timpson], review of *Catalogue of the Miocene Shells of the Atlantic Slope*, by T. A. Conrad, *AJS* 35 (1863): 428–30. Other followers of Klein included Otto Mörch of Denmark and Henry and Arthur Adams.

62. Stimpson, "On Botanical and Zoological Nomenclature," *AJS* 29 (1860): 292. Four years later Carpenter called for a congress of malacologists, crediting Stimpson with the idea. Carpenter, "On the Present State of Malacological Nomenclature," reprinted in *AJC* 1 (1865): 89–92.

63. Philip P. Carpenter wrote in 1856, "As human life is so short, and those who have the inclination for scientific pursuits have generally so little leisure, it is a serious evil when so large a proportion of that little has to be devoted to the labour of making out the errors of predecessors." Carpenter, "Report on the Present State of our Knowledge with Regard to the Mollusca of the West Coast of North America," *Report of the British Association for the Advancement of Science for 1856* (1857), 165. Malacology was not the only discipline that was reexamining what constituted an acceptable scientific nomenclature. In terms of professionalism, a comparison can be made to American entomology, which as Conner Sorensen has shown was going through a maturation period of "increasing scientific sophistication" through its own reevaluation of nomenclature in the 1860s and 1870s. See Sorensen, *Brethren of the Net* (1995), 242–52.

64. Stimpson, "On Botanical and Zoological Nomenclature," 289, 291, 293.

65. Louis Agassiz, "On the Origin of Species," *AJS* 30 (1860): 142–54.

66. Ronald L. Numbers, *Darwinism Comes to America* (Cambridge, MA: Harvard University Press, 1998), 28; Ronald L. Numbers, *The Creationists: From Scientific Creationism to Intelligent Design*, expanded edition (Cambridge, MA: Harvard University Press, 2006), 19–20, 30. Agassiz's own words make it clear that the geographic distribution of species had a supernatural explanation: "the geographical distribution of organized beings displays more fully the direct intervention of a Supreme Intelligence in the plan of Creation, than any other adaptation in the physical world." Cited in Lurie, *Louis Agassiz*, 151.

67. Quoted in Edmund B. Bolles, *The Ice Finders: How a Poet, a Professor, and a Politician Discovered the Ice Age* (Washington, DC: Counterpoint, 1999), 215–16.

68. From a notice of the book *On the Origins of Species*, in the *AJS* 29 (1860), 149.

69. Peter Bowler, quoted in E. Alison Kay, "Darwin's Biogeography and the Oceanic Islands of the Central Pacific, 1859–1909," in *Darwin's Laboratory*, ed. Roy Macleod and Philip F. Rehbock (Honolulu: University of Hawaii Press, 1994), 61. For Darwinism and birth order, see Ronald L. Numbers, *Darwinism Comes to America* (Cambridge, MA: Harvard University Press, 1998), 24–48, and Frank J. Sulloway, *Born to Rebel: Birth Order, Family Dynamics, and Creative Lives* (New York: Pantheon Books, 1996).

70. Dirk Struik, *Yankee Science in the Making* (New York: Collier Books, 1962), 391–92. Struik argued that Agassiz's opposition to evolution impeded the acceptance of the theory among New

England naturalists. It is perhaps just as likely that by the mid-1860s, Agassiz's declining reputation among American naturalists and his dictatorial treatment of his student assistants may have actually contributed to a pro-evolution stance (at least in private) among his former students.

71. Mayer, "William Stimpson" (1918), 422.

72. Louis Agassiz, *Contributions to the Natural History of the United States of America*, vol. 3 (Boston: Little, Brown, 1860), 91. An excerpt from this work containing the remarks on the *Lingula* was published in the *AJS* 30 (1860): 145.

73. Stimpson to Baird, telegraphic dispatch, dated by internal evidence to March 23, 1860, SIA RU 305.

74. William Stimpson, "A trip to Beaufort N. Carolina," *AJS* 29 (1860): 442–45. This species is now known as *Glottidia pyramidata*.

75. A. E. Verrill's journal, September 5, 1860, Harvard University Archives. Agassiz was also one of the editors of the *American Journal of Science*.

76. William Stimpson, "Notes on North American Crustacea, no. 2," *Annals of the Lyceum of Natural History of New York* 7 (1862): 177–78.

77. Nathaniel Shaler, *The Autobiography of Nathaniel Southgate Shaler* (Boston: Houghton and Mifflin, 1909), 128–29. Mary P. Winsor, *Reading the Shape of Nature: Comparative Zoology at the Agassiz Museum* (Chicago: University of Chicago Press, 1991), 38, states that in 1860 Agassiz encouraged his students to read Darwin.

78. Paul Jerome Croce, "Probabilistic Darwinism: Louis Agassiz vs. Asa Gray on Science, Religion, and Certainty," *Journal of Religious History* 22 (1998): 44. Croce wrote (42), "Where religiously inspired scientists saw the certainty and purposefulness of divine contrivance, Darwin postulated the impersonal operation of naturalistic patterns." Some of Agassiz's resistance to evolution was that it seemed to take away a need for the "Divine Intelligence," a phrase he used often in his "Essay on Classification."

79. Stimpson fragment, May 26, 1863, probably from a letter to Robert E. C. Stearns, Stimpson Papers, Shay Collection, CAS Archives; Stimpson to Hayden, June 13, 1857, NA RG 57. For natural theology, see Herbert Hovenkamp, *Science and Religion in America, 1800–1860* (Philadelphia: University of Pennsylvania Press, 1978), 57–145; Philip F. Rehbock, *The Philosophical Naturalists: Themes in Early Nineteenth-Century British Biology* (Madison: University of Wisconsin Press, 1983).

80. This may be due to the fragmentary nature of his papers. Stimpson to Baird, March 11, 1860, SIA RU 305; Stimpson to Baird, May 19, 1862, SIA RU 7002. For the Broad Church, see Frederick Gregory, "The Impact of Darwinian Evolution on Protestant Theology in the Nineteenth Century," in David C. Lindberg and Ronald L. Numbers, eds., *God and Nature: Historical Essays on the Encounter between Christianity and Science* (Berkeley: University of California Press, 1986), 369–90.

81. W.[illiam] S.[timpson], review of *The Kjökkenmöddings: Recent Geologico-Archaeological Researches in Denmark*, *AJS* 33 (1862): 297–98. Stimpson quoted from the book of Genesis, ch. 2, verse 7. Stimpson's words are similar to those used by Darwin in 1859—"light will be thrown on the origin of man and his history." Cited in Martin J. S. Rudwick, *Earth's Deep History: How it Was Discovered and Why it Matters* (Chicago: University of Chicago Press, 2014), 203.

82. See Edward J. Pfeifer, in Thomas F. Glick, ed., *The Comparative Reception of Darwinism* (Chicago: University of Chicago Press, 1988), 192–93; Numbers, *Darwinism Comes to America* (1998), 24–48.

83. Stimpson, review of "The Northern Buccinums, and Remarks on some other Northern Marine Mollusks," part 1, *Canadian Naturalist and Geologist*, new series 2 (1865): 381; W. S., review of *Catalogue of the Miocene Shells of the Atlantic Slope*, *AJS* 35 (1863): 429.

84. Stimpson to Meek, June 16, August 29, 1868, SIA RU 7062.

85. Dall Diary, January 22, 1865, SIA RU 7073; Dall, "Some American Conchologists," *Proceedings of the Biological Society of Washington* 4 (1888): 97; Mayer, "William Stimpson" (1918), 424; Shaler, *Autobiography* (1909), 128–29.

86. Winsor, *Reading the Shape of Nature*, 37–38. In a note in the Smithsonian Archives, RU 7253, Stimpson wrote, "Chief objection to Darwin's theory—that specific dif.[ferences] are frequently of ornamentation which is of no benefit on life pertaining to survival."

87. Bruce, *The Launching of Modern American Science*, 342.

Chapter 5

1. John Steinbeck and Edward F. Ricketts, *The Log from the Sea of Cortez* (New York : Penguin Books, 1995 [1951]).

2. Elliott Coues and D. Webster Prentiss, "Avifauna Columbiana," *Bulletin of the United States National Museum*, no. 26 (1883): 8.

3. "The Smithsonian Museum," *Washington Evening Star*, January 13, 1860; "Highway Robbery in Smithsonian Grounds," *Washington Evening Star*, June 18, 1860.

4. Robert V. Bruce, *The Launching of Modern American Science 1846–1876* (Ithaca, NY: Cornell University Press, 1987), 187.

5. No one missed the irony that the funding for "America's museum" came from an Englishman, James Smithson. For the Smithsonian in a national context, see A. Hunter Dupree, *Science in the Federal Government: A History of Policies and Activities* (Baltimore: Johns Hopkins University Press, 1986 [1957]): 66–90. On intellectual life in Washington, see Wilcomb E. Washburn, "The Influence of the Smithsonian Institution on Intellectual Life in Mid-Nineteenth-Century Washington," *Records of the Columbia Historical Society of Washington, D.C. for 1963–1965* (Washington, DC: Columbia Historical Society, 1966), 96–121.

6. William A. Deiss, "Spencer F. Baird and his Collectors," *J. Soc. Bibliography of Natural History* 9, no. 4 (1980): 638.

7. For the full story, see Pamela M. Henson, "Spencer Baird's Dream: A U.S. National Museum," in Michael T. Ghiselin and Alan E. Leviton, *Cultures and Institutions of Natural History* (San Francisco: California Academy of Sciences, 2000), 101–26.

8. For the Smithsonian's museum during the years 1850–1870, see Joel Orosz, *Curators and Culture: The Museum Movement in America, 1740–1870* (Tuscaloosa: University of Alabama Press, 1990), 201–12.

9. James G. Cooper to Fan, March 22, 1857, Bancroft Library, University of California at Berkeley. For Cooper, see Eugene Coan, *James Graham Cooper* (Moscow: University Press of Idaho, 1981).

10. Stimpson to Hayden, June 13, 1857, NA RG 57.

11. The phrase is mentioned in letters from John L. LeConte to Ferdinand Hayden, April 10, 15, 1858, NA RG 57.

12. Robert Carter, *A Summer Cruise on the Coast of New England* (Boston: Crosby and Nichols, 1864), 2–3.

13. Cooper to Fan, January 22, 1858, Bancroft Library.

14. Stimpson, "Notes on the American oyster," unpublished notes, SIA RU 7093.

15. The earliest mention of the Megatherium Club is from a letter from Stimpson to Hayden, June 13, 1857, NA RG 57. Very little has been written on the Megatherium Club. For a brief account, see Vasile, "The Megatherium Club," in James Conaway, *The Smithsonian: 150 Years of Adventure, Discovery, and Wonder* (New York: Knopf, 1995), 86–87. There were precedents for such a club of brash young naturalists. As an admirer of Edward Forbes, Stimpson had no doubt heard of the dinners of the Red Lions, a similarly irreverent group of British naturalists that coalesced around Forbes. See Adrian Desmond, *Huxley: From Devil's Disciple to Evolution's High Priest* (Reading, MA: Helix Books, 1997 [1994]), 157. For Forbes and the Red Lions, see Brian G. Gardiner, "Edward Forbes, Richard Owen and the Red Lions," *Archives of Natural History* 20, no. 3 (1993): 349–72.

16. Robert Kennicott to the folks at home, February 17, 1863, GNHL.

17. Bettie G. R. to Robert Kennicott, June 1, 1858, GNHL.

18. James G. Cassidy, *Ferdinand Hayden, Entrepreneur of Science* (Lincoln: University of Nebraska Press, 2000); Mike Foster, *Strange Genius: The Life of Ferdinand Vandeveer Hayden* (Niwot, CO: Roberts Rinehart Publishers, 1994).

19. James G. Cooper to Fan, January 22, 1858, Bancroft Library. For Kennicott, see Ronald S. Vasile, "Robert Kennicott," *ANB* 12 (1999): 583–85.

20. "Biography of Robert Kennicott," *Transactions of the Chicago Academy of Sciences* 1, part 2 (1869): 140.

21. Kennicott to home, February 17, 1863, GNHL; William H. Dall, "Theodore Nicholas Gill," *National Academy of Sciences Biographical Memoirs* 8 (1916): 318.

22. Bettie G. R. to Kennicott, June 1, 1858, GNHL. See Charles A. White, "Biographical Memoir of Fielding Bradford Meek," *National Academy of Sciences Biographical Memoirs* 4 (Washington, DC: National Academy of Sciences, 1902): 77–80; Mike W. Foster, "Fielding Bradford Meek," *ANB* 14 (1999): 240–42.

23. William H. Dall, "Baird, the Man," *Science*, new series 57, no. 1468 (February 16, 1923): 196; Kennicott to his mother, December 20, 1857, GNHL.

24. Stimpson to Frederic Putnam, September 15, 1858, Harvard University Archives.

25. Stimpson to Hayden, September 9, 1857, NA RG 57.

26. Verrill journal, March 9, 17, 1861, Harvard University Archives.

27. Stimpson to Hayden, June 13, 1857, NA RG 57; Stimpson to Hayden, February 28, 1858, SIA RU 7230 (page 1) and RU 7177 (page 2).

28. For Stimpson and Mary Henry, see Bettie G. R. to Kennicott, June 1, 1858, GNHL.

29. Stimpson to Hayden, November 23, September 9, 1857, NA RG 57.

30. Stimpson to Hayden, December 19, 1859, February 26, 1860, NA RG 57. For the Henry family, see Sarah R. Riedman, *Trailblazer of American Science: The Life of Joseph Henry* (Chicago: Rand McNally, 1961).

31. Stimpson to Hayden, August 13, 1859, NA RG 57.

32. William Turner to Baird, September 20, 1858, SIA RU 7002, box 34.

33. Stimpson to Putnam, September 15, 1858, Harvard University Archives; Stimpson to Hayden, December 19, 1859, NA RG 57.

34. Stimpson to Hayden, February 26, 1860, NA RG 57.

35. Stimpson to Hayden, June 13, 1857, September 14, 1858, NA RG 57.

36. Stimpson to Putnam, January 27, 1859, Harvard University Archives.

37. Stimpson to William Turner, September 13, 1859, Stimpson Papers, Shay Collection, CAS Archives.

38. Stimpson to Hayden, September 9, 1857, NA RG 57.

39. William H. Dall, *Spencer Fullerton Baird* (Philadelphia: J. B. Lippincott, 1915), 231–32.

40. For Congress, see the *Congressional Globe*, January 29, 1858, 482–91. Graft and corruption in the publication of government reports also explained Congress's hostility. See Ron Tyler, "Illustrated Government Publications of the American West, 1848–1863," in Edward C. Carter II, ed., *Surveying the Record: North American Scientific Exploration to 1930* (Philadelphia: American Philosophical Society, 1999), esp. 158–60; Kennicott to Major, February 4, 1858, GNHL.

41. For the Patent Office, see Robert C. Post, "'Liberalizers' versus 'Scientific Men' in the Antebellum Patent Office," *Technology and Culture* 17 (1976): 24–54. In all, six Patent Office employees became members of the club. The average age of Stimpson, Kennicott, Hayden, and Cooper was 26. The average age of the other founders was 47.

42. Phillip Drennon Thomas, Titian Ramsey Peale, *ANB* 17 (1999): 203–5.

43. For Foreman, see Curtis M. Hinsley, *The Smithsonian and the American Indian: Making a Moral Anthropology in Victorian America* (Washington, DC: Smithsonian Institution Press, 1994 [1981]), 71–73.

44. See "Regulations of the Potomac Side Naturalists Club," SIA RU 7210.

45. J. W. Chickering, "The Potomac-Side Naturalists' Club," *Science* 23 (1906): 264–65. Two notebooks containing accounts of each meeting were known to exist as late as 1906 but have not resurfaced.

46. Cooper to "Fan," March 22, 1857, Bancroft Library; Stimpson to Hayden, June 13, 1857, NA RG 57; William D. Haley, ed., *Philp's Washington Described. A Complete View of the American Capital, and the District of Columbia; with many notices Historical, Topographical, and Scientific, of the Seat of Government* (New York: Rudd and Carleton, 1861), 33–34.

47. Others nominated by Stimpson were William E. Jillson, F. W. Vaughan, Henry Engelmann, and W. H. W. Campbell. For explorers and the West, see William H. Goetzmann, *Exploration and Empire: The Explorer and the Scientist in the Winning of the American West* (New York: W. W. Norton, 1978 [1966]).

48. "Annual Commencement of the National Medical College," *Washington Evening Star*, March 2, 1860. For Lincoln, see George Crossette, *Founders of the Cosmos Club of Washington 1878* (Cosmos Club, 1966), 106–8.

49. Stimpson to Hayden, December 19, 1859, February 26, 1860, NA RG 57; Stimpson, "Obituary of William Turner," *AJS* 29 (1860): 152.

50. Stimpson to Hayden, September 29, December 19, 1859. For Jane Turner, see Caroline H. Dall, "In Memoriam: Susan Wadden Turner, William Wadden Turner, Jane Wadden Turner" (1898), 19, http://archive.org/stream/inmemoriamsusanw00dall#page/n5/mode/2up.

51. 1860 federal census, Washington, DC. Stimpson's name is misspelled (Stimson) and his personal estate was $500.

52. James G. Cooper to Fan, October 25, 1858, Bancroft Library.

53. It is ironic that the longest contemporary account of the expedition was published in German and has not been translated into English. See Heine, *Die Expedition in die Seen von China, Japan, Ochotsk unter Commando von Commodore Calw. Ringgold und Commodore John Rodgers, im Auftrage der Regierung der Vereinigten Staaten unternommen in den Jahren 1853 bis 1856* (Leipzig: Purfurst, 1858–1859).

54. The best source on Heine in English is *With Perry to Japan: A Memoir by William Heine*, ed. and trans. Frederic Trautmann (Honolulu: University of Hawaii Press, 1990), 9-22. For Schönborn, see http://www.prospecthillcemetery.org/Biographies.html.

55. For Ulke, see Arnold Mallis, *American Entomologists* (New Brunswick, NJ: Rutgers University Press, (1971), 258–60. Ulke painted portraits of Kennicott and Stimpson, both of which survive at the Chicago Academy of Sciences.

56. For Osten Sacken, see Mallis, *American Entomologists* (1971), 381–84; Keir B. Sterling, "Carl Robert Romanovich von Osten Sacken," *ANB* 16 (1999): 808–9.

57. "Medical Interview," *Washington Evening Star*, May 19, 1860. For more, see Tadashi Aruga, "The First Japanese Mission to the United States," in Marc Pachter and Frances Wein, eds., *Abroad in America: Visitors to the New Nation 1776–1914*, National Portrait Gallery (1976), 134–43; Kae Takarabe, "Samurai at the Smithsonian: First Japanese Visitors to Western Museum in the U.S.," in Michael Ghiselin and Alan E. Leviton, eds., *Cultures and Institutions of Natural History* (San Francisco: California Academy of Sciences, 2000), 161–82.

58. Stimpson to Putnam, January 27, 1859, Harvard University Archives.

59. Kennicott to the folks at home, February 17, 1863; for Cassin as vulgar-looking, see Kennicott to his mother, December 26, 1857, GNHL. For Cassin, see Keir B. Sterling, "John Cassin," *ANB* 4 (1999): 556–57.

60. For LeConte, see Mallis, *American Entomologists*, 242–48.

61. Cope did not much care for Washington either, calling it a "second-rate place." Cope to his father, February 1, 1861, Library of the American Museum of Natural History. For Cope, see H. F. Osborn, *Cope: Master Naturalist* (Princeton, NJ: Princeton University Press, 1931).

62. Verrill recorded in his journal that on the matter of distributing specimens, there appeared to be a misunderstanding between Henry and Agassiz. See Verrill journal, February 6 and 7, 1861. Cope, who arrived in Washington two weeks before the Cambridge naturalists, noting how generous Baird had been to the Philadelphia Academy, gave his own critique of Agassiz's museum: "The facilities which Prof.'s Baird & Henry have put us in possession of, have surprised me, in view of their supposed great friendship for Cambridge, but the men at the latter place have not earned sufficient respect,

by industry, or confidence, by perfect good faith, to place them in competition with the Academy at present." Cope to his father, January 23, 1861, American Museum of Natural History.

63. Verrill journal, February 10–12, 1861, Harvard University Archives.

64. Verrill journal, February 14, 1861, Harvard University Archives.

65. Dall, *Spencer Fullerton Baird*, 224.

66. Stimpson to Hayden, September 9, November 23, June 13, 1857, NA RG 57. The *Oxford English Dictionary* defines *papilionaceous* as relating to butterflies, but other meanings include showy, frivolous, erratic, and capricious, http://www.oed.com/view/Entry/137156?redirectedFrom=papilionaceous#eid.

67. Stimpson to Hayden, November 23, 1857, NA RG 57. This is very reminiscent of Edward Forbes's "The Universal Brotherhood of Friends of Truth," a "Union of the Searchers after Truth" which demanded "Love for the good and the beautiful. . . . Co-operation in research [and] . . . advice and firm friendship." See "The Brotherhood of Truth," *The British Controversialist and Literary Magazine* (London, 1869), 150–51.

Chapter 6

1. "The Dredging Song" by Edward Forbes, first and third verses, from Philip F. Rehbock, "The Early Dredgers: 'Naturalizing' in British seas, 1830–1850," *Journal of the History of Biology* 12, no. 2 (Fall 1979): 326–27.

2. Stimpson to Meek, August 17, 1869, SIA RU 7062, box 6.

3. Janet Browne, *The Secular Ark: Studies in the History of Biogeography* (New Haven, CT: Yale University Press, 1983), 75.

4. Stimpson, "On a new form of parasitic gasteropodous mollusca, *Cochliolepis parasiticus*," *Proc. BSNH* 6 (1856–1859), 307–9.

5. See Stimpson to Kane, February 4 and 15, 1853, American Philosophical Society Library, Kane papers, BK 132. For Tufts, see Burchsted, J. C. A., and F. Burchsted, "Samuel Tufts Jr. (1817–1902), a Massachusetts shell collector and aquarium stocker," *Archives of Natural History* 34 (2007): 229–34.

6. Stimpson to Hayden, April 23, 1863, NA RG 57.

7. Stimpson to Ravenel, December 21, 1857, Charleston Museum.

8. Ravenel to Stimpson, March 1858, and Stimpson to Ravenel, April 29, 1858, Charleston Museum. Ravenel to Gibbes, April 2, 1858, Charleston Museum.

9. Stimpson to Ravenel, December 24, 1860, January 30, 1861, Charleston Museum.

10. [W.] [S.], "Catalogue of the Recent Marine Shells found on the coasts of North and South Carolina," *AJS* 29 (1860): 294–95.

11. See Dorothy Wayman, *Edward Sylvester Morse: A Biography* (Cambridge, MA: Harvard University Press, 1942), 35–37.

12. W. F. Ganong, "John Robert Willis, the First Nova Scotian Conchologist: A Memorial," *Transactions of the Nova Scotian Institute of Natural Science* 7, pt. 4 (1889–1890): 407–8, http://archive.org/stream/cihm_12507#page/n3/mode/2up.

13. *Smithsonian Annual Report (1859)*, 68.

14. Stimpson, "On the Collection and Preservation of Marine Invertebrates," in "Directions for Collecting, Preserving, and Transporting Specimens of Natural History prepared for the use of the Smithsonian Institution," *Smithsonian Miscellaneous Collections*, 3rd edition (March 1859): 37–40.

15. Russell Lant Carpenter, *Memoirs of the Life and Work of Philip Pearsall Carpenter* (London: C. Kegan Paul, 1880), 235, 256.

16. Stimpson to Hayden, April 22, 1860, NA RG 57. For Carpenter, see Russell Lant Carpenter, *Memoirs of the Life and Work of Philip Pearsall Carpenter* (London: C. Kegan Paul, 1880); for British naturalists focusing on the shell and not the animals, see Anne Larsen, "Equipment for the Field," 359–60, in *Cultures of Natural History*, ed. N. Jardine, J. A. Secord, and E. C. Spary (Cambridge: Cambridge University Press, 1996).

17. Carpenter, "Lectures on the Shells of the Gulf of California," *Smithsonian Annual Report* (1859), 195–219, and "Lectures on Mollusca; or "Shell-Fish" and their Allies," *Smithsonian Annual Report* (1860), 151–283.

18. Carpenter, "Memoirs," 275–76. When Carpenter returned to England he took with him many of the Smithsonian's West Coast mollusks in order to compare with specimens in Hugh Cuming's collection, thus ensuring that "there would be uniformity between Eng. and Am. naturalists." See Carpenter to Ravenel, November 23, 1860, Charleston Museum.

19. Smithsonian Annual Report (1862), 21–22.

20. Check Lists of the Shells of North America, by Isaac Lea, P. P. Carpenter, William Stimpson, William G. Binney, and Temple Prime, *Smithsonian Misc. Collections* (1860).

21. See Daniel Goldstein, "Yours for Science: The Smithsonian Institution's Correspondents and the Shape of Scientific Community in Nineteenth-Century America," *Isis* 85, no. 4 (1994): 573–99. For collectors, see Matthew Laubacher, "Cultures of Collection in late Nineteenth Century American Natural History" (PhD diss., Arizona State University, 2011), http://repository.asu.edu/attachments/56624/content/Laubacher_asu_0010E_10662.pdf; for shell collectors, see S. Peter Dance, *A History of Shell Collecting* (Leiden: E. J. Brill, 1986).

22. W. Conner Sorensen, *Brethren of the Net: American Entomology, 1840–1880* (Tuscaloosa: University of Alabama Press, 1995), also expands on the idea of information networks, 28–32.

23. Robert E. Kohler, *All Creatures: Naturalists, Collectors, and Biodiversity, 1850–1950* (Princeton, NJ: Princeton University Press, 2006), 4–16; Robert V. Bruce, *The Launching of Modern American Science 1846–1876* (Ithaca, NY: Cornell University Press, 1987), 66.

24. Kohler, *All Creatures* (2006), 255. Juan Ilerbaig, "Pride in Place: Fieldwork, Geography, and American Field Zoology, 1850–1920" (PhD diss., University of Minnesota, 2002), noted Baird's use of tables and measurements, 55–56. Ilerbaig argued that Baird and Agassiz trained their followers to become philosophical naturalists who expounded generalizations, if not theories, from their work.

25. Stimpson, "On the question concerning the identity of the common species Buccinum of the European with the North American Coast," undated manuscript, but probably from 1864 to 1865. SIA RU 7093.

26. For debates on Baconians and Humboldtians, see Lynn K. Nyhart, *Biology Takes Form: Animal Morphology and the German Universities, 1800–1900* (Chicago: University of Chicago Press, 1995), 91–93; Susan Faye Cannon, Science in Culture: *The Early Victorian Period* (New York: Dawson and Science History Publications, 1978), esp. 73–110; Janet Browne, *The Secular Ark*, 80–85; Hugh Richard Slotten, *Patronage, Practice, and the Culture of American Science: Alexander Dallas Bache and the U.S. Coast Survey* (Cambridge: Cambridge University Press, 1994), 112–46.

27. Bacon quoted in Loren Eiseley, *The Man Who Saw Through Time*, revised and enlarged edition of *Francis Bacon and the Modern Dilemma* (New York: Charles Scribner's Sons, 1973), 40; Theodore Gill, "The Doctrine of Darwin," *Proceedings of the Biological Society of Washington* 1 (1880–1881): 55. From the German perspective, Lynn K. Nyhart, *Biology Takes Form* (1995), 90–102, sees zoology becoming more important in the late 1840s, as natural history shifted to morphology through an emphasis on comparative anatomy and development and generation. This is exactly the approach that Louis Agassiz inculcated in his American students. For the transformation from natural history to biology, also see Lynn L. Merrill, *The Romance of Victorian Natural History* (New York: Oxford University Press, 1989), 12–13, 75–102; Lynn K. Nyhart, "Natural history and the 'new' biology," in N. Jardine, J. A. Secord, and E. C. Spary, *Cultures of Natural History* (Cambridge: Cambridge University Press, 1996), 426–43; Keith R. Benson, "From Museum Research to Laboratory Research: The Transformation of Natural History into Academic Biology," in Ronald Rainger, Keith R. Benson; and Jane Maienschein, *The American Development of Biology* (Philadelphia: University of Pennsylvania Press, 1988), 49–86.

28. S. Peter Dance, *A History of Shell Collecting*, 199. Dance noted the confusion surrounding the use of both terms. *Malacology* has the upper hand today, at least judging by natural history museums, which overwhelmingly use the term to designate this department of science.

29. Dall, review of "Notes on Lingual Dentition of Mollusca," by W. G. Binney and Thomas Bland, *AJC* 6 (1870): 169–71. Dall argued that the character of dentition today "is accepted by the best authorities as one of the most important and reliable aids to the proper classification of species and genera."

30. Many of them also did not have access to the works in which shells were described. Philip P. Carpenter concluded in 1856 that the latter fact alone "almost limits the satisfactory production of original works to those who have frequent access to the capital," by which he meant the scientific literature. Carpenter, "Report on the Present State of our Knowledge with Regard to the Mollusca of the West Coast of North America," *Report of the British Association for the Advancement of Science for 1856* (1857): 164.

31. R. J. Cleevely, in "Some Malacological Pioneers and their links with the transition of shell-collecting to conchology during the first half of the nineteenth century," *Archives of Natural History* 22, no. 3 (1995): 385–418, summarizes the literature on the subject from a British and Continental perspective. He notes (408) that the transition from shell-collecting to a science was not a simple progression but instead occurred through a series of separate events, concluding that "the substantial achievements of the period were in the continued refinement of classification that was the outcome of much deeper and broader research into the anatomy, morphology and occurrence of the Mollusca."

32. Amos Binney, *The Terrestrial Air-Breathing Mollusks of the United States, and the Adjacent Territories of North America*, vol. 1 (Boston: Little and Brown, 1851), 68–69.

33. C. B. Adams, "On the Value of the Shells of Mollusca for the purpose of distinguishing Species and Higher Groups," *Contributions to Conchology*, vol. 1 (New York: H. Bailliere Brothers, 1852), 195.

34. For Tryon, see S. Peter Dance, *A History of Shell Collecting* (Leiden: E. J. Brill, 1986), 144–46, 176–77; Leonard Warren, "George Washington Tryon Jr.," *ANB* 21 (1999): 883–84. Historians of science have weighed in with various theories regarding the professionalization of science and the amateur/professional divide. In 1950 Marston Bates noted that the distinction between amateur and professional is not useful in natural history, preferring to focus instead on full- or part-time. See Bates, *The Nature of Natural History* (New York: Scribner's, 1950), 263. Tryon is an example of someone who worked full-time and made lasting contributions although he had no formal training. Perhaps most pertinent to this debate is the judgment of Curtis Hinsley: "In the middle of the nineteenth century, being a scientist in America was still as much a matter of character and integrity as one of specific academic or laboratory training." Hinsley, *The Smithsonian and the American Indian* (1994), 34–35.

35. George W. Tryon, "Monograph of the Terrestrial Mollusca of the United States," *AJC* 2 (1866): 218–20. This echoes the amateur vs. professional divide in ornithology in the 1880s, as described by *The Feathery Tribe: Robert Ridgway and the Modern Study of Birds* (New Haven, CT: Yale University Press, 2012), esp. 181–84, 213–16.

36. George W. Tryon, "Observations on the Family Strepomatidae," *AJC* 1 (1865): 100.

37. Stimpson, "Researches on the Hydrobiinae and allied forms," *Smithsonian Misc. Coll.*, vol. 7 (1866), 6, 20, 38.

38. Stimpson, "On the structural characters of the so-called Melanians of North America," *AJS* 38 (1864): 41.

39. Stimpson's descriptions of the family Ptychatractidae and the genus Ptychatractus are from his paper, "On Certain Genera and Families of Zoophagous Gasteropods," *AJC* 1 (1865): 59.

40. Kohler, *All Creatures* (2006), 229.

41. Stimpson to Tryon, June 16, 1863, Academy of Natural Sciences of Philadelphia, no. 98.

42. Stimpson, "Researches on the Hydrobiinae," 48–49. Tryonia is still a valid genus.

43. George W. Tryon Jr., *American Marine Conchology* (Philadelphia: published by the author, 1873), vi.

44. Dall, "Some American Conchologists," 97.

45. Stimpson to George W. Tryon, June 16, 1863, Academy of Natural Sciences of Philadelphia, no. 98; Stimpson to Baird, September 8, 1863, SIA RU 7002.

46. J. Henry, *Smithsonian Annual Report* (1864), 21; *Smithsonian Annual Report* (1865), 40.

47. Stimpson, "On Certain Genera and Families of Zoophagous Gasteropods," *AJC* 1 (1865): 60.

48. This undated note is in the Stimpson Papers, SIA RU 7093.

49. Rehbock, "The Early Dredgers" (1979), 312–16. For Dawson, see Susan Sheets-Pyenson, *John William Dawson: Faith, Hope, and Science* (Montreal & Kingston: McGill-Queen's University Press, 1996).

50. William Stimpson, "Review of the Northern Buccinums and Remarks on some other Northern Marine Mollusks, Part 1," *Canadian Naturalist*, new series 2 (1865): 364–89. Part 2 was never published.

51. Dall to Baird, December 16, 1865, SIA RU 7213.

52. Stimpson, "Malacozoological Notices, no. 1: On the Genus Gundlachia," *Proc. BSNH* 9 (1863): 249–52. A second Malacozoological Notices was presented at the June 15, 1864, meeting of the Boston Society of Natural History but never published.

53. Carpenter, "Lectures on Mollusca," *Smithsonian Annual Report* (1860), 206; Stimpson, "On the structural characters of the so-called Melanians of North America," *AJS* 38 (1864): 41–53. Carpenter later praised Stimpson's Melanian paper as "interesting and important." Carpenter, "A Supplementary Report on the Present State of Our Knowledge with regard to the Mollusca of the West Coast of North America," *Report of the British Association for the Advancement of Science for 1863* (1864), 677.

54. Stimpson, "On the structural characters of the so-called Melanians of North America," 53.

55. Stimpson to Haldeman, February 17, 1866, Academy of Natural Sciences of Philadelphia, no. 73; Tryon, review of *On the structural characters of the so-called Melanians of North America, AJC* 1 (1865): 78–79.

56. Stimpson, "Researches on the Hydrobiinae," *Smithsonian Misc. Coll.*, vol. 7 (1866), 6.

57. Stimpson, "Researches Upon the Hydrobiinae," 1866, 15.

58. Alfred Mayer to Stimpson, April 11, 1864, Stimpson Papers, Shay Collection, CAS Archives.

59. *Smithsonian Report for 1865*, 42–43; Elmer G. Berry, "The Amnicolidae of Michigan: Distribution, Ecology, and Taxonomy," *Miscellaneous Publications, Museum of Zoology, University of Michigan*, no. 57 (1943), 11–12.

60. See Lynn K. Nyhart, "Natural History and the 'New' Biology," in *Cultures of Natural History*, ed. N. Jardine, J. A. Secord, and E. C. Spary (Cambridge: Cambridge University Press, 1996), 430–32.

61. George W. Tryon, "A Sketch of the History of Conchology in the United States," *AJS* 33 (1862): 161–80.

62. Eric L. Mills, "Edward Forbes, John Gwyn Jeffreys, and British dredging before the *Challenger* expedition," *Journal of the Society for the Bibliography of Natural History* 8, no. 4 (1978): 512. For dredging, see Rehbock, "The Early Dredgers" (1979), 295–98; Eric L. Mills, "Problems of Deep-Sea Biology: An Historical Perspective," in *The Sea*, vol. 8: *Deep Sea Biology*, ed. G. T. Rowe (1983), 1–18; Susan Schlee, *The Edge of an Unfamiliar World: A History of Oceanography* (New York: E. P. Dutton, 1973), 80–98; and Helen M. Rozwadowski, *Fathoming the Ocean: The Discovery and Exploration of the Deep Sea* (Cambridge, MA: Belknap Press of Harvard University Press, 2005), 117–36.

63. Stimpson to Sars, October 10, 1857, University of Oslo Library; Stimpson to Baird, August 8, 1857, SIA RU 7002.

64. Stimpson to Lewis R. Gibbes, December 25, 1859, Gibbes Papers, Library of Congress; Stimpson, "On an oceanic Isopod found near the southeastern shore of Massachusetts," *Proc. ANSP* (1862): 133–34.

65. Theodore Gill to Spencer F. Baird, March 11, 1860, SIA RU 7002.

66. Stimpson to Baird, March 11, 20, 1860, SIA RU 7002.

67. Gill to Baird, March 11, 1860, SIA RU 7002. Stimpson added an addendum to Gill's letter.

68. Stimpson to Baird, March 20, 1860, SIA RU 7002.

69. Nathaniel Shaler, *The Autobiography of Nathaniel Southgate Shaler* (Boston: Houghton and Mifflin, 1909), 132–33.

70. "Dr. William Stimpson," *The Prairie Farmer*, June 29, 1872, vol. 43, no. 26, 203.

71. Stimpson to Baird, September 25, 1861, SIA RU 7002; Stimpson to Meek, October 7, 1861, SIA RU 7062.

72. Shaler, *Autobiography*, 164.

73. The book has also been published as *Carter's Coast of New England*. For Carter, see James R. Simmons Jr., "Robert Carter," *ANB* 4 (1999): 494–95. For contemporary reviews, see the *Atlantic Monthly* 14, no. 84 (October 1864), 515, Cornell University Making of America website; *New York Times*, August 1, 1864.

74. The book mentions that the Professor was the naturalist on the North Pacific Exploring Expedition. Stimpson wrote to Baird from Marblehead, MA on July 11, 1858 (SIA RU 7002) at the same time the book has them there.

Chapter 7

1. Robert Kennicott to Roderick McFarlane, April 15, 1864, SIA RU 7213.

2. "The inauguration of the President Elect," *Washington Daily National Intelligencer*, March 4, 1861.

3. For southern science, see Lester D. Stephens, *Science, Race and Religion in the American South: John Bachman and the Charleston Circle of Naturalists* (Chapel Hill and London: University of North Carolina Press, 2000), especially 101–26 on Gibbes.

4. Stimpson to Gibbes, October 31, December 8, 25, 1859, Library of Congress.

5. W.[illiam] S.[timpson], reviews of "Descriptions of Oceania (Turritopsis) nutricula, n. s. and the embryological history of a singular Medusan Larva found in the cavity of its bell"; "The Gymnophthalmata of Charleston Harbor"; "On the Zoological Affinities of the Graptolites," *AJS* 29 (1860): 130–32. For McCrady, see Lester D. Stephens and Dale R. Calder, "John McCrady of South Carolina: Pioneer Student of North American Hydrozoa," *Archives of Natural History* 19 (1992): 39–54; also see Stephens, *Science, Race and Religion*, 146–64.

6. "Letter from John McCrady, Esq., Charleston, on the Lingula pyramidata described by Mr. W. Stimpson," *AJS* 30 (1860): 157–58.

7. Stimpson to Ravenel, December 24, 1860, Charleston Museum. Robert V. Bruce cited this letter in *The Launching of American Science* (272): "A Northern zoologist wrote a wholly nonpolitical letter to a fellow scientist at Charleston, promising (without any evident sense of double meaning) to send a 'fine collection of Northern shells.'" Bruce was unaware of Stimpson's recent trip to Grand Manan.

8. Edmund Ravenel to Stimpson, January 1861, Charleston Museum.

9. Stimpson to Ravenel, January 30, 1861, Charleston Museum. Stimpson wrote the date as 1851, but the context of the letter makes clear that it was 1861; Edmund Ravenel, "Descriptions of New Recent Shells from the Coast of South Carolina," *Proc. ANSP* (1861): 41–44. For Ravenel and McCrady, see Stephens, *Science, Race, and Religion*. Unaware that his paper had already been published, on June 10, 1861, Ravenel wrote Lewis R. Gibbes that he doubted that a paper from a Rebel would be published in a Northern journal. Ravenel to Doctor, June 10, 1861, Charleston Museum.

10. See Stimpson to Ravenel, June 1, 1860, and April 14, 1861, Charleston Museum.

11. Stimpson to Ravenel, April 14, 1861, Charleston Museum.

12. Stimpson to John William Dawson, April 17, 1861, McGill University. The reference to McPeake is in a letter from Meek to Hayden, June 27, 1861, SIA RU 7062.

13. Marc Rothenberg, *The Papers of Joseph Henry*, Vol. 10 (Washington, DC: Smithsonian Institution Press, 2004), 206.

14. Stimpson to Dawson, April 17, 1861, McGill University Archives, Montreal. For a gripping account of this attack, see David Detzer, *Dissonance: The Turbulent Days Between Fort Sumter and Bull Run* (Orlando, FL: Harcourt et al., 2006), 104–24.

15. William Schouler, *A History of Massachusetts in the Civil War*, vol. 1 (Boston: E. P. Dutton, 1868), 298–99. The best account of wartime Washington is still Margaret Leech, *Reveille in Washington* (New York: Carroll and Graf Publishers, 1991 [1941]).

16. Stimpson to Baird, July 12, 1861, SIA RU 7002.

17. Stimpson to Baird, September 25, 1861, SIA RU 7002; Stimpson to Meek, September 26, 1861, SIA RU 7062. (Some of his other names for crustaceans include crusty-shins, milligrubs, pollywogs, and side-walkers.)

18. Bruce, *The Launching of Modern American Science* (1987), 300.

19. E. B. Long with Barbara Long, *The Civil War Day by Day: An Almanac 1861–1865* (New York: Da Capo Press, 1971), 704–5.

20. E. F. Rivinus and E. M. Youssef, *Spencer Baird of the Smithsonian* (Washington, DC: Smithsonian Institution Press, 1992), 108. Bruce, *The Launching of Modern American Science* (279–80), noted that about one-quarter of the naturalists served, with ninety-eight members of the BSNH enrolled.

21. Meek to Hayden, April 18, 1861, NA RG 57.

22. Bruce, *The Launching of Modern American* Science, 280.

23. Stimpson to Morse, December 17, 1861, Morse Papers, Peabody Essex Museum.

24. See Dorothy Wayman, *Edward Sylvester Morse: A Biography* (Cambridge, MA: Harvard University Press, 1942), 183–87.

25. Mary P. Winsor, *Reading the Shape of Nature: Comparative Zoology at the Agassiz Museum* (Chicago: University of Chicago Press, 1991), 46.

26. Stimpson to Baird, May 28, 1862, SIA RU 7002. Also see Stimpson to Putnam, February 9, 1862, Harvard University Archives; Verrill journal, January 17 and 22, 1862.

27. Stimpson to Putnam, February 9, 1862, Harvard University Archives.

28. Stimpson to Baird, March 14, 1862, SIA RU 7002.

29. Stimpson to Baird, May 19, 1862, SIA RU 7002.

30. Stimpson to Baird, April 5, 1862, SIA RU 7002.

31. Stimpson to Baird, May 2, 19, 1862, SIA RU 7002. Latin translation courtesy of Calvin Byre. Bickmore later founded the American Museum of Natural History. See Douglas J. Preston, *Dinosaurs in the Attic* (New York: Ballantine Books, 1986), 19.

32. Stimpson to Baird, May 2, 1862, SIA RU 7002.

33. William mentioned that Jim had been a mate aboard the *Norway* in a letter to Ferdinand Hayden, April 22, 1860, NA RG 57. The mutiny is described by Edgar Holton, "A Chapter on the Coolie Trade," *Harper's New Monthly Magazine* 29 (1864), 1–11, Cornell University Making of America website.

34. Stimpson to Baird, May 19, 28, 1862, SIA RU 7002.

35. Stimpson to Baird, April 5, 11, and 30, 1862, and Baird to Stimpson, April 12 and 13, 1862, SIA RU 7002; Stimpson to Hayden, May 22, 1862, NA RG 57. At the beginning of the Crimean War, Herbert Stimpson bought up the available Springfield rifles and sold them to the British government for a tidy profit. See undated transcript on Herbert H. Stimpson, Stimpson Papers, Shay Collection, CAS Archives.

36. Stimpson to Baird, July 1, 1862, SIA RU 7002. For more on the Verreaux brothers, see Erwin Stresemann, *Ornithology: From Aristotle to the Present*, ed. G. William Cottrell (Cambridge, MA: Harvard University Press, 1975), 162–63.

37. Stimpson sent Mary Baird a posy from the ancient castle of Rolandseck near Bonn. See Stimpson to Baird, November 12, 1862, SIA RU 7002.

38. Stimpson to Baird, October 21, 1862, SIA RU 7002; for details on Stimpson's collections, see Louis Agassiz, "Fifth Annual Report of the Director of the Museum of Comparative Zoology" (January 1864), 25–54.

39. Stimpson to Baird, October 21, 1862, SIA RU 7002.

40. Stimpson to Baird, October 27, 1862, SIA RU 7002.

41. Verrill journal, December 17, 1862, January 13, 15, 1863, Harvard University Archives; Verrill, "Revision of the Polypi of the eastern coast of the United States," *Memoirs of the Boston Society of Natural History* 1 (1863), 1–45.

42. Henry James Clark, "Prodromus of the History, Structure, and Physiology of the order Lucernariae," *Boston Journal of Natural History* 7 (1859–1863), 531–67.

43. For the Clark controversy, see Winsor, *Reading the Shape of Nature*, 47–65, and Christoph Irmscher, *Louis Agassiz: Creator of American Science* (Boston: Houghton Mifflin Harcourt, 2013), 168–95. While Winsor believes it "most unlikely" that Clark worked on the paper in secret (51), it seems clear that many MCZ students prepared papers for publication without telling Agassiz.

44. Stimpson to Baird, October 27 and December 22, 1862, SIA RU 7002.

45. Stimpson to Baird, October 21, 1862, SIA RU 7002.

46. Janet Browne, *Charles Darwin, Voyaging* (Princeton, NJ: Princeton University Press, 1995), 207–8.

47. Stimpson to Baird, October 27, 1862, SIA RU 7002; A. Agassiz to Stimpson, January 22, 1864, MCZ Archives.

48. See Stimpson to Baird, June 17, 1863, SIA RU 7002.

49. Kennicott to Home, undated but probably January 1863, GNHL.

50. Kennicott to Folks at Home, February 17, 1863, GNHL.

51. Stimpson to Hayden, February 15, 1863, NA RG 57.

52. Kennicott to the Dear Folk at Home, December 22, 1862, GNHL.

53. Robert Kennicott to John Kennicott, March 11, 1863, GNHL.

54. Kennicott to Charlie Kennicott, February 17, 1863, GNHL.

55. George Gibbs to Kennicott, February 22 and April 16, 1863, GNHL.

56. Ira (Bruno) Kennicott to his father, April 16, 1863, GNHL.

57. Ira (Bruno) Kennicott to James Redfield, April 14, 1863, GNHL.

58. Kennicott to the folks at home, February 17, 1863, GNHL.

59. L. E. Chittenden, *Recollections of Abraham Lincoln and His Administration* (New York: Harper and Brothers, 1891), 239.

60. Kennicott to home, February 17, 1863, GNHL.

61. Kennicott to folks at home, February 17, 1863; Kennicott to little daughter, January 28, 1863, GNHL.

62. Kennicott to Home, February 17, 1863, GNHL.

63. Kennicott to the folks at home, February 17, 1863, GNHL; Chittenden, *Recollections* (1891), 241–45.

64. Kennicott to Home, undated but probably January 1863, GNHL.

65. Henry to Baird, August 24, 1863, SIA RU 7002.

66. Stimpson to Baird, September 4, 1863, SIA RU 7002. Henry was not opposed to natural history and recognized the value of what Baird and his assistants were doing. At the same time that he was evicting them from the building, Henry complained to James D. Dana that he did not feel "properly appreciated by naturalists" at the Smithsonian. Henry to Dana, August 21, 1863, quoted in Henry papers, vol. 10, 333–34. That same year Philip P. Carpenter responded to one of Henry's letters in which the Smithsonian chief had clearly been complaining about naturalists. Carpenter wrote, "I fully agree with what you say about naturalists and their specimens. But don't judge them all. . . . Our science is as though you wanted to deduce laws of meteorology and electricity from a vast mass of unsorted and *badly observed* facts. Either you must throw Natural History overboard, or you must give us *time.*" Russell Lant Carpenter, *Memoirs of the Life and Work of Philip Pearsall Carpenter* (London: C. Kegan Paul, 1880), 275–76.

67. Theodore Gill, "The History of Classification in Zoology," 22nd meeting, January 20, 1882, *Proceedings of the Biological Society of Washington* 1 (1880–1882): 36.

68. Stimpson to Baird, July 17, 1863, SIA RU 7002; Stimpson sent some of the birds to Louis Agassiz. *MCZ Annual Report for 1864*, 23; *Smithsonian Report for 1863*, 59–61.

69. "Deaths," *Cambridge Chronicle*, August 8, 1863, Cambridge Public Library.

70. For Francis Stimpson, see F. B. Heitman, *Historical Register of the United States Army, from its Organization, September 29, 1789, to September 29, 1889*, The National Tribune (Washington, DC, 1890), 618; for James, see *Official Records of the Union and Confederate Navies in the War of Rebellion*, Series 1, vol. 13, *South Atlantic Blockading Squadron, May 14, 1862–April 7, 1863* (Washington, DC: Government Printing Office, 1901), Cornell University Making of America website.

71. Stimpson to William Turner, September 13, 1859, Stimpson Papers, Shay Collection, CAS Archives.

72. Stimpson to Baird, September 4 and 8, 1863, SIA RU 7002; Dean C. Allard, *Spencer Fullerton Baird and the U.S. Fish Commission* (New York: Arno Press, 1978).

73. Edward Lurie, *Louis Agassiz, A Life in Science* (Baltimore: Johns Hopkins University Press, 1988 (314–15.

74. Ralph W. Dexter, "The 'Salem Secession' of Agassiz Zoologists," *Essex Institute Hist. Collections* 101, no. 1 (1965): 29, 34.

75. Dexter, "The 'Salem Secession,'" 34.

76. Stimpson to Putnam, January 4, 1864, Harvard University Archives; *Fifth Annual Report of the MCZ, January 1864*, p. 9.

77. A. E. Verrill to Joseph Henry, April 28, 1864, cited in Marc Rothenberg, *The Papers of Joseph Henry,* vol. 10 (2004), 368–71; Baird to Agassiz, December 13, 1869, cited in Elmer C. Herber, ed., *Correspondence between Spencer Fullerton Baird and Louis Agassiz* (Washington, DC: Smithsonian Institution Press, 1963), 189–94.

78. Lurie, *Louis Agassiz,* 312; Winsor, *Reading the Shape of Nature* (1991), 54.

79. William James to Henry James, May 3, 1865, cited in Ignas K. Skrupskelis and Elizabeth M. Berkeley, eds., *The Correspondence of William James,* vol. 1: *William and Henry 1861–1884* (Charlottesville: University Press of Virginia, 1992), 5–8.

80. Herber, *Correspondence,* 1963. See 77–78 for just one example.

81. Joseph Henry to Asa Gray, August 14, 1860, cited in Marc Rothenberg, *The Papers of Joseph Henry,* vol. 10: *January 1858-December 1865, the Smithsonian Years,* 163–65; Rivinus and Youssef, *Spencer Baird* (1992), 103–5. This biography devotes a chapter (98–105) to the Baird-Agassiz relationship.

82. Stimpson to Putnam, April 6, 1865, Harvard University Archives.

83. Dall, *Spencer Baird* (1915), 154.

Chapter 8

1. *Smithsonian Annual Report for 1862,* 14.

2. Robert V. Bruce, *The Launching of Modern American Science 1846–1876* (Ithaca, NY: Cornell University Press, 1987), 279.

3. Stimpson, "Notes on North American Crustacea," nos. 1 and 2, *Annals of New York Lyceum* 7 (1862), 49–93, 176–246.

4. Stimpson, "Notes on North American Crustacea," no. 2, 197; Stimpson to Riise, May 30, 1861, January 6, 1862, American Philosophical Society Library, Letters to Scientists. Riise created a commercial empire that continues to this day, as A. H. Riise is well known in the Virgin Islands for its duty-free stores.

5. Xantus to Baird, July 5, 1859, cited in Ann H. Zwinger, ed., *Xantus: The Letters of John Xantus to Spencer Fullerton Baird from San Francisco and Cabo San Lucas 1859–1861* (Los Angeles: Dawson's Book Shop, 1986), 120–21; Zwinger, "John Xantus," *ANB* 24 (1999): 97–98. For a very funny story about Xantus, see John Steinbeck, *The Log from the Sea of Cortez* (New York: Penguin Books, 1995 [1951]), 52.

6. Stimpson's comment is contained in a footnote in a paper by Baird, "Notes on a collection of Birds made by Mr. John Xantus, at Cape St. Lucas, Lower California, and now in the Museum of the Smithsonian Institution," *Proc. ANSP* 11 (1859): 302.

7. Zwinger, *The Letters of John Xantus,* 220, 254–56.

8. Albert Ordway, "Monograph of the Genus Callinectes," *Boston Journal of Natural History* 7, no. 4 (1863): 3–18.

9. Stimpson, "On the classification of the Brachyura and on the homologies of the antennary joints in decapod crustacea," *AJS* 35 (1863): 139–43. This article was reprinted in the *Annals and Magazine of Natural History* 11: 233–37. Stimpson also believed that E. Percival Wright had erred in describing an annelid from having worked with preserved specimens only. "On the genus Bipaliura," *AJS* 31 (1861): 135.

10. Stimpson, "On the fossil crab (Archaeoplax signifera) of Gay Head," *Journal of the Boston Society of Natural History* 7 (1863): 583–89. Stimpson originally concluded that the deposits were from the Cretaceous period but admitted his error. Stimpson, "Cretaceous Strata at Gay Head, Mass," *AJS* 29 (1860): 145.

11. Stimpson, "Note on the 'Glass Coral' of Japan," *AJS* 35 (1863): 458–59.

12. Baird, "Review of American Birds, in the Museum of the Smithsonian Institution," *Smithsonian Miscellaneous Collections*, no. 181 (1864–1866). Stimpson's note and drawings are on pp. 162–64.

13. Draft of a letter from Stimpson to Knorr, no date but probably late 1861 or early 1862, SIA RU 7093.

14. Stimpson, "On the question of whether Diatoms live on the sea-bottom at great depths," *AJS* 35 (1863): 454–55.

15. G. C. Wallich, "Do Diatoms live on the Sea-Bottom at Great Depths?," *Annals and Magazine of Natural History* 12, 3rd series (1863): 166. For a good view of Wallich, see A. L. Rice, Harold L. Burstyn, and A. G. E. Jones, "G. C. Wallich M. D.—megalomaniac or mis-used oceanographic genius?," *Journal of the Society for the Bibliography of Natural History* 7, no. 4 (1976): 423–50.

16. *Stimpson Journal*, July 19, 1855, 93.

17. See Mills, "Problems of Deep-Sea Biology: An Historical Perspective," in *The Sea*, vol. 8: *Deep-sea Biology*, ed. G. T. Rowe (New York: John Wiley and Sons, 1983), 1–79.

18. Stimpson to Meek, March 14, 1867, SIA RU 7062. In his last credited review for the *American Journal of Science*, appearing shortly after Wallich's rebuke, Stimpson ridiculed Wallich's claim that he had dredged a vertebrate jaw that was 1/100th of an inch in diameter. Several naturalists immediately discounted this assertion, with Stimpson writing a short rebuttal that included two figures, showing that the jaws of certain annelids were pseudomorphs of vertebrate jaws. W. S., "On the 'Minute Vertebrate Lower Jaw,'" *AJS* 36 (1863): 299–300.

19. Stimpson, "Descriptions of new species of marine invertebrates from Puget Sound," *Proc. ANSP* 16 (1864): 153–61. The full report was never published. Stimpson and Verrill's manuscript, minus the drawings, is in the Stimpson Papers at the Smithsonian Institution Archives, RU 7093.

20. *Report of the Secretary of War, communicating, in compliance with a resolution of the Senate, Lieutenant Michler's report of his survey for an interoceanic ship canal near the Isthmus of Darien* (US War Dept., 1861), 264, University of Michigan Making of America website.

21. Stimpson, "Synopsis of the Marine Invertebrata collected by the late Arctic expedition, under I. I. Hayes," *Proc. ANSP* 15 (1863): 138–42.

22. Stimpson mentions the Jessup Fund in a letter to Joseph Leidy, July 27, 1864, College of Physicians of Philadelphia.

23. Stimpson to Frederic Putnam, January 4 and March 1, 1864, Harvard University Archives.

24. Birth record (June 13, 1836) for Ann L. Gordon, International Genealogical Index, www.familysearchg.org. No photographs of her are known to exist.

25. Mayer, "William Stimpson" (1918), 423. For James Frisby Gordon, see Michael Owen Bourne, "Tracing an Architectural Pedigree," *Washington College Magazine* 47, no. 1 (Fall 1998): 24–26; James Frisby Gordon, http://worldconnect.rootsweb.com/cgi-bin/igm.cgi?op=REG&db=brucen&id=I57.

26. Stimpson to Putnam, March 1, 1864, Harvard University Archives.

27. For Couthouy's death, see Gary D. Joiner, ed., *Little to Eat and Thin Mud to Drink: Letters, Diaries, and Memoirs from the Red River Campaigns, 1863–1864* (Knoxville: University of Tennessee Press, 2007), 219–29. Stimpson's notes on Couthouy are in his private North Pacific journal, Stimpson Papers, Shay Collection, CAS Archives.

28. E. B. Long editor, *Personal Memoirs of U. S. Grant* (New York: Da Capo Press, 1982), 408.

29. David M. Jordan, *"Happiness is Not My Companion": The Life of General G. K. Warren* (Bloomington: Indiana University Press, 2001), 150–52.

30. Information on Francis Stimpson is from Francis B. Heitman, *Historical Register and Dictionary of the U.S. Army*, vol. 1 (Washington, DC: Government Printing Office, 1903; reprint, Urbana: University of Illinois Press, 1965), 926.

31. For James H. Stimpson's military record, see Stimpson and Price, *A Stimpson Family in America* (2004), 39–46.

32. Kennicott to Baird, July 8, 1864, SIA RU 7002.

33. See Kennicott to Baird, July 11, 1864, SIA RU 7002; and Bruce Catton, *Never Call Retreat* (New York: Doubleday, 1965), 379–80.

34. Kennicott to Baird, July 25, 1864, SIA RU 7002.

35. Molly W. Berger, *Hotel Dreams: Luxury, Technology, and Urban Ambition in America, 1829–1929* (Baltimore: Johns Hopkins University Press, 2011), 83–110.

36. Patricia Tyson Stroud, *Thomas Say: New World Naturalist* (Philadelphia: University of Pennsylvania Press, 1992).

37. Stimpson to Baird, June 7, 1864, SIA RU 7002.

38. Stimpson to Meek, July 31 and August 7, 1864, SIA RU 7062.

39. Stimpson to Meek, August 7, 1864, SIA RU 7062; Stimpson to Putnam, November 5, 1864, Harvard University Archives.

40. Baird's career was significantly enhanced by his wife's family's social and political connections. See Debra Lindsay, "Intimate Inmates: Wives, Households, and Science in Nineteenth-Century America," *Isis* 89, no. 4 (1998): 631–52.

41. Kennicott to "The folks at home," February 17, 1863, GNHL.

42. Kennicott to Baird, December 11, 1864, SIA RU 7002.

43. "CAS Board of Trustees Minutes 1864–1890," December 30, 1864, 36, CAS Archives.

Chapter 9

1. Quoted from Marc Pachter and Frances Wein, eds., *Abroad in America: Visitors to the New Nation, 1776–1914* (Reading, MA: Addison-Wesley, and National Portrait Gallery, Smithsonian Institution, 1976), 251.

2. For the prefire history of Chicago, see Bessie Louise Pierce, *A History of Chicago*, vol. 2, *From Town to City 1848–1871* (Chicago: University of Chicago Press, 1940); William Cronon, *Nature's Metropolis: Chicago and the Great West* (W. W. Norton, 1991); Perry R. Duis, *Challenging Chicago: Coping with Everyday Life, 1837–1920* (Chicago: University of Illinois Press, 1998); Harold M. Mayer and Richard C. Wade, *Chicago: Growth of a Metropolis* (Chicago: University of Chicago Press, 1969); Donald L. Miller, *City of the Century: The Epic of Chicago and the Making of America* (New York: Simon and Schuster, 1997).

3. Robert Collyer, "Talks about Life," *The Western Monthly* 3 (1871): 389; "A New Englander Estray," *Springfield (Mass.) Republican*, July 7, 1869.

4. Lloyd Lewis and Henry Justin Smith, *Chicago: A History of its Reputation* (New York: Harcourt Brace, 1929), 111.

5. For the early history of the Chicago Academy, see Edmund Andrews, "Historical Sketch of the Chicago Academy of Sciences," in *Act of Incorporation, Constitution, By-Laws, and the Lists of Officers and Members of the Chicago Academy of Sciences; with a Historical Sketch of the Association, and Reports on the Museum and Library* (Chicago: Brewster & Hanscom, 1865); Walter B. Hendrickson and William J. Beecher, "In the Service of Science: The History of the Chicago Academy of Sciences," *Bulletin of the Chicago Academy of Sciences* 11, no. 7 (1972); William Kerr Higley, "Historical Sketch of the Chicago Academy of Sciences," *Special Publication of the Chicago Academy of Sciences*, 1902.

6. Kennicott to Baird, November 18, 1863, February 22, 1864, SIA RU 7002. Ralph S. Bates has noted the rise of scientific societies in the United States and stressed that they were more democratic than similar European societies. Bates, *Scientific Societies in the United States*, 3rd edition (Cambridge, MA: MIT Press, 1965), 28.

7. "CAS Board of Trustees Minutes," CAS Archives, 7–9.

8. Quoted in David Lowenthal, *George Perkins Marsh: Prophet of Conservation* (Seattle: University of Washington Press, 2000), 180.

9. Quote from Ronald S. Vasile, "The Early Career of Robert Kennicott, Illinois' Pioneering Naturalist," *Illinois Historical Journal* 87, no. 3 (1994): 162.

10. Stimpson to Frederic Putnam, April 6, 1865, Harvard University Archives.

11. "The Smithsonian Institution," *Chicago Tribune*, January 25, 1865.

12. Fortuitously, just before leaving for Chicago he had sent two papers for publication to the new *American Journal of Conchology*. See Stimpson to George W. Tryon, February 9, 1865, Academy of Natural Sciences of Philadelphia, no. 98.

13. Joseph Henry diary, January 25, 1865, SIA RU 7001. For the fire, see Kenneth Hafertepe, *America's Castle: The Evolution of the Smithsonian Building and its Institution, 1840–1878* (Washington, DC: Smithsonian Institution Press, 1984), 132–35. Stimpson suspected that Smithsonian employee William De Beust and his son were responsible for the theft of his property. See Stimpson to Meek, April 5, 1865, SIA RU 7062.

14. "Academy of Sciences," *Chicago Tribune*, March 15, 1865.

15. "Academy of Sciences," *Chicago Times*, March 15, 1865.

16. Stimpson to Baird, March 21, June 16, 1865, SIA RU 52.

17. Stimpson to Putnam, April 6, 1865, Harvard University Archives; Stimpson to Baird, March 15, 1865, SIA RU 52.

18. Stimpson to Baird, March 15, 1865, SIA RU 52; the quote on hellbenders is from Roger Conant, *A Field Guide to Reptiles and Amphibians of Eastern/Central America*, 2nd edition (Boston: Houghton Mifflin, 1975), 240.

19. *Smithsonian Annual Report for 1864*, 51.

20. The classic account is Herbert Asbury's *Gem of the Prairie: An Informal History of Chicago's Underworld* (New York: Knopf, 1940).

21. Mary Baird to Annie Stimpson, March 29, 1869, Stimpson letterbook, CAS Archives.

22. Stimpson to Frederic Putnam, March 21, 1865, transcribed copy, Peabody Essex Museum.

23. Stimpson to William S. Vaux, September 20, 1865, Academy of Natural Sciences of Philadelphia, no. 567.

24. Walter B. Hendrickson, "Science and Culture in the American Middle West," *Isis* 64 (1973): 326–40.

25. Safford's lectures at the Academy are noted in the *Chicago Tribune*, "Academy of Sciences," June 12, 1867, July 10, 1867, April 16, 1868. See Philip Fox, "General Account of the Dearborn Observatory," in *Annals of the Dearborn Observatory of Northwestern University*, vol. 1 (1915), 1–20. I am indebted to Phyllis Pitluga for this reference. For Safford see Karen Hunger Parshall, "Truman Henry Safford," *ANB* 19 (1999), 190–191. For early science in Chicago see Austin H. Clark and Leila G. Forbes, "Science in Chicago," *The Scientific Monthly*, vol. 36, no. 6 (1933), 556–567.

26. For cultural philanthropy in Chicago, see Kathleen D. McCarthy, *Noblesse Oblige: Charity and Cultural Philanthropy in Chicago, 1849–1929* (Chicago: University of Chicago Press, 1982); Helen Lefkowitz Horowitz, *Culture and the City: Cultural Philanthropy in Chicago from the 1880s to 1917* (Chicago: University of Chicago Press, 1976).

27. Stimpson to Meek, April 5, 1865, SIA RU 7062.

28. Stimpson to Meek, April 5, 1865, SIA RU 7062.

29. Stimpson to Meek, March 12, 1865, SIA RU 7062. For Worthen, see Daniel Goldstein, "Amos Henry Worthen," *ANB* 23 (1999): 885–86.

30. Stimpson to Meek, March 12, 1865, SIA RU 7062.

31. Meek to Ferdinand Hayden, June 30, 1865, NA RG 57.

32. Stimpson to Meek, July 25, 1865, SIA RU 7062.

33. For example, see Stimpson to Meek, March 21, 1867, SIA RU 7062, where Stimpson wrote, "Why don't you go in as chief on the Nebraska Survey?—don't let Hayden get ahead of you this time— you furnishing the brains & he getting the pay."

34. Meek to Hayden, July 29, 1865, NA RG 57.

35. Stimpson to Meek, August 13, 1865, SIA RU 7062.

36. Andrews, *Act of Incorporation* (1865), xxvi–xli; Joseph Henry to Stimpson, June 13, 1865, SIA RU 33, Smithsonian Institution Archives.

37. Stimpson to Joseph Henry, March 1, 1866, SIA RU 26. For more on the Smithsonian, see Daniel Goldstein, "Yours for Science: The Smithsonian Institution's Correspondents and the Shape of Scientific Community in Nineteenth-Century America," *Isis* 85, no. 4 (1994): 573–99.

38. For a view of the Smithsonian's close ties with another organization, see Michele L. Aldrich and Alan E. Leviton, "West and East: The California Academy of Sciences and the Smithsonian Institution 1852–1906," in Michael T. Ghiselin and Alan E. Leviton, eds., *Cultures and Institutions of Natural History* (San Francisco: California Academy of Sciences, 2000), 183–202.

39. Andrews, *Act of Incorporation* (1865), xxvi–xliv.

40. "Chicago Academy of Sciences," *AJS* 41 (1866): 281. Fielding B. Meek wrote the review; see Stimpson to Meek, January 18, 1866, SIA RU 7062.

41. "Academy of Sciences," *Chicago Tribune*, March 9, July 14, 1864. Ties between the Chicago and St. Louis academies seem to have been sparse. For comparisons between these two academies, see Hendrickson, "Science and Culture in the American Middle West," *Isis* 64 (1973): 326–40.

42. George Engelmann, "President's Annual Address, 1868," *Transactions of the Academy of Science of St. Louis,* vol. 2, 1861–1868, 581; George Engelmann, "President's Annual Address, 1867," *Transactions of the Academy of Science of St. Louis,* vol. 2, 1861–1868, 569. For the St. Louis Academy, see Daniel Goldstein, "Midwestern Naturalists: Academies of Science in the Mississippi Valley, 1850–1900" (PhD diss., Yale University, 1989), 106–54, University Microfilms International.

43. Many naturalists of Stimpson's day began their careers with medical training. Edmund Andrews was a leading Chicago physician who made important contributions as a geologist and was a founder of the Chicago Academy of Sciences. Thomas Neville Bonner, *Medicine in Chicago 1850–1950: A Chapter in the Development of the Social and Scientific Development of a City,* 2nd edition (Urbana: University of Illinois Press, 1991), 42–43.

44. Stimpson to Frederic Putnam, October 26, 1865, Peabody Essex Museum. The donations are listed in the Academy's *Proceedings* and in the *Chicago Times*, "Academy of Sciences," February 14, March 14, April 11, and May 9, 1866.

45. "Academy of Sciences," *Chicago Tribune*, February 17, 1866.

46. Stimpson to Baird, October 12, 1865, SIA RU 52; "CAS Board of Trustees Minutes," 44–45.

47. "The Academy of Sciences," *Chicago Times,* April 11, 1866; "CAS Board of Trustees Minutes," 47.

48. Stimpson to Dall, February 9, 1866, SIA RU 7073.

49. Stimpson to Baird, February 6, 1866, SIA RU 52.

50. Stimpson to Baird, February 6, 1866, SIA RU 52.

51. A. T. Andreas, *History of Chicago,* vol. 2 (1885), 550–53; Charles E. Rosenberg, *The Cholera Years: The United States in 1832, 1849, and 1866* (Chicago: University of Chicago Press, 1987).

52. "Academy of Sciences," *Chicago Times*, May 9, 1866. Bonner, *Medicine in Chicago*, notes that the 1866 cholera epidemic led to a new Board of Health but does not mention the Academy's connection to its creation.

53. For Chicago and pork, see Cronon, *Nature's Metropolis*, 225–35.

54. *Proceedings of the Chicago Academy of Sciences* (1866): 43–45. Joseph Leidy had reported in 1846 the existence of the trichina in a hog from America, but his report was overlooked. See Leonard Warren, *Joseph Leidy, The Last Man Who Knew Everything* (New Haven, CT: Yale University Press, 1998), 65–67.

55. Stimpson to Baird, March 24, 1866, SIA RU 52. Neither Swift nor Armour ever joined the Academy.

56. "The Trichinae Question," *Chicago Republican*, April 13, 1866, reprinted in the *Chicago Medical Examiner*, May 1866, 7, 5; APS Online, 300.

57. *The New York Times*, April 17, 1866. The *Daily Cleveland Herald*, April 14, 1866, reprinted the report, and the *American Naturalist, Scientific American*, and *The Nation* all commented on it; Stimpson to Baird, April 16, 1866, SIA RU 52.

58. "Academy of Sciences," *Chicago Times*, July 11, 1866. The Academy's role in investigating early public health issues has not been previously documented. Two Academy members served on the new Board of Health.

59. "Academy of Sciences," *Chicago Times*, March 14, 1866.

60. Theodore J. Karamanski, *Rally 'Round the Flag: Chicago and the Civil War* (Chicago: Nelson-Hall Publishers, 1993), 164; Stimpson to Baird, February 8, 1867, SIA RU 7002.

61. Robert E. Kohler argues that such landscapes were a dominant feature of the American environment from the 1870s to the 1920s. See *All Creatures: Naturalists, Collectors, and Biodiversity, 1850–1950* (Princeton, NJ: Princeton University Press, 2006), esp. 17–46.

62. Stimpson to Lapham, April 30, 1866, State Historical Society of Wisconsin.

63. Eliphalet Blatchford, "Sketch of the Chicago Academy of Sciences," Supplement B of Reminiscences of Eliphalet Wickes Blatchford, p. B-2, Blatchford papers, Newberry Library, Chicago.

64. Stimpson to Baird, May 29, 1866, SIA RU 52.

65. For the Ottawa Academy, see *The Past and Present of LaSalle County, Illinois* (Chicago: H. F. Kett, 1877), 289–91.

66. Stimpson to Frederic Putnam, May 31, 1866, Peabody Essex Museum.

Chapter 10

1. Ezra B. McCagg, "Beneath the Dust of a Generation," typescript copy, McCagg Papers, Chicago History Museum.

2. For fires in Chicago, see A. T. Andreas, *History of Chicago*, vol. 2 (A. T. Andreas, 1885), 102; statistic on New York fires is from H. A. Musham, "The Great Chicago Fire, October 8–10, 1871," *Papers in Illinois History and Transactions for the Year 1940* (Illinois State Historical Society, 1940): 86.

3. Herman Leroy Fairchild, *A History of the New York Academy of Sciences, 1887* (published by the author, New York), 50; L. M. Eastman, "The Portland Society of Natural History: The Rise and Fall of a Venerable Institution," *Northeastern Naturalist* 13, monograph 1 (2006): 1–38.

4. "Destruction of Scientific Museums by Fire," *AJS* 42 (1866): 135.

5. This account of the fire is based on articles in the *Chicago Tribune*, "The Conflagration," and in the *Chicago Times*, June 8, 1866. While rushing to the scene, the steam engine *Economy* collided with a train, killing fireman Daniel Heartt and seriously injuring two others. Heartt was the third Chicago fireman killed since the professionalization of the department in 1858.

6. "Circular on CAS fire losses," June 7, 1866, CAS Archives.

7. *Chicago Times*, June 8, 1866.

8. Stimpson to Baird, June 19, 1866, SIA RU 52; Baird to Stimpson, June 26, 1866, SIA RU 53.

9. Stimpson to Baird, June 19, 1866, SIA RU 52.

10. Ezra B. McCagg, "Address on Presentation of a Portrait," no date, typescript copy, McCagg Papers, Chicago History Museum.

11. Stimpson to Baird, July 12, 1866, SIA RU 52.

12. Stimpson to Baird, June 19, 1866, SIA RU 52.

13. Mark V. Barrow Jr., *A Passion for Birds: American Ornithology after Audubon* (Princeton, NJ: Princeton University Press, 1998), 39; for Kennicott's and Baird's interest, see Debra Lindsay, *Science in the Subarctic: Trappers, Traders, and the Smithsonian Institution* (Washington, DC: Smithsonian Institution Press, 1993), 24–25.

14. Baird to Stimpson, July 25, 1866, SIA RU 53.

15. For Baird's revision, see Stimpson to Baird, August 6, 1866, SIA RU 52; for the insurance money, see Stimpson to Meek, August 7, 1866, SIA RU 7062; Stimpson to Samuel Scudder, October 5, 1866, Boston Museum of Science.

16. Stimpson to Baird, June 19, 1866, June 25, 1866, SIA RU 52; Baird to Stimpson, June 21, 1866, SIA RU 53.

17. Stimpson to Meek, June 19, 1866, SIA RU 7062.

18. Stimpson to Dall, March 14, 1867, SIA RU 7073.

19. George Walker to Baird, October 22, 1866, SIA RU 52.

20. "The Late Major Kennicott," *Chicago Tribune,* November 15, 1866.

21. "CAS Board of Trustees Minutes," November 20, 1866, 56.

22. Stimpson to Meek, January 4, 1867, SIA RU 7062.

23. Stimpson to Meek, January 4, January 16, 1867, SIA RU 7062. See Eugene H. Cropsey, *Crosby's Opera House: Symbol of Chicago's Cultural Awakening* (London: Fairleigh Dickinson University Press, Associated University Presses, 1999).

24. Stimpson to Baird, January 30, 1867, SIA RU 7002.

25. For Stimpson's neighbors, see 1870 federal census, Cicero Township. For Oak Park, see Everett Chamberlin, *Chicago and its Suburbs* (Chicago: T. A. Hungerford, 1874), 426–27; Jean Guarino, *Oak Park, A Pictorial History* (St. Louis: G. Bradley Publishing, 1988), 8–12.

26. Stimpson to Meek, September 4, 1867, SIA RU 7062. The 1860 census lists Herbert Stimpson's personal estate at $25,000 and his real estate at $10,000. For William Stimpson, see the 1870 federal census, Cicero Township, M 593, National Archives.

27. Stimpson to Baird, August 27, 1867, SIA RU 7002.

28. Baird to Stimpson, September 10, 1867, SIA RU 53.

29. Louis Agassiz to Stimpson, April 29, 1867, MCZ Archives; Stimpson to Meek, May 6, 1867, SIA RU 7062. In January 1867, Philip R. Uhler, one of the last of Agassiz's American students, had resigned.

30. Stimpson to Putnam, March 14, 1867, Peabody Essex Museum.

31. See Ralph W. Dexter, "The Early American Naturalist as Revealed by Letters to the Founders," *The American Naturalist* 90, no. 853 (July–August 1956): 209–25. Margaret Welch, *The Book of Nature: Natural History in the United States 1825–1875* (Boston: Northeastern University Press, 1998), 96, notes that this and other journals helped provide an outlet for the contribution of amateur naturalists as well.

32. F. B. Meek, "Notes on the affinities of the Bellerophontidae," *Proceedings of the Chicago Academy of Sciences* 1 (1866): 9–11; F. B. Meek and A. H. Worthen, "Descriptions of Palaeozoic Fossils from the Silurian, Devonian, and Carboniferous rocks of Illinois, and other Western States," *Proceedings of the Chicago Academy of Sciences* 1 (1866): 11–23; Theodore Gill, "On a New Species of the Genus Macrorhinus," *Proceedings of the Chicago Academy of Sciences* 1 (1866): 33–34.

33. Stimpson, "Descriptions of new Genera and Species of Macrurous Crustacea from the Coasts of North America," *Proceedings of the Chicago Academy of Sciences* 1 (1866): 46–48. The completed version of the paper appeared as Stimpson's "Notes on North American Crustacea in the Museum of the Smithsonian Institution, no. III," *Annals of the New York Lyceum of Natural History* 10 (1873): 92–136.

34. Dall to Robert Kennicott, June 5, 1866, SIA RU 7213.

35. For McChesney, see "Prof. Joseph H. McChesney Dies," *Chicago Tribune,* March 14, 1895; John William McLure, "A History of the Illinois State Geological Survey, 1851–1875" (MA thesis, University of Illinois, 1962), 124–36; Stimpson to Meek, May 6, 1867, SIA RU 7062.

36. Stimpson to Baird, March 27, 1865, SIA RU 52.

37. Stimpson to Baird, October 19, 1865, SIA RU 52.

38. Stimpson to Baird, June 25, 1866, SIA RU 52; Ann Shelby Blum, *Picturing Nature* (Princeton, NJ: Princeton University Press, 1993), 194.

39. Blum, *Picturing Nature.*

40. Stimpson to Baird, March 27, April 5, 1865, SIA RU 52.

41. Stimpson to Baird, October 29, 1867, SIA RU 7002; Stimpson, "Illustrations of North American birds in the Museum of the Chicago Academy of Sciences," *Transactions of the Chicago Academy of Sciences* 1 (1867–1869): 128–29.

42. For Lapham, see Graham Parker Hawks, "Increase A. Lapham, Wisconsin's First Scientist" (PhD diss., University of Wisconsin, 1960), University Microfilms International.

43. Stimpson to Meek, November 8, 1867, SIA RU 7062.

44. Stimpson to Baird, January 10, 1868, SIA RU 7002.

45. "Transactions of the Chicago Academy," *American Naturalist* 2, no. 3 (1868): 158. The regular misspelling of Stimpson's name evokes the quote attributed to Lord Byron—"Fame consists in falling on the field of glory, and having your name spelled wrong in the dispatches," quoted from Donald Culross Peattie, *The Road of a Naturalist* (Boston: Houghton Mifflin, 1941), 261.

46. *AJC* 1 (1865): 96; *AJC* 2 (1866): 384.

47. Stimpson to Baird, March 9, 1867, SIA RU 7002; "Academy of Sciences," *Chicago Tribune*, March 14, 1867, *Chicago Times*, July 10, 1867.

48. Stimpson to Baird, April 12, March 9, 1867, SIA RU 7002.

49. Baird to Stimpson, May 12, 1867, SIA RU 53. Robert Kennicott had begun collecting ethnological specimens for the Academy in 1859 and encouraged Hudson's Bay Company employees to do the same. The Smithsonian donation to the Academy is mentioned in Jane MacLaren Walsh, "Collections as Currency," in William L. Merrill and Ives Goddard, eds., *Anthropology, History, and American Indians: Essays in Honor of William Curtis Sturtevant*, Smithsonian Contributions to Anthropology, no. 44 (Smithsonian Institution Press, 2002), 204.

50. Undated fragment, but probably 1864 or 1865, SIA RU 7213.

51. For the natural history of the region, see Joel Greenberg, *A Natural History of the Chicago Region* (Chicago: University of Chicago Press, 2002).

52. Kennicott to Baird, February 26, 1864, SIA RU 7002. For Kate Doggett, see Rima Lunin Schultz, "Kate Newell Doggett," in *Women Building Chicago 1790–1990, A Biographical Dictionary*, edited by Rima L. Schultz and Adele Hast (Bloomington: Indiana University Press, 2001), 224–29.

53. Stimpson to Baird, November 8, 1870, SIA RU 26. Stimpson wrote Baird that Borcherdt was farming near Denver. Could Borcherdt be the mysterious taxidermist who inspired Martha Maxwell? Maxine Benson's book *Martha Maxwell, Rocky Mountain Naturalist* (Lincoln: University of Nebraska Press, 1986), 68–70, mentions a similarly unscrupulous German taxidermist living near Denver.

54. Stimpson to Baird, June 6, 1867, SIA RU 7002.

55. "To the Exposition of 1867 and Return," *Prairie Farmer*, June 1, 1867; Stimpson to Baird, February 17, 1868, SIA RU 7002.

56. Stimpson to Joel A. Allen, March 21, June 18, 1867, American Museum of Natural History, Ornithology Dept.

57. Donald Worster, *A River Running West: The Life of John Wesley Powell* (New York: Oxford University Press, 2001), 117. Stimpson's letter of introduction for Powell is mentioned in Stimpson to Baird, April 23, 1867, SIA RU 7002, but I failed to find the letter itself.

58. Stimpson to Baird, May 20, 27, 31, June 6, 1867, SIA RU 7002.

59. "Academy of Sciences," October 10, 1867. The find is also mentioned in *Scientific American*, vol. 17, October 19, 1867, Cornell University Making of America website.

60. Stimpson to Meek, October 22, 1867, SIA RU 7062; Stimpson to Baird, October 12, 1867, SIA RU 7002. For vertebrate paleontology in this era, see Keith Thomson, *The Legacy of the Mastodon*, 2008.

61. For a discussion of the systematic approach to collecting, see Robert Kohler, *All Creatures: Naturalists, Collectors, and Biodiversity, 1850–1950* (Princeton, NJ: Princeton University Press, 2006), esp. 111–17; for the Smithsonian, see Debra Lindsay, *Science in the Subarctic: Trappers, Traders, and the Smithsonian Institution* (Washington, DC: Smithsonian Institution Press, 1993).

62. *Annual Report of the Board of Regents of the Smithsonian Institution for 1867*, 42–43; Baird to Stimpson, April 22, 1867, SIA RU 53.

63. F. M. Brown, "Itineraries of the Wheeler Survey Naturalists 1871—Ferdinand Bischoff," *Journal of the New York Entomological Society* 65 (1957): 219–34.

64. Kennicott to Baird, December 11, 1864, January 7, 1865.

65. Baird to Stimpson, February 27, 1867, SIA RU 53.

66. Baird to Stimpson, February 20, 1867, SIA RU 53; George Walker to Baird, March 12, 1867, SIA RU 7002. For the Boston Society's offer, see Baird to George Walker, March 14, 1867, quoted in "CAS Board of Trustees Minutes," 60–61.

67. Stimpson to Baird, April 12, 1867, SIA RU 7002.

68. Baird to Stimpson, June 30, 1867, SIA RU 53; Stimpson to Baird, September 2, October 12, 1867, SIA RU 7002.

69. Stimpson to Baird, November 9, 1867, SIA RU 7002. The bird is now known as *Otus kennicottii*, the western screech owl.

70. Stimpson to Baird, November 9, 1867, SIA RU 7002. Elliot would later become the first curator of zoology at Chicago's Field Museum.

71. Stimpson to Baird, February 6, June 29, 1868, SIA RU 7002. The Bischoff story has a sad ending. In 1869 an army officer at Sitka, Alaska described him as "a little dried up old man who had spent 40 years of his life in the wilds of America collecting insects, birds &c for scientific institutions. . . . He barely makes a living and works for the love of his Science." See "An Army Officer's Trip to Alaska in 1869," ed. Robert G. Athearn, *Pacific Northwest Quarterly* 40 (1949): 44–64, quote on 51–52. Two years later, Bischoff joined a government expedition to Nevada and Arizona, and it is likely that on this trip he wandered off into the desert and was never heard from again, a story related by Dall, *Spencer Fullerton Baird* (1915), 377.

72. Stimpson to Baird, November 25, 1867, SIA RU 7002.

Chapter 11

1. G. M. Kellogg, "Self-Made Men," *Western Monthly* 2 (1869): 37.

2. Duis, *Challenging Chicago* (1998), 206–7; for one example see "Amusements," *Chicago Tribune*, June 20, 1871.

3. Paul M. Angle, *The Chicago Historical Society 1856–1956: An Unconventional Chronicle* (New York: Rand McNally, 1956), 15–60.

4. "The American Museum of Natural History," *American Naturalist* 4, no. 7 (September 1870): 436–37. The American Museum had been founded in 1869 but lacked its own building. The article harshly criticized those in charge for not having a qualified naturalist involved in the management of the museum.

5. For Boyington, see Thomas L. Sloan, "William W. Boyington," in Adolf K. Placzek, ed., *Macmillan Encyclopedia of Architects*, vol. 1 (London: The Free Press, 1982), 267.

6. Stimpson to Baird, December 29, 1867, SIA RU 7002; Stimpson to Meek, January 15, 1868, SIA RU 7062.

7. "Academy of Sciences," *Chicago Tribune*, January 22, 29, 1868. The building lacked one feature that might have helped bring more public attention to the Academy. The Smithsonian and other institutions had lecture halls or auditoriums capable of attracting large numbers of people for popular public lectures. The monthly Academy meetings, with attendance hovering around twenty people, were held in the library.

8. For Franklin Scammon, see "Death of Prof. Franklin Scammon," *Chicago Tribune*, February 12, 1864.

9. Stimpson took full advantage of the Smithsonian's offer to distribute books and specimens to Europe. In 1870 alone, the Smithsonian shipped fifty boxes to Europe and received fifty-nine on behalf of the Chicago Academy.

10. Stimpson to Dall, May 6, 1868, SIA RU 7073. The account of the Academy opening is from "Academy of Sciences," *Chicago Tribune*, January 29, 1868. One inspiration for the Academy building may have come from the Wagner Free Institute of Science in Philadelphia. Built between 1859 and 1865 by a wealthy merchant, like the Academy it had galleries lit by skylights around a central hall. See Sally Gregory Kohlstedt and Paul Brinkman, "Framing Nature: The Formative Years of Natural History Museum Development in the United States," *Proceedings of the California Academy of Sciences* 55, supp. 1, no. 2 (2004): 15.

11. James Parton, "Chicago," *Atlantic Monthly* 19 (March 1867), 341, Cornell University Making of America website, www.cornell.edu.

12. Joel J. Orosz, *Curators and Culture, the Museum Movement in America, 1740–1870* (Tuscaloosa: University of Alabama Press, 1990); Kohlstedt and Brinkman, "Framing Nature" (2004), 7–33; Steven Conn, *Museums and American Intellectual Life, 1876–1926* (Chicago: University of Chicago Press, 1998), 8–45.

13. Stimpson to Baird, February 6, 1868, SIA RU 7002.

14. For the AAAS, see Kohlstedt, *The Formation of the American Scientific Community: The American Association for the Advancement of Science 1848–1860* (Urbana: University of Illinois Press, 1976); Kohlstedt, "Creating a Forum for Science: AAAS in the Nineteenth Century," in *The Establishment of Science in America: 150 Years of the American Association for the Advancement of Science*, ed. Sally G. Kohlstedt, Michael M. Sokal, and Bruce V. Lowenstein (New Brunswick: Rutgers University Press, 1999), 7–49. Robert V. Bruce noted that the "AAAS did not loom as large in American science as it had before the war." Robert V. Bruce, *The Launching of Modern American Science 1846–1876* (Ithaca, NY: Cornell University Press, 1987), 289, 313–14; "The Scientific Convention," *Chicago Tribune*, June 25, 1868.

15. *Proceedings of Association for the Advancement of Science, Sixteenth Meeting, Held at Burlington, Vermont, August 1867* (Cambridge: Joseph Lovering, 1868): 169.

16. Local committee circular, attached to a letter from Stimpson to Meek, July 25, 1868, SIA RU 7062; Stimpson to Gibbes, July 13, 1868, Library of Congress #4451–4452. I am indebted to Lester D. Stephens for his notes on this collection. Charles U. Shepard Jr. of South Carolina did attend the meeting briefly.

17. Stimpson to Baird, July 13 and 17, 1868, SIA RU 7002; Stimpson to Joseph Henry, July 28, 1868, SIA RU 26.

18. "Science," *Chicago Tribune*, August 6, 1868.

19. *Proceedings of the AAAS for 1868, 17th meeting* (published by Joseph Lovering, Cambridge, 1869): 355–59.

20. See "Science," *Chicago Tribune*, August 7, 1868, and the *American Naturalist*, (September 1868): 389; F. B. Meek and A. H. Worthen, "Notice of some New Types of Organic Remains, from the Coal Measures of Illinois," *Proc. ANSP* 17 (1865): 41–53; for Cope's comments, see the *American Naturalist* 2 (September 1868): 388.

21. "Lake Excursion," *Chicago Tribune*, August 9, 1868.

22. *AAAS Proceedings for 1868*. For Atwater, see Mary Clemmer, *Memorial Sketch of Elizabeth Emerson Atwater, Written for Her Friends* (Buffalo, NY: The Courier Company, 1879), 33.

23. *American Naturalist*, September 1868, 384–85; "Western Correspondence," *The Independent*, August 27, 1868, APS Online, 6.

24. Stimpson to Baird, August 22, 1868, SIA RU 7002; CAS Board of Trustees book, 1864–1893, 73; *Proceedings of the AAAS for 1868*, 361–62. For the complaint, see "The American Association for the Advancement of Science," *Manufacturer and Builder* 5, no. 10 (October 1873): 234. Cornell University Making of America website.

25. "Academy of Sciences," *Chicago Times*, November 11, 1868. The Academy of Natural Sciences of Philadelphia was also open to the public, and the display of the first articulated dinosaur in 1868 was so popular that in 1869 the Academy began to charge an admission fee. Robert McCracken Peck and Patricia Tyson Stroud, *A Glorious Enterprise: The Academy of Natural Sciences of Philadelphia and the Making of American Science* (Philadelphia: University of Pennsylvania Press, 2012), 139.

26. Kate Doggett was the first female member, elected in late 1869. The full story of the Chicago Academy of Sciences between 1865 and 1871 is beyond the scope of this book. The author hopes to publish a separate article dealing with other aspects of the Academy's work.

27. E. F. Rivinus and E. M. Youssef, *Spencer Baird of the Smithsonian* (Washington, DC: Smithsonian Institution Press, 1992), 104.

28. A. Hunter Dupree, *Science in the Federal Government: A History of Policies and Activities* (Baltimore: Johns Hopkins University Press, 1986 [1957]), 135–48; Dupree, "The National Academy of Sciences and the American Definition of Science," in *The Organization of Knowledge in Modern*

America, 1860–1920, ed. Alexandra Oleson and John Voss (Baltimore: Johns Hopkins University Press, 1979), 342–63.

29. Henry to Baird, August 12, 1866, SIA RU 7002, box 25, folder 10.

30. See Edward Lurie, *Louis Agassiz, A Life in Science* (Baltimore: Johns Hopkins University Press, 1988), for Agassiz's involvement in the Lazzaroni.

31. Kenneth Silverman, *Edgar A. Poe: Mournful and Never-ending Remembrance* (New York: Harper Perennial, 1991), 215; for the number of deaths, see Sheila M. Rothman, *Living in the Shadow of Death: Tuberculosis and the Social Experience of Illness in American History* (Baltimore: Johns Hopkins University Press, 1994), 2.

32. Stimpson to Mörch, December 10, 1869, American Philosophical Society Library.

33. George Walker to Baird, September 22 and 24, 1868, SIA RU 7002; Blatchford to Baird, November 3, 1868, SIA RU 26.

34. Harry Harris, "Robert Ridgway with a Bibliography of his Published Writings," *The Condor* 30, January 1928, 31–32.

35. Stimpson to Lewis R. Gibbes, November 7, 1868, January 13, 1869, Library of Congress Mss. Division, Gibbes Collection.

36. For Stearns, see Pamela M. Henson, "Robert Edwards Carter Stearns," *ANB* 20 (1999): 596–97. My account is based on Robert E. C. Stearns, "Rambles in Florida," *American Naturalist* 3 (1869): 281–88, 349–356, 397–404, 455–70. In a letter from Tampa, dated January 27, 1869, Stimpson told Baird, "Stearns & Col. J. send you their regards." SIA RU 26.

37. See "Remarks of Robert E. C. Stearns on the Death of Colonel Ezekiel Jewett, before the California Academy of Sciences, June 18th, 1877"; Edwin H. Colbert, *The Great Dinosaur Hunters and Their Discoveries* (New York: Dover, 1984), 67.

38. Robert E. C Stearns, "Some remarks on the death of William Stimpson," *Proceedings California Academy of Sciences* 4, series 1 (1872): 230–32.

39. Stimpson to Baird, January 27, 1869, SIA RU 26.

40. Stimpson, "Some remarks upon the shell-mounds of West Florida," *American Naturalist* 3 (December 1869): 558–60.

41. Stimpson, "Distribution of the Marine Shells of Florida," *American Naturalist* 4 (1871): 586–87.

42. Stimpson to Baird, June 29, 1869, SIA RU 7002; Baird to Stimpson, July 16, 1869, SIA RU 7002.

43. Baird to Stimpson, April 3, 1869, SIA RU 7002.

44. Baird to Stimpson, April 2, 1868, SIA RU 53; Stimpson to Baird, April 6, 1868, SIA RU 7002. Mary P. Winsor, in *Reading the Shape of Nature: Comparative Zoology at the Agassiz Museum* (Chicago: University of Chicago Press, 1991), notes Hagen's groundbreaking discoveries on the crayfish, 92–110.

45. Stimpson to Leidy, June 22, 1859, Academy of Natural Sciences of Philadelphia, no. 1. Edmund Ravenel also complained that Agassiz did not return shells. See Ravenel to Stimpson, January 1861, Charleston Museum.

46. Stimpson to Baird, March 23, 1867, SIA RU 7002; Stimpson to Baird, January 15, 1868, SIA RU 7002.

47. See Henry to Stimpson, April 22, 1869, SIA RU 33, vol. 11, p. 179. Wilcomb E. Washburn, "The Influence of the Smithsonian Institution on Intellectual Life in Mid-Nineteenth-Century Washington," *Records of the Columbia Historical Society of Washington, D. C. for 1963–1965* (Washington, DC: Columbia Historical Society, 1966), 108–13, relates how between 1865 and 1870 Henry gave the bulk of the Smithsonian library to the Library of Congress and the fine arts collection to William Corcoran. Orosz, *Curators and Culture* (210), notes that Henry had the herbarium transferred to the Department of Agriculture in 1868.

48. Stimpson to Joseph Henry, July 13 and 27, 1869, SIA RU 26; Joseph Henry to Stimpson, July 23, 1869, SIA RU 33.

49. Stimpson to George Walker, June 18, 1869, in CAS Board of Trustees Minutes, 78.

50. Stimpson to Henry, November 19, 1869, SIA RU 26; Stimpson to Dall, July 21, 1870, SIA RU 7073. Brown is listed as the clerk in charge of transportation in the 1869 Smithsonian Annual Report.

See Terrica M. Gibson, "Solomon G. Brown," in James Conaway, *The Smithsonian: 150 Years of Adventure, Discovery, and Wonder* (Washington: Smithsonian Books, Knopf, 1996), 94–95.

51. Stimpson to Baird, June 29, 1869, SIA RU 7002; Stimpson to Meek, August 17, 1869, SIA RU 7062.

52. Stimpson, "Experiments upon a solution of carbolic acid as a substitute for alcohol in the preservation of wet specimens," *American Naturalist* 3 (December 1869): 557–58. Also see Stimpson to Baird, September 20, 1869, SIA RU 7002.

53. Stimpson to Baird, October 9, 1869, SIA RU 7002.

54. See Ralph W. Dexter, "The 'Salem Secession' of Agassiz Zoologists," *Essex Institute Historical Collections* 101, no. 1 (1965): 37; Stimpson to Edward Morse, September 15, 1869, Peabody Essex Museum, Morse papers.

55. Stimpson to Meek, October 19, 1869, SIA RU 7062.

56. "Scientific," *Chicago Republican*, January 12, 1870.

57. Stimpson to Joseph Henry, August 1, 1868, SIA RU 26.

58. For Bridge's analysis, see "Academy of Sciences," *Chicago Tribune*, February 10, 1869.

59. E.[dmund] Andrews, "Military Surgery Among the Apache Indians," *Chicago Medical Examiner* 10, no. 10 (October 1869): 599–601.

60. See *Richardson's Guide to the Fossil Fauna of Mazon Creek*, ed. Charles W. Shabica and Andrew A. Hay (Chicago: Northeastern Illinois University, 1997). Stimpson to Meek, March 27, 1868, SIA RU 7062, mentions the Joseph Evans collection that had been destroyed.

61. Stimpson to Meek, September 11, 1869, SIA RU 7062. The fire is noted in the *American Naturalist* 5, no. 5 (1871): 301, 321.

62. W. Conner Sorensen, *Brethren of the Net: American Entomology, 1840–1880* (Tuscaloosa: University of Alabama Press, 1995), 199–200.

63. Quoted in F. Garvin Davenport, "Natural Scientists and the Farmers of Illinois, 1865–1900," *Journal of the Illinois State Historical Society* 51 (1958): 361.

64. "Academy of Sciences," *Chicago Times*, October 12, 1870; Carol A. Sheppard, "Benjamin Dann Walsh: Pioneer Entomologist and Proponent of Darwinian Theory," *Annual Review of Entomology* 49 (2004): 1–25.

Chapter 12

1. "Chicago," quoted in Paul M. Angle, ed., *Prairie State: Impressions of Illinois, 1673–1967, by Travelers and Other Observers* (Chicago: University of Chicago Press, 1968), 362–69 (quote on pages 363, 367).

2. On the Academy's aid, see "A Party About to Start for a Two Years' Survey of the Rocky Mountains," *New York Times*, June 27, 1868; "The Powell Expedition," *Chicago Tribune*, May 29, 1869. For Powell and Stimpson's lecture, see "Academy of Sciences," *Chicago Tribune*, October 13, 1869.

3. *Chicago Times*, November 9, 1869.

4. *Chicago Times*, November 9, 1869; "Interesting Discovery," *Chicago Evening Journal*, November 9, 1869. *Appleton's Journal* reported that the Cardiff Giant had been made by a German in Chicago. *Appleton's Journal* 3, no. 50 (March 12, 1870): 307, University of Michigan Making of America website.

5. See "John Wells Foster," *Appleton's Cyclopedia of American Biography*, vol. 2, ed. James Grant Wilson and John Fiske (D. Appleton, 1887 [1873]), 512; "The Late Col. Foster," *Chicago Tribune*, July 9, 1873.

6. Robert Silverberg, *Mound Builders of Ancient America: The Archaeology of a Myth* (Greenwich, CT: New York Graphic Society, 1968), 1–7, 157–61. Silverberg writes that Foster's depiction of Native Americans is "eerily reminiscent" of German scientists' views of the Jews in the 1930s.

7. Stimpson to Joseph Henry, December 21, 1869, SIA RU 26.

8. Sally G. Kohlstedt, *The Formation of the American Scientific Community: The American Association for the Advancement of Science 1848–1860* (Urbana: University of Illinois Press, 1976), 139.

9. Spencer F. Baird, "On Additions to the Bird-Fauna of North America, made by the Scientific Corps of the Russo-American Telegraph Expedition," *Transactions of the Chicago Academy of Sciences* 1 (1869): 314; Chicago Academy of Sciences, "Biography of Robert Kennicott", *Transactions of the Chicago Academy of Sciences* 1 (1869): 133–226.

10. William H. Dall and Henry M. Bannister, "List of the Birds of Alaska, with Biographical Notes," *Transactions of the Chicago Academy of Sciences* 1 (1869): 267–310; *American Naturalist* 4, no. 6 (1870): 367–71.

11. Stimpson to Baird, September 20, October 7, 1869, SIA RU 7002.

12. Stimpson to Baird, November 15, December 22, 1869, January 31, 1870, SIA RU 7002.

13. *Western Monthly*, February 1870, 160–61. George Engelmann, president of the Academy of Sciences at St. Louis, pointed out in 1871 that, "Chicago has by its splendid volumes far surpassed us." See Engelmann's Presidential Address, *Transactions of the Academy of Sciences of St. Louis* 3 (1868–1877): xxxix.

14. "Scientific," *Chicago Republican*, January 12, 1870.

15. Stimpson to Joseph Henry, January 8, 1870, SIA RU 26; Stimpson to Baird, August 25, 1870, SIA RU 26.

16. Stimpson to Packard, November 3, 1869, letter quoted in Ralph Dexter, "The Early American Naturalist as Revealed by Letters to the Founders," *American Naturalist* 90 (1956): 214, 209–25.

17. Stimpson to Meek, March 27 and June 16, 1868; for the dispute with Billings, see Stimpson to Meek, August 17, October 19, 1869, SIA RU 7062.

18. Stimpson to Dall, July 21, 1870, SIA RU 7073.

19. Stimpson to Baird, April 9, 1870, SIA RU 7002.

20. Stimpson, "On the deep-water fauna of Lake Michigan," *American Naturalist* 4 (September 1870): 403–5.

21. Stimpson to Baird, May 23, 1870, SIA RU 7002; "Academy of Sciences," *Chicago Tribune*, June 15, 1870.

22. A. W. Schorger, "Philo Romayne Hoy: Wisconsin's Greatest Pioneer Zoologist," *The Passenger Pigeon* 6, no. 3 (July 1944): 55–59; "Philo Romyne [sic] Hoy, M.D.," *The United States Biographical Dictionary, Wisconsin Volume* (Chicago, Cincinnati, and New York: American Biographical Publishing Company, 1877), 78–79.

23. Stimpson, "On the Deep-Water Fauna (1870), 403–5; P. R. Hoy, "Deep-Water Fauna of Lake Michigan," *Transactions of the Wisconsin Academy of Sciences, Arts, and Letters* 1 (1872), 98–101.

24. Hoy to Baird, June 29, 1870, SIA RU 7002; Stimpson to Baird, June 29, 1870, SIA RU 7002.

25. "Academy of Sciences," *Chicago Evening Journal*, June 9, 1869; for the Wheaton mastodon, see "Academy of Sciences," *Chicago Republican*, December 15, 1869, and "Academy of Sciences," *Chicago Times*, December 16, 1869, and October 12, 1870.

26. See "Academy of Sciences," *Chicago Times*, October 12 and November 9, 1870; Stimpson to Baird, November 9 and 19, 1870, SIA RU 7002; Robert E. Warren and Russell W. Graham, "*Cervalces*: An Ice Age Discovery," *Living Museum* 50, no. 3 (1988): 38–41. For Milner, see "Obituary, Prof. James W. Milner," *New York Times*, January 18, 1880.

27. Stimpson to Baird, October 8, 1870, SIA RU 26. Inspired by the Academy's success in obtaining the Smithsonian marine invertebrates in alcohol, Louis Agassiz unsuccessfully attempted to forge a similar agreement with Baird by asking him to send the Smithsonian's fish collection to the MCZ under the same conditions that applied to the Chicago Academy's loan. Agassiz to Baird, December 12, 1870, cited in Elmer C. Herber, ed., *Correspondence between Spencer Fullerton Baird and Louis Agassiz* (Washington, DC: Smithsonian Institution Press, 1963), 200–1.

28. *Annual Report of the National Academy of Sciences* (1879), 11; *Proceedings of the National Academy of Sciences* (1870): 75, 77; Stimpson to Baird, June 15 and June 29, 1870, SIA RU 7002.

29. Baird to Stimpson, November 15, 1870, SIA RU 7002; Stimpson to Baird, November 17, 1870, SIA RU 7002. For Hall, see Chauncey C. Loomis, *Weird and Tragic Shores: The Story of Charles Francis Hall, Explorer* (Lincoln: University of Nebraska Press, 1991).

30. Helen M. Rozwadowski, *Fathoming the Ocean: The Discovery and Exploration of the Deep Sea* (Cambridge, MA: Belknap Press of Harvard University Press, 2005); Susan Schlee, *The Edge of an Unfamiliar World: A History of Oceanography* (New York: E. P. Dutton, 1973), 92–106.

31. Alexander Agassiz to Stimpson, December 24, 1868, MCZ Archives, Agassiz collection, vol. 5, 60–61. For Pourtales's work, see Schlee, *The Edge of an Unfamiliar World*, 103–4; Rudolf S. Scheltema and Amelie H. Scheltema, "Deep-sea Biological Studies in America, 1846 to 1872—their contribution to the *Challenger* Expedition," *Proceedings of the Royal Society of Edinburgh* (B) 72 (1971–1972): 133–44; Richard Rathbun, "The American Initiative in Methods of Deep-Sea Dredging," *Science* 4, no. 76 (July 18, 1884): 54–57.

32. He named two of the new genera for Florida caciques, or Native American chiefs mentioned in de Soto's accounts of his explorations in Florida. Stimpson, "Preliminary Report on the Crustacea dredged in the Gulf Stream in the Straits of Florida," by L. F. de Pourtales, Assist. US Coast Survey. Part 1. Brachyura; *Bulletin of the Museum of Comparative Zoology* 2 (1871): 109–60; "Scientific Intelligence," *Harper's Weekly*, March 11, March 25, 1871, www.harpweek.com; Gilberto Rodriguez, "From Oviedo to Rathbun: The development of brachyuran crab taxonomy in the Neotropics (1535–1937)," in *Crustacean Issues: History of Carcinology* 8, ed. Frank Truesdale, 109–18 (Rotterdam: A. A. Balkema, 1993), 54.

33. Louis Agassiz, "Report upon the Deep-Sea Dredgings in the Gulf Stream, during the Third cruise of the U.S. Steamer *Bibb*," addressed to Professor Benjamin Peirce, Superintendent U.S. Coast Survey," *Bulletin of the Museum of Comparative Zoology* 1, no. 13 (1869), quotes on 363, 363–86; Louis Agassiz to Stimpson, November 22 and December 2, 1870, MCZ Archives.

34. CAS Board of Trustee Minutes book, 1864–1890, p. 88.

35. "Reminiscences of Eliphalet Wickes Blatchford," ed. Charles Hammond Blatchford Jr.; "Sketch of the Chicago Academy of Sciences," Supplement B, p. B-3, copy in CAS archives, original in the Newberry Library, E. W. Blatchford Papers.

36. Quoted in Sheila M. Rothman, *Living in the Shadow of Death: Tuberculosis and the Social Experience of Illness in American History* (Baltimore: Johns Hopkins University Press, 1994), 47.

37. J. B. Holder to Stimpson, October 22, 1870, Blatchford papers, Newberry Library.

38. Stimpson to Baird, December 18, 1870, SIA RU 7002; Stimpson to Joseph Henry, January 12, 1871, SIA RU 26.

39. The following account is based on Eliphalet Blatchford, "Sketch of the Chicago Academy of Sciences": Supplement B of Reminiscences of Eliphalet Wickes Blatchford, Newberry Library, pp. B-5–6.

40. Robert Kohler, *All Creatures: Naturalists, Collectors, and Biodiversity, 1850–1950* (Princeton, NJ: Princeton University Press, 2006), esp. 10–16.

41. Kohler, *All Creatures*, notes (91) that museum curators collected on unpaid working vacations. Stimpson, however, was paid by the Academy to collect as part of his duties as director of the museum.

42. Paula M. Mikkelsen and Rüdiger Bieler, *Seashells of Southern Florida: Living Marine Mollusks of the Florida Keys and Adjacent Regions: Bivalves*, vol. 1 (Princeton, NJ: Princeton University Press, 2007), 5.

43. The reference to the fyke net is in Stimpson to Baird, March 28, 1871, SIA RU 52; Blatchford, "Reminiscences," B-5.

44. Blatchford, "Reminiscences," B-6.

45. Stimpson to Baird, March 28, 1871, SIA RU 52; Blatchford, Supplement B of the "Reminiscences of Eliphalet Wickes Blatchford," Newberry Library.

46. "Scientific Intelligence," *Harper's Weekly*, May 13, 1871, www.harpweek.

47. Stimpson to Baird, May 6, 1871, SIA RU 52.

48. "CAS Board of Trustees Minutes," 87.

49. L. Agassiz to Stimpson, June 1, 1871, MCZ Archives, vol. 6, 200.

50. Stimpson to Baird, May 14, 29, 1871, SIA RU 52; Stimpson to Dall, July 21, 1870, SIA RU 7073.

51. Edmund Andrews, "The North American Lakes considered as Chronometers of Post-Glacial Time," copy in CAS bound *Transactions*. For the other article, see Stimpson to Packard, June 16, 1871,

Peabody Essex Museum; Alpheus Packard, "Notice of Hymenoptera & Nocturnal Lepidoptera, Collected in Alaska, by W. H. Dall, Director Sci. Corp, W. U. T. Exp., with a list of Neuroptera, by P. R. Uhler and Dr. H. Hagen," copy in CAS Archives, original in Museum of Comparative Zoology.

52. Stimpson to Joseph Henry, May 15, 1871, SIA RU 7260.

53. Stimpson to Baird, May 6, 14, 1871, SIA RU 52.

Chapter 13

1. Quoted in W. S. W. Ruschenberger, "A Sketch of the Life of Joseph Leidy," *Proceedings of the American Philosophical Society* 30 (1892): 135–84, quote on 147.

2. Richard Fortey, *Dry Storeroom No. 1: The Secret Life of the Natural History Museum* (New York: Knopf, 2008), 206.

3. "To the Friends and Correspondents of the Chicago Academy of Sciences, October 30, 1871."

4. Stimpson to Meek, July 31, 1871, SIA RU 7062.

5. When Joseph Henry asked Stimpson in 1870 to send some of the unpublished drawings of the expedition's crustacea to Alphonse Milne Edwards, Stimpson demurred, claiming that he had "nearly completed arrangements for their publication in this country." Whether this was merely a dodge to prevent a European from getting his hands on them or whether Stimpson had really made progress on the publication is unknown. Stimpson to Henry, March 16, 1870, SIA RU 26.

6. The Murex and North-West Boundary report are in the Stimpson Papers at the SIA RU 7093.

7. See Stimpson to Chesbrough, May 25, 1871, Stimpson Papers, CAS Archives; Academy trustee book, 95. The collection was ready to be sent when word arrived of the Chicago Fire. According to Herman Hagen, the "collection is now far more important than it ever was before, as it contains a part of the published types of W. Stimpson." See *Annual Report of the Trustees of the Museum of Comparative Zoology, at Harvard College, in Cambridge; together with the Report of the Director for 1872* (Boston: Wright and Potter, 1873), 28.

8. For Jeffreys, see Eric L. Mills, "Edward Forbes, John Gwyn Jeffreys, and British dredging before the *Challenger* expedition," *Journal for the Society of the Bibliography of Natural History* 8 (1978): 507–36. Also see Philip F. Rehbock, "The Early Dredgers: "Naturalizing" in British Seas, 1830–1850," *Journal of the History of Biology* 12 (1979): 293–368.

9. Stimpson to Baird, October 3, 1871, SIA RU 52; Jeffreys to Stimpson, October 6, 1871, CAS Archives, Stimpson letterbook.

10. See Stimpson to Baird, October 3, 1871, SIA RU 52.

11. Stimpson, "Notes on North American Crustacea, in the Museum of the Smithsonian Institution, no. 3," *Annals of the Lyceum of Natural History of New York* 10 (1873): 92–136.

12. Unless otherwise noted, the following account of the fire is from H. A. Musham, "The Great Chicago Fire, October 8–10, 1871," *Transactions of the Illinois State Historical Society* (1940): 69–189.

13. Musham, "The Great Chicago Fire," 93.

14. Stephen J. Pyne, *Fire in America: A Cultural History of Wildland and Rural Fire* (Princeton, NJ: Princeton University Press, 1982).

15. Herbert Asbury, *Gem of the Prairie: An Informal History of Chicago's Underworld* (New York: Knopf, 1940), 86.

16. William Kerr Higley, *Historical Sketch of the Academy*, Special Publication no. 1, Chicago Academy of Sciences 23 (1902).

17. Robert Cromie, *The Great Chicago Fire, Illustrated Edition* (Nashville, TN: Rutledge Hill Press, 1994 [1958]), 165.

18. Blatchford, Reminiscences, B8. Transcript in CAS Archives, original in Newberry Library.

19. James Milner to Henry Bannister, October 14, 1871, Northwestern University Archives, Bannister collection.

20. "Sketch of the Chicago Academy of Sciences, Supplement B of Reminiscences of Eliphalet Wickes Blatchford." Copy in CAS archives.

21. Stimpson to Henry, October 10, 1871, SIA RU 26, box 41.

22. Baird to Stimpson, October 18, 1871, Stimpson letterbook, CAS Archives. Baird noted the destruction of the Academy museum in "Scientific Intelligence," *Harper's Weekly*, November 18, 1871, www.harpweek.com.

23. Louis Agassiz to Stimpson, October 12, 1871, SIA RU 7260.

24. Verrill to Stimpson, October 19, 1871, CAS Archives, Stimpson letterbook.

25. Meek to Stimpson, December 15, 1871, CAS Archives, Stimpson letterbook.

26. Lea to J. W. Foster, December 19, 1871, CAS Archives, Stimpson letterbook.

27. Carpenter to Stimpson, December 31, 1871, CAS Archives, Stimpson letterbook. Richard Rathbun of the Smithsonian would later write that the fire and Stimpson's death had "deprived our country of a most important chapter in the history of submarine exploration." Rathbun, "The American Initiative in Methods of Deep-Sea Dredging," *Science* 4, no. 76 (July 18, 1884): 55.

28. Paul Bartsch, Harald Rehder, and Beulah Shields, *A Bibliography and Short Biographical Sketch of William Healey Dall*, Smithsonian Miscellaneous Collections 104, no. 15 (1946): 11–12.

29. Frederic Putnam to Stimpson, October 19, 1871, Stimpson Papers, Shay Collection, CAS Archives.

30. Ezekiel Jewett to Stimpson, October 16, 1871, Stimpson Papers, Chicago Academy of Sciences.

31. Dana to Foster, January 18, 1872, Stimpson letterbook, CAS Archives; Jeffreys to Stimpson, October 10, 1871, SIA RU 7260; Alexander Agassiz to Stimpson, October 10, 1871, Stimpson letterbook, CAS Archives.

32. Silliman to Foster, December 18, 1871, Stimpson letterbook, CAS Archives; "Destruction of the Museum of the Chicago Academy of Sciences," *AJS* 2, 3rd series (1871): 387–88.

33. *The Nation*, November 9, 1871, 302–5; November 16, 1871, 321.

34. Stimpson to Baird, October 25, 1871, SIA RU 52.

35. "CAS Minutes, 1857–93," 78.

36. "CAS Board of Trustees Minutes," 115.

37. Stimpson to Gentlemen, undated but probably December 1871, Stimpson letterbook, CAS Archives.

38. "CAS Board of Trustees Minutes," 96.

39. "To the Friends and Correspondents of the Chicago Academy of Sciences, October 30th, 1871."

40. "A Great Loss to Science," *Appleton's Journal* 6, no. 141 (December 9, 1871): 669, University of Michigan Making of America website.

41. "Dr. William Stimpson," *The Prairie Farmer* 43, no. 26 (June 29, 1872): 203.

Chapter 14

1. Samuel T. Coleridge's "The Dungeon," quoted by Philip P. Carpenter, "Lectures on the Shells of the Gulf of California," *Smithsonian Annual Report for 1859*, 195–219. Quote is from 219. For the entire poem, see I. A. Richards, *The Portable Coleridge* (New York: Penguin Books, 1986), 79–80.

2. Stimpson to Baird, November 14, 1871, SIA RU 52.

3. Louis Agassiz to Stimpson, October 14, 1871, CAS Archives, Stimpson letterbook.

4. Louis Agassiz to Stimpson, October 22, 1871, CAS Archives, Stimpson letterbook. As it turned out, the urgency was not necessary: Agassiz misjudged his departure by over a month.

5. Baird to Stimpson, October 18, 1871; Benjamin Peirce to Stimpson, November 10, 1871, Stimpson letterbook, CAS Archives.

6. Pourtales to Stimpson, November 29, 1871, CAS Archives, Stimpson letterbook.

7. "Remarks of Robert E. C. Stearns on the Death of Dr. William Stimpson before the California Academy of Sciences, June 17, 1872, *Proceedings of the California Academy of Sciences* 4, series 1 (1868–1873): 230–32.

8. Stimpson to Baird, December 7 and 12, 1871, SIA RU 52; Stimpson to George Tryon, December 18, 1871, Academy of Natural Sciences of Philadelphia, no. 98.

9. For Velie, see A. T. Andreas, *History of Chicago*, vol. 2 (A. T. Andreas, 1885), 429–31.

10. Agassiz, "A Letter concerning Deep-Sea Dredgings, addressed to Professor Benjamin Peirce, Superintendent United States Coast Survey," *Bulletin of the MCZ* 3 (1871–1876): 49–53.

11. For Platt, see http://www.history.noaa.gov/stories_tales/platt.html.

12. Platt to Peirce, March 23, 1872, NA RG 23; Stimpson to Baird, April 19, 1872, SIA RU 52.

13. Platt to Peirce, March 23, 1872, NA RG 23.

14. Stimpson to Baird, April 19, 1872, SIA RU 52.

15. Platt to Peirce, February 22, 1872, NA RG 23.

16. Stimpson to Willie Stimpson, March 31, 1872, Stimpson Papers, Shay Collection, CAS Archives.

17. *Report of the Superintendent of the United States Coast Survey, 1872* (Washington, DC: Government Printing Office, 1873), 32–33. Five of the crabs dredged from the *Bache* were later described as new by Alphonse Milne Edwards. See Gilberto Rodriguez, "From Oviedo to Rathbun: The development of brachyuran crab taxonomy in the Neotropics (1535–1937)," in *Crustacean Issues: History of Carcinology* 8, ed. Frank Truesdale, 109–18 (Rotterdam: A. A. Balkema, 1993), 54.

18. Stimpson to Baird, April 19, 1872, SIA RU 52; "Remarks of Robert E. C. Stearns on the Death of Dr. William Stimpson"; *Annual Report of the Museum of Comparative Zoology for 1872*, 20–21.

19. Stimpson to Baird, April 19, 1872, SIA RU 52. For more on the history of crinoid research, see Mills, "Problems of Deep-Sea Biology" (1983), 10–18. Baird reported on Stimpson's trip in "Scientific Intelligence," *Harper's Weekly*, May 25, 1872, www.harpweek.com. He also summarized the expedition in Spencer F. Baird, ed., *Annual Record of Science and Industry for 1872* (New York: Harper and Brothers, 1873), 154–55.

20. "In Memoriam, The Late Dr. William Stimpson, of Chicago," *Chicago Tribune*, June 26, 1872. After Stimpson's death, Velie became the leading scientific presence at the Academy.

21. "In Memoriam, The Late Dr. William Stimpson," *Chicago Tribune*, June 12, 1872.

22. Blatchford to Stimpson, April 12 and May 22, 1872, Stimpson letterbook, CAS Archives.

23. See Sheila M. Rothman, *Living in the Shadow of Death: Tuberculosis and the Social Experience of Illness in American History* (Baltimore: Johns Hopkins University Press, 1994), 16–17.

24. "Remarks of Robert E. C. Stearns on the Death of Dr. William Stimpson before the California Academy of Sciences." Obituary notices of Stimpson appeared in the *American Journal of Science*, the *American Naturalist*, *Nature*, and the *Journal de Conchyliologie*.

25. "In Memoriam—William Stimpson, M. D.," *Overland Monthly* 9, no. 4 (October 1872): 381, University of Michigan Making of America website.

26. William W. Everts, "The Church and Science," *The Chicago Pulpit* 1, no. 24 (1872): 221–28, CHS library.

27. "CAS Minutes," June 11, 1872, 94–97.

28. Dall, "Some American Conchologists," presidential address before the Biological Society of Washington, 1888, pp. 95–134.

29. "In Memoriam, The Late Dr. William Stimpson," *Chicago Tribune*.

30. Nathaniel Shaler, *The Autobiography of Nathaniel Southgate Shaler* (Boston: Houghton and Mifflin, 1909), 128.

31. Herbert H. Stimpson to William B. Rogers, February 15, 1882; Rogers to Stimpson, April 1, 1882; Stimpson to Rogers, April 24, 1882, Institute Archives, MIT; "Mrs. Anne Louise Gordon Stimpson," *The Transcript*, September 10, 1904, http://mdhistory.net/msa_sc6000/msa_sc6000_42_1/pdf/msa_sc6000_42_1-0296.pdf.

32. Blatchford to Baird, February 20, 1880, SIA RU 7002; William Gordon Stimpson, *Who Was Who in America,* vol. 1, 1897–1942 (Chicago: Marquis Who's Who, 1943), 1188.

33. Herbert Baird Stimpson, *Who Was Who in America,* vol. 5, 1969–1973 (Chicago: Marquis Who's Who, 1973), 696.

34. Baird published a brief note on Stimpson, "Scientific Intelligence," *Harper's Weekly,* June 15, 1872, www.harpweek.com.

35. William H. Dall to A. G. Mayer, December 27, 1916, Dall papers, SIA RU 7073.

36. It was written by marine biologist Alfred Goldsborough Mayor, the son of Stimpson's friend Alfred Marshall Mayer. A. G. Mayer changed his name to Mayor in 1918. Like Stimpson, Mayor did fieldwork in the Florida Keys and died of tuberculosis. See Lester D. Stephens and Dale R. Calder, *Seafaring Scientist: Alfred Goldsborough Mayor, Pioneer in Marine Biology* (Columbia: University of South Carolina Press, 2006).

37. Theodore Lyman, "Zoological Results of the Hassler Expedition, Ophiuridae and Astrophytidae," *Illustrated Catalogue of the Museum of Comparative Zoology,* no. 8 (1874).

Epilogue

1. Loren Eiseley, *The Man Who Saw Through Time,* revised and enlarged edition of *Francis Bacon and the Modern Dilemma* (New York: Charles Scribner's Sons, 1973), 34.

2. Andrew J. Lewis, *A Democracy of Facts: Natural History in the Early Republic* (Philadelphia: University of Pennsylvania Press, 2011); D. Graham Burnett, *Trying Leviathan: The Nineteenth-Century New York Court case that Put the Whale on Trial and Challenged the Order of Nature* (Princeton, NJ: Princeton University Press, 2007); Lee Alan Dugatkin, *Mr. Jefferson and the Giant Moose: Natural History in Early America* (Chicago: University of Chicago Press, 2009); Richard W. Judd, *The Untilled Garden: Natural History and the Spirit of Conservation in America, 1740–1840* (New York: Cambridge University Press, 2009); Amy R. W. Meyers, ed., *Knowing Nature: Art and Science in Philadelphia, 1740–1840* (New Haven, CT: Yale University Press, 2012).

3. Bruce, *The Launching of Modern American Science* (1987); Keith Thomson, *The Legacy of the Mastodon: The Golden Age of Fossils in America* (New Haven, CT: Yale University Press, 2008); W. Conner Sorensen, *Brethren of the Net: American Entomology, 1840–1880* (Tuscaloosa: University of Alabama Press, 1995); Mark V. Barrow Jr., *A Passion for Birds: American Ornithology after Audubon* (Princeton, NJ: Princeton University Press, 1998).

4. Lester D. Stephens, *Science, Race and Religion in the American South: John Bachman and the Charleston Circle of Naturalists* (Chapel Hill and London: University of North Carolina Press, 2000); Richard G. Beidelman, *California's Frontier Naturalists* (Berkeley: University of California Press, 2006).

5. E. F. Rivinus and E. M. Youssef, *Spencer Baird of the Smithsonian* (Washington, DC: Smithsonian Institution Press, 1992); A. Hunter Dupree, *Asa Gray: American Botanist, Friend of Darwin* (Cambridge, MA: Harvard University Press, 1959); Edward Lurie, *Louis Agassiz, A Life in Science* (Baltimore: Johns Hopkins University Press, 1988; Christoph Irmscher, *Louis Agassiz: Creator of American Science* (Boston: Houghton Mifflin Harcourt, 2013); James G. Cassidy, *Ferdinand Hayden, Entrepreneur of Science* (Lincoln: University of Nebraska Press, 2000).

6. Bruce, *The Launching of Modern American Science.*

7. Robert E. Kohler, "Reflections on the History of Systematics," in Andrew Hamilton, ed., *The Evolution of Phylogenetic Systematics* (Berkeley: University of California Press, 2014), 17–46.

8. Robert Kohler, *All Creatures: Naturalists, Collectors, and Biodiversity, 1850–1950* (Princeton, NJ: Princeton University Press, 2006), 227.

9. Shaler, *Autobiography,* 128. For Shaler, see David Livingstone, *Nathaniel Southgate Shaler and the Culture of American Science* (Tuscaloosa: University of Alabama Press, 1987).

10. Kohler, *All Creatures,* 227. Kohler extends the argument of David E. Allen that later biologists derided the work of taxonomists as a way to advance their own lab-based careers. David

E. Allen, *The Naturalist in Britain: A Social History* (London: Allen Lane, 1976), 176–94. Carol Kaesuk Yoon criticized taxonomy for making people more disconnected from nature, a claim that seems ridiculous given the myriad other ways that people have lost connection with the natural world. Carol Kaesuk Yoon, *Naming Nature: The Clash Between Instinct and Science* (New York: W. W. Norton, 2009).

11. John Steinbeck and Edward F. Ricketts, *Sea of Cortez* (Mamaroneck, NY: Paul A. Appel, 1971 [1941]), 286.

12. Peter J. Bowler, *The Norton History of the Environmental Sciences* (New York: W. W. Norton, 1993). Bowler asserted (5) that "the transition from natural history to biology and geology was a complex process, and some areas have remained largely untouched by explanatory schemes such as the theory of evolution. . . . Yet 'mere' classification is itself a theoretical process." Lynn K. Nyhart, "Natural history and the 'new' biology," in N. Jardine, J. A. Secord, and E. C. Spary, eds., *Cultures of Natural History* (Cambridge: Cambridge University Press, 1996), 42; Keith R. Benson, "From Museum Research to Laboratory Research: The Transformation of Natural History into Academic Biology," in Ronald Rainger, Keith R. Benson, and Jane Maienschein, eds., *The American Development of Biology* (Philadelphia: University of Pennsylvania Press, 1988), 49–86.

13. Robert Kohler, *All Creatures*, 2; Anne L. Larsen, "Not Since Noah: The English Scientific Zoologists and the Craft of Collecting, 1800–1840" (PhD diss., Princeton University, 1993).

14. Judd, *The Untilled Garden*, 55, 118, 121–22.

15. Lynn Barber, *The Heyday of Natural History* (Garden City, NY: Doubleday, 1980), 40, 60. Lynn Merrill in *The Romance of Natural History* stated (81), "The closet-field division is an important feature of nineteenth-century natural history. . . . What is important is not so much whether a given writer is a field or closet observer, but the fact that a tension always exists between the broad view—natural ecology, the landscape as a whole, objects within their setting—and the narrow view—anatomical details, microscopic focus, the object as isolated." Merrill argues that this "macroscopic-microscopic tension" is crucial to understanding nineteenth-century natural history.

16. William Leach, *Butterfly People: An American Encounter with the Beauty of the World* (New York: Pantheon Books, 2013), 129.

17. "Is there a specifically *American* natural history?" wrote Joyce E. Chaplin in "Nature and Nation," in Sue Ann Prince, ed., *Stuffing Birds, Pressing Plants, Shaping Knowledge: Natural History in North America, 1730–1860* (Philadelphia: American Philosophical Society, 2003), 75–95. Chaplin sees early American naturalists highlighting the distinctiveness of American animals and plants after the Revolution in order to overcome the effects of British colonialism.

18. Daniel Lewis's book *The Feathery Tribe: Robert Ridgway and the Modern Study of Birds* (New Haven, CT: Yale University Press, 2012), 122–44, notes that by the last quarter of the nineteenth century American ornithologists relied mainly on collectors to do their fieldwork for them. Ornithology was one of the more popular disciplines and had many amateur collectors and dealers. Birds are also relatively easy to collect and preserve compared to crustaceans and echinoderms. Each zoological discipline has its own unique pattern of support and growth.

19. Addison E. Verrill, William H. Dall, Alexander Agassiz, Louis Francois de Pourtales, J. Gywn Jeffries, and the *Challenger* expedition are examples of the field tradition, which continues today. Sorensen, in *Brethren of the Net* (1995), notes (45) that many taxonomic entomologists were in the field collecting in the West and elsewhere.

20. Susan Schlee, *The Edge of an Unfamiliar World: A History of Oceanography* (New York: E. P. Dutton, 1973), 67.

21. Frederick R. Schram, preface to *History of Carcinology, Crustacean Issues 8*, ed. Frank Truesdale (Rotterdam/Brookfield: A. A. Balkema, 1993), vii.

22. Austin B. Williams, "Reflections on crab research in North America since 1758," in Truesdale, *History of Carcinology*, 260–61.

Appendix

1. http://www.marinespecies.org/index.php, accessed August 21, 2016. These numbers fluctuate as the nomenclature is revised.

2. https://scholar.google.com/citations?user=SVdRgCMAAAAJ&hl=en&oi=ao, accessed August 9, 2016.

3. Elizabeth C. Agassiz and Alexander Agassiz, *Seaside Studies in Natural History: Marine Animals of Massachusetts Bay, Radiates* (Boston: James R. Osgood, 1871), 25; Kohler, *All Creatures* (2006), 3.

BIBLIOGRAPHY

MANUSCRIPT SOURCES

(Major collections are noted by asterisks.)

ACADEMY OF NATURAL SCIENCES OF PHILADELPHIA

Collection #1
#98 to Tryon
#73 to Haldeman
#567 to Vaux

LIBRARY OF THE AMERICAN MUSEUM OF NATURAL HISTORY

Joel A. Allen Papers
Cope Papers
Ornithology Department

AMERICAN PHILOSOPHICAL SOCIETY

Joseph Henry Papers
Elisha Kane Papers
John L. LeConte Papers
Letters to Scientists

BANCROFT LIBRARY, UNIVERSITY OF CALIFORNIA AT BERKELEY

James G. Cooper Papers

BOSTON MUSEUM OF SCIENCE

Samuel Scudder Papers

CHARLESTON MUSEUM

Ravenel Papers

CHICAGO ACADEMY OF SCIENCES

*Board of Trustees minutes book, 1864–1890
CAS Minutes of regular meetings, 1857–1893
*William Stimpson Letterbook

CHICAGO HISTORICAL SOCIETY

Ezra B. McCagg Papers
Colonial Williamsburg Foundation Library
Robert Carter Papers, Shirley Plantation Collection

COLLEGE OF PHYSICIANS OF PHILADELPHIA

Joseph Leidy Papers

FRANCIS A. COUNTWAY LIBRARY, HARVARD MEDICAL LIBRARY

Jeffries Wyman Papers

DET KONGELIGE BIBLIOTEK, COPENHAGEN

*Christian Lütken Papers
*Johannes Steenstrup Papers

GRAY HERBARIUM LIBRARY, HARVARD UNIVERSITY

*Charles Wright Papers

THE GROVE NATIONAL HISTORIC LANDMARK, GLENVIEW, IL

*Robert Kennicott Papers

HARVARD UNIVERSITY ARCHIVES

Putnam Papers
*Addison E. Verrill Journal

LIBRARY OF CONGRESS

Lewis R. Gibbes Papers

McGILL UNIVERSITY ARCHIVES

John W. Dawson Papers

MIT ARCHIVES

William H. Rogers Papers

MUSEUM OF COMPARATIVE ZOOLOGY

Alex Agassiz Papers
Louis Agassiz Papers

NATIONAL ARCHIVES

*Naval Records Collection of the Office of Naval Records and Library—Record Group 45
*Records of the Coast and Geodetic Survey—Record Group 23
*Records of the US Geological Survey—Record Group 57

NEWBERRY LIBRARY, CHICAGO

Eliphalet Blatchford Papers

NORTHWESTERN UNIVERSITY

Henry Bannister Papers

PEABODY ESSEX MUSEUM OF SALEM

Edward Morse Papers
Frederic Putnam Letters

ENOCH PRATT LIBRARY

John P. Kennedy Papers

SMITHSONIAN INSTITUTION ARCHIVES

Accession Records RU 305
*Spencer F. Baird Papers RU 7002
*William H. Dall Papers RU 7073
Theodore Gill Papers
*Joseph Henry Papers
* Fielding Meek Papers RU 7062
National Anthropological Archives
*North Pacific Exploring Expedition Papers RU 7253
*Office of the Secretary RU 26
*Potomac Side Naturalists Club RU 7210
*William Stimpson Papers RU 7093

STATE HISTORICAL SOCIETY OF WISCONSIN

Increase Lapham Papers

UNIVERSITY OF OSLO LIBRARY

Sars Papers

YALE UNIVERSITY ARCHIVES

James D. Dana Papers

NEWSPAPERS

Boston Daily Atlas 1842, 1848, 1853
Boston Evening Transcript 1847
Brooklyn Daily Eagle 1853
Cambridge Chronicle 1863
Chicago Evening Journal 1869
Chicago Medical Examiner 1866
Chicago Republican 1866, 1869–1870
Chicago Times 1865–1872
Chicago Tribune 1864–1872, 1895
Daily National Intelligencer (Washington, DC) 1852–1853, 1860–1861
Friend's Intelligencer 1855
Macon Weekly Telegraph 1868
New York Daily Tribune 1858
New York Times 1854, 1866–1867
Prairie Farmer 1867, 1872
Sacramento Daily Union 1855
Springfield (MA) *Republican* 1869
The Times of London 1853
Washington Evening Star 1860

PUBLICATIONS OF WILLIAM STIMPSON

(listed chronologically)

"Description of a new species of *Helix.*" *Proc. BSNH* 3 (1848–1851): 175.

"Descriptions of two new species of *Philine* obtained in Boston Harbor." *Proc. BSNH* 3 (1848–1851): 333–34.

"On two new species of shells and a holothurian from Massachusetts Bay." *Proc. BSNH* 4 (1851–1854): 7–9.

"List of fossils found in the post-Pliocene deposit, in Chelsea Mass., at Point Shirley." *Proc. BSNH* 4 (1851–1854): 9–10.

"Notices of several new species of Testaceous Mollusca new to Massachusetts Bay." *Proc. BSNH* 4 (1851–1854): 12–18.

"Observations on the identity of *Nucula navicularis* and *N. thraciaeformis.*" *Proc. BSNH* 4 (1851–1854): 26–27.

"Remarks on an Ascidian found in Massachusetts Bay." *Proc. BSNH* 4 (1851–1854): 49.

"Observations on the Stations of the Echinodermata." *Proc. BSNH* 4 (1851–1854): 55.

"A New species of *Pentacta.*" *Proc. BSNH* 4 (1851–1854): 66–67.

"Observations on the fauna of the islands at the mouth of the Bay of Fundy and of the extreme northeast coast of Maine." *Proc. BSNH* 4 (1851–1854): 95–100.

"Monograph of the genus *Caecum* in the United States." *Proc. BSNH* 4 (1851–1854): 112–13.

"Descriptions of several new species of shells from the northern coast of New England." *Proc. BSNH* 4 (1851–1854): 113–14.

Kurtz, J. D., and William Stimpson. "Descriptions of several new species of shells from the Southern coast." *Proc. BSNH* 4 (1851–1854): 114–15.

"Description of a new crustacean belonging to the genus *Axius.*" *Proc. BSNH* 4 (1851–1854): 222–23.

"Descriptions of two new species of *Ophiolepis*, from the Southern coast of the United States." *Proc. BSNH* 4 (1851–1854): 224–26.

"Several new Ascidians from the coast of the United States." *Proc. BSNH* 4 (1851–1854): 228–32.

Revision of the Synonymy of the Testaceous Mollusks of New England, with notes on their structure, and their geographical and bathymetrical distribution. Boston: Phillips and Sampson, 1851.

"Synopsis of the Marine Invertebrata of Grand Manan; or the region about the mouth of the Bay of Fundy." *Smithsonian Contributions to Knowledge* 6, article 5 (1854): 1–68.

"Descriptions of some of the new Marine Invertebrata from the Chinese and Japanese Seas." *Proc. ANSP* 7 (1855): 375–84.

"Descriptions of some new Marine Invertebrata." *Proc. ANSP* 7 (1855): 385–95.

"On some remarkable marine invertebrata inhabiting the shores of South Carolina." *Proc. BSNH* 5 (1854–1856): 110–17.

"On some California Crustacea." *Proceedings of the California Academy of Natural Sciences* 1 (1856): 95–99.

"Notices of New species of Crustacea of Western North America." *Proc. BSNH* 6 (1856–1859): 84–89.

"On a new form of parasitic gasteropodous mollusca, *Cochliolepis parasiticus.*" *Proc. BSNH* 6 (1856–1859): 307–9.

"Notice of the Scientific Results of the Expedition to the North Pacific Ocean, under the command of Com. John Rodgers." *American Journal of Science* 23 (1857): 136–38.

"Prodromus descriptionis animalium evertebratorum quae in Expeditione ad Oceanum Pacificum Septentrionalem, a Republica Federata missa, Cadwalader Ringgold et Johanne Rodgers Duce, observavit et descripsit." Part 1, Turbellaria Dendrocoela. *Proc. ANSP* 9 (1857): 19–31. Part 2, Turbellarieorum Nemertineorum. *Proc. ANSP* 9 (1857): 159–66. Part 3, Crustacea Maioidea. *Proc. ANSP* 9 (1857): 216–21. Part 4, Crustacea Cancroidea et Corystoidea. *Proc. ANSP* 10 (1858): 31–40. Part 5, Crustacea Ocypodoidea. *Proc. ANSP* 10 (1858): 93–110. Part 6, Crustacea Oxystomata. *Proc. ANSP* 10 (1858): 159–63. Part 7, Crustacea Anomoura. *Proc. ANSP* 10 (1858): 225–52. Part 8, Crustacea Macrura. *Proc. ANSP* 12 (1860): 22–47.

"On the crustacea and echinodermata of the Pacific shores of North America." *Journal of the Boston Society of Natural History* 6 (1857): 444–532.

"[Description of *Scutella interlineata*]." In "Geological Report," *From Reports of explorations and surveys, to ascertain the most practicable and economical route for a railroad from the Mississippi River to the Pacific Ocean,* edited by William P. Blake, 153–54. Washington, 1857. Aka Pacific Railroad Reports.

[*Hapalocarcinus marsupialis,* a remarkable new form of brachyurous crustacean on the coral reefs at Hawaii]. *Proc. BSNH* 6 (1859): 412–13.

[Note on crustacea collected by John Xantus]. In "Notes on a collection of Birds made by Mr. John Xantus, at Cape St. Lucas, Lower California, and now in the Museum of the Smithsonian Institution," edited by Spencer F. Baird. *Proc. ANSP* 11 (1859): 302.

"On the Collection and Preservation of Marine Invertebrates." In "Directions for Collecting, Preserving, and Transporting Specimens of Natural History prepared for the use of the Smithsonian Institution," 37–40. *Smithsonian Miscellaneous Collections,* 3rd edition. Washington, DC, 1859.

"Sketch of a revision of the genus *Mithracidae.*" *American Journal of Science* 29 (1860): 132–33.

"Cretaceous Strata at Gay Head, Mass." *American Journal of Science* 29 (1860): 145.

"Obituary of William Turner." *American Journal of Science* 29 (1860): 152.

"On Botanical and Zoological Nomenclature." *American Journal of Science* 29 (1860): 289–93.

"A trip to Beaufort N. Carolina." *American Journal of Science* 29 (1860): 442–45.

"On the marine shells brought by Mr. Drexler from Hudson's Bay, and the occurrence of a Pleistocene deposit on the southern shore of James Bay." *Proc. ANSP* 13 (1861): 97–98.

"Notes on certain decapod crustacea." *Proc. ANSP* 13 (1861): 372–73.

"On new genera and species of star-fishes of the family Pycnopodidae (*Astercanthion* Müll. and Trosch.)" *Proc. BSNH* 8 (1861): 261–73.

"On the genus Bipaliura." *American Journal of Science* 31 (1861): 134–35.

"Description of a new *Cardium* from the Pleistocene Hudson's Bay." *Proc. ANSP* 14 (1862): 58–59.

"List of crustacea found in the Bay of Cartajena and its vicinity westward." In *Report of the Secretary of War, communicating, in compliance with a resolution of the Senate, Lieutenant Michler's report of his survey for an interoceanic ship canal near the Isthmus of Darien,* 264. Washington, DC: US War Dept, 1861. 36th Congress, 2nd session, Senate exec. Document no. 9. University of Michigan, Making of America website, http://name.umdl.umich.edu/AKR5089.0001.001.

"Check Lists of the Shells of North America." With Isaac Lea, P. P. Carpenter, William Stimpson, William G. Binney, and Temple Prime. Stimpson contributed the section "East Coast: Arctic Seas to Georgia." *Smithsonian Misc. Collections* 2 article 6, 1862, n. p.

"On an oceanic Isopod found near the southeastern shore of Massachusetts." *Proc. ANSP* 14 (1862): 133–34.

"Notes on North American Crustacea, No. 1." *Annals of the Lyceum of Natural History of New York* 7 (1862): 49–93.

"Notes on North American Crustacea, No. 2." *Annals of the Lyceum of Natural History of New York* 7 (1862): 175–246.

"On the fossil crab (Archaeoplax signifera) of Gay Head." *Journal of the Boston Society of Natural History* 7 (1863): 583–89.

"Malacozoological Notices, No. 1: On the Genus Gundlachia." *Proc. BSNH* 9 (1863): 249–52.

"Synopsis of the Marine Invertebrata collected by the late Arctic expedition, under I. I. Hayes." *Proc. ANSP* 15 (1863): 138–42.

"On the classification of the Brachyura and on the homologies of the antennary joints in decapod crustacea." *American Journal of Science* 35 (1863): 139–43.

"On the question of whether Diatoms live on the sea-bottom at great depths." *American Journal of Science* 35 (1863): 454–55.

"Note on the 'Glass Coral' of Japan." *American Journal of Science* 35 (1863): 458–59.

"On the structural characters of the so-called Melanians of North America." *American Journal of Science* 38 (1864): 41–53.

"Descriptions of new species of marine invertebrates from Puget Sound." *Proc. ANSP* 16 (1864): 153–61.

[Notes on bird tongues] in Spencer F. Baird. "Review of American Birds, in the Museum of the Smithsonian Institution." *Smithsonian Miscellaneous Collections,* no. 181 (1864–1866): 162–64.

"Diagnosis of newly discovered genera of gasteropods belonging to the subfamily Hydrobiinae, of the family Rissoidea." *American Journal of Conchology* 1 (1865): 52–54.

"On Certain Genera and Families of Zoophagous Gasteropods." *American Journal of Conchology* 1 (1865): 55–64.

"Review of the Northern Buccinums, and Remarks on some other Northern Marine Mollusks. Part 1." *Canadian Naturalist and Geologist* 2 (1865): 364–89.

"Researches on the Hydrobiinae and allied forms." *Smithsonian Misc. Coll* 7 article 4 (1866): 1–59.

"Descriptions of new Genera and Species of Macrurous Crustacea from the Coasts of North America." *Proceedings of the Chicago Academy of Sciences* 1 (1866): 46–48.

"Illustrations of North American birds in the Museum of the Chicago Academy of Sciences." *Transactions of the Chicago Academy of Sciences* 1 (1867–1869): 128–29.

"Experiments upon a solution of carbolic acid as a substitute for alcohol in the preservation of wet specimens." *American Naturalist* 3 (1869): 557–58.

"Some remarks upon the shell-mounds of West Florida, particularly those of Tampa Bay." *American Naturalist* 3 (1869): 558–60.

"On the deep-water fauna of Lake Michigan." *American Naturalist* 4 (1870): 403–5.

"Preliminary Report on the Crustacea dredged in the Gulf Stream in the Straits of Florida, by L. F. de Pourtales, Assist. U.S. Coast Survey. Part 1. Brachyura." *Bulletin of the Museum of Comparative Zoology* 2 (1871): 109–60.

"Distribution of the Marine Shells of Florida." *American Naturalist* (1871): 586–87.

"Notes on North American Crustacea in the Museum of the Smithsonian Institution, No. III." *Annals of the Lyceum of Natural History of New York* 10 (1873): 92–136.

Report on the Crustacea (Brachyura and Anomura) Collected by the North Pacific Exploring Expedition, 1853–1856. Washington: Smithsonian Misc. Collections, 49, 1907.

REVIEWS BY WILLIAM STIMPSON IN THE *AMERICAN JOURNAL OF SCIENCE AND ARTS*

Only reviews in which Stimpson added significant comments are listed.

"Naturhistoriske Bidrag til en Belkrivelle af Grönland," by J. Reinhardt, J. C. Schiödte, O. A. L. Mörch, C. F. Lütken, J. Lange, and H. Rink. *AJS* 25 (1858): 124–26.

"*Leptosiagion* Trask, nov. gen.," by John B. Trask. *AJS* 25 (1858): 295.

"Life Beneath the Waters, or, The Aquarium in America," by A. M. Edwards. *AJS* 26 (1858): 284–85.

"Memoires pour servir a l' Histoire Naturelle du Mexique, des Antilles et des Etats-Unis," by Henri de Saussure. *AJS* 27 (1859): 445–47.

Bidrag till Spitsbergens Molluskfauna: jemte en allman öfversigt af Artiska Regionens, naturforhallanden och forntida utbredning," by Otto Torell. *AJS* 28 (1859): 444–45.

"Bidrag till Kännedomen om Skandinaviens Amphipoda Gammaridea," by R. M. Bruzelius. *AJS* 28 (1859): 445.

"Die Klassen und Ordnungen des Thier-Reichs, wissenschaftlich dur gestellt in Wort und Bild," by H. G. Bronn. *AJS* 29 (1860): 130.

"Descriptions of Oceania (*Turritopsis*) *nutricula*, n. s. and the embryological history of a singular Medusan Larva found in the cavity of its bell," and "The Gymnophthalmata of Charleston Harbor," by John McCrady. *AJS* 29 (1860): 130–31.

"On the Zoological Affinities of the Graptolites," by John McCrady. *AJS* 29 (1860): 131–32.

"On the genus Synapta," by Woodward and Barrett. *AJS* 29 (1860): 134.

"Geology for Teachers, Classes, and Private Students," by Sanborn Tenney. *AJS* 29 (1860): 288.

"Les genres Loriope et Peltogaster," by H. Rathke and W. Liljeborg. *AJS* 29 (1860): 293.

"Neue Wirbeilose Thiere, beobachtet und gesammelt auf einer Reise am die Erde," by Ludwig K. Schmarda. *AJS* 29 (1860): 293–94.

"A Supplement to the 'Terrestrial Air-breathing Mollusks of the United States and the Adjacent Territories of North America,'" by W. G. Binney. *AJS* 29 (1860): 294.

"Catalogue of the Recent Marine Shells found on the coasts of North and South Carolina," by J. D. Kurtz. *AJS* 29 (1860): 294–95.

"Catalogue of Acanthopterygian Fishes in the collection of the British Museum," by A. Gunther. *AJS* 30 (1860): 140.

"On the genus Peasia," by W. H. Pease. *AJS* 31 (1861): 134.

"Mollusca, or the Shell-fish and their Allies," by P. P. Carpenter, and "The Kjökkenmöddings: Recent Geologico-Archaeological Researches in Denmark," by John Lubbock. *AJS* 33 (1862): 297–98.

"On the Ornithology of Labrador," by Elliott Coues. *AJS* 33 (1862): 298.

"Post-pliocene Fossils of South Carolina," by Francis S. Holmes. *AJS* 33 (1862): 298–300.

"Catalogue of the Miocene Shells of the Atlantic Slope," by T. A. Conrad. *AJS* 35 (1863): 428–30.

"On the 'genus Diplothyra,'" *AJS* 35 (1863): 455.

"On the 'Minute Vertebrate Lower Jaw,'" by Dr. Wallich. *AJS* 36 (1863): 299–300.

References

Abbott, R. T., and M. E. Young, eds. *American Malacologists: A national register of professional and amateur malacologists and private shell collectors and biographies of early American mollusk workers born between 1618 and 1900*. Falls Church, VA: American Malacologists, 1973.

Adams, Charles B. "On the Value of the Shells of Mollusca for the purpose of distinguishing Species and Higher Groups." *Contributions to Conchology*, vol. 1. New York: H. Bailliere Brothers, 1852.

Agassiz, Elizabeth, ed. *Louis Agassiz, His Life and Correspondence*. 2 vols. Boston: Houghton, Mifflin, 1886.

Agassiz, Elizabeth C., and Alexander Agassiz. *Seaside Studies in Natural History: Marine Animals of Massachusetts Bay, Radiates*. Boston: James R. Osgood, 1871.

Agassiz, Louis. *Contributions to the Natural History of the United States of America*, vol. 3. Boston: Little, Brown, 1860.

——. "A Letter Concerning Deep-Sea Dredgings, Addressed to Professor Benjamin Peirce, Superintendent United States Coast Survey." *Bulletin of the MCZ* 3 (1871–1876): 49–53.

——. "On the Origin of Species." *American Journal of Science* 30 (1860): 142–54.

——. "Report upon the Deep-Sea Dredgings in the Gulf Stream, during the Third Cruise of the U.S. Steamer *Bibb*, Addressed to Professor Benjamin Peirce, Superintendent U.S. Coast Survey." *Bulletin of the Museum of Comparative Zoology* 1, no. 13 (1869): 363–86.

Allard, Dean C. *Spencer Fullerton Baird and the U.S. Fish Commission*. New York: Arno Press, 1978.

Allen, David. E. *The Naturalist in Britain: A Social History*. London: Allen Lane, 1976.

"The American Museum of Natural History." *American Naturalist* 4, no. 7 (September 1870): 436–37.

Andreas, A. T. *History of Chicago*, vol. 2. Chicago: A. T. Andreas, 1885.

——. *History of Chicago*, vol. 3. Chicago: A. T. Andreas, 1886.

Andrews, Edmund. "Historical Sketch of the Chicago Academy of Sciences." In *Act of Incorporation, Constitution, By-Laws, and the Lists of Officers and Members of the Chicago Academy of Sciences; with a Historical Sketch of the Association, and Reports on the Museum and Library*. Chicago: Brewster & Hanscom, 1865.

——. "Military Surgery Among the Apache Indians." *Chicago Medical Examiner* 10 (1869): 599–601.

——. "The North American Lakes considered as Chronometers of Post-Glacial Time." *Transactions of the Chicago Academy of Sciences*, 1871.

Angle, Paul M. *The Chicago Historical Society 1856–1956: An Unconventional Chronicle*. New York: Rand McNally, 1956.

Angle, Paul M., ed. *Prairie State: Impressions of Illinois, 1673–1967, by Travelers and Other Observers*. Chicago: University of Chicago Press, 1968.

Aruga, Tadashi. "The First Japanese Mission to the United States." In *Abroad in America: Visitors to the New Nation 1776–1914*, edited by Marc Pachter and Frances Wein, 134–43. National Portrait Gallery, 1976.

Asbury, Herbert. *Gem of the Prairie: An Informal History of Chicago's Underworld*. New York: Knopf, 1940.

Athearn, Robert G. "An Army Officer's Trip to Alaska in 1869." *Pacific Northwest Quarterly* 40 (1949): 44–64.

Bailey, Jacob M. "On some specimens of deep-sea bottom, from the sea of Kamschatka, collected by Lieut. Brooke, U.S.N." *American Journal of Science* 21 (1855): 284–85.

Baird, Spencer F. "Notes on a collection of Birds made by Mr. John Xantus, at Cape St. Lucas, Lower California, and now in the Museum of the Smithsonian Institution." *Proc. ANSP* 11 (1859): 299–306.

——. "On Additions to the Bird-Fauna of North America, made by the Scientific Corps of the Russo-American Telegraph Expedition." *Transactions of the Chicago Academy of Sciences* 1 (1869): 311–25.

——. "Review of American Birds, in the Museum of the Smithsonian Institution." *Smithsonian Miscellaneous Collections*, no. 181 (1864–1866).

Baird, Spencer F., ed. *Annual Record of Science and Industry for 1872*. New York: Harper and Brothers, 1873.

Barber, Lynn. *The Heyday of Natural History*. Garden City, NY: Doubleday, 1980.

Barrow, Mark V. Jr. *Nature's Ghosts: Confronting Extinction from the Age of Jefferson to the Age of Ecology*. Chicago: University of Chicago Press, 2009.

——. *A Passion for Birds: American Ornithology after Audubon*. Princeton, NJ: Princeton University Press, 1998.

Bartsch, Paul, Harald Rehder, and Beulah Shields. "A Bibliography and Short Biographical Sketch of William Healey Dall." *Smithsonian Miscellaneous Collections* 104, no. 15 (1946).

Bates, Marston. *The Nature of Natural History*. New York: Scribner's, 1950.

Bates, Ralph S. *Scientific Societies in the United States*. 3rd edition. Cambridge, MA: MIT Press, 1965.

Beidelman, Richard G. *California's Frontier Naturalists*. Berkeley: University of California Press, 2006.

Benson, Keith R. "From Museum Research to Laboratory Research: The Transformation of Natural History into Academic Biology." In *The American Development of Biology*, edited by Ronald Rainger, Keith R. Benson, and Jane Maienschein, 49–86. Philadelphia: University of Pennsylvania Press, 1988.

Benson, Maxine. *Martha Maxwell, Rocky Mountain Naturalist*. Lincoln: University of Nebraska Press, 1986.

Berger, Molly W. *Hotel Dreams: Luxury, Technology, and Urban Ambition in America, 1829–1929*. Baltimore: Johns Hopkins University Press, 2011.

Berry, Elmer G. "The Amnicolidae of Michigan: Distribution, Ecology, and Taxonomy." *Miscellaneous Publications, Museum of Zoology, University of Michigan*. Misc. Publications no. 57 (1943).

Binney, Amos. *The Terrestrial air breathing mollusks of the U.S.*, edited by Augustus Gould. Boston: Little and Brown, 1851.

Blum, Ann S. *Picturing Nature*. Princeton, NJ: Princeton University Press, 1993.

Bolles, Edmund B. *The Ice Finders: How a Poet, a Professor, and a Politician Discovered the Ice Age*. Washington, DC: Counterpoint, 1999.

Bonner, Thomas N. *Medicine in Chicago 1850–1950: A Chapter in the Development of the Social and Scientific Development of a City*. 2nd edition. Urbana: University of Illinois Press, 1991.

Bourne, Michael O. "Tracing an Architectural Pedigree." *Washington College Magazine* 48, no. 1 (1998), 24–26.

Bouve, Tomas T. "Historical Sketch of the Boston Society of Natural History, with a Notice of the Linnaean Society which Preceded It." In *Anniversary Memoirs of the Boston Society of Natural History published in celebration of the Fiftieth Anniversary of the Society's Foundation, 1830–1880*. Boston: Boston Society of Natural History, 1880.

Bowler, Peter J. *The Norton History of the Environmental Sciences*. New York: W. W. Norton, 1993.

Brooke, George M. Jr. *John M. Brooke, Naval Scientist and Educator*. Charlottesville: University Press of Virginia, 1980.

——. "John Mercer Brooke, Naval Scientist." PhD diss., University of North Carolina, 1955.

Brown, Chandos M. *Benjamin Silliman: A Life in the Young Republic*. Princeton, NJ: Princeton University Press, 1989.

Brown, F. M. "Itineraries of the Wheeler Survey Naturalists 1871—Ferdinand Bischoff." *Journal of the New York Entomological Society* 65 (1957): 219–34.

Browne, Janet. *Charles Darwin, Voyaging*. Princeton, NJ: Princeton University Press, 1995.

——. *The Secular Ark: Studies in the History of Biogeography*. New Haven, CT: Yale University Press, 1983.

Bruce, Robert V. *The Launching of Modern American Science 1846–1876*. Ithaca, NY: Cornell University Press, 1987.

Burchsted, J. C. A., and F. Burchsted. "Samuel Tufts Jr. (1817–1902), a Massachusetts Shell Collector and Aquarium Stocker." *Archives of Natural History* 34 (2007): 229–34.

Burkhardt, Frederick, and Sydney Smith, eds. *The Correspondence of Charles Darwin*, vol. 6, 1856–1857. Cambridge: Cambridge University Press, 1990.

Cannon, Susan F. *Cultures of Science: The Early Victorian Period*. New York: Dawson and Science History Publications, 1978.

Carpenter, Philip P. "Lectures on Mollusca; or 'Shell-Fish' and their Allies." *Smithsonian Annual Report for 1860* (1861): 151–283.

———. "Lectures on the Shells of the Gulf of California." *Smithsonian Annual Report for 1859* (1860): 195–219.

———. "Report on the Present State of our Knowledge with Regard to the Mollusca of the West Coast of North America." *Report of the British Association for the Advancement of Science for 1856* (1857): 159–368.

———. "A Supplementary report of the Present State of Our Knowledge with Regard to the Mollusca of the West Coast of North America." *Report of the British Association for the Advancement of Science for 1863* (1864): 517–86.

Carpenter, Russell L. *Memoirs of the Life and Work of Philip Pearsall Carpenter*. London: C. Kegan Paul, 1880.

Carter, Robert. *A Summer Cruise on the Coast of New England*. Boston: Crosby and Nichols, 1864.

Cassidy, James G. *Ferdinand Hayden, Entrepreneur of Science*. Lincoln: University of Nebraska Press, 2000.

Catton, Bruce. *Never Call Retreat*. New York: Doubleday, 1965.

Chamberlin, Everett. *Chicago and its Suburbs*. Chicago: T. A. Hungerford, 1874.

Chaplin, Joyce E. "Nature and Nation." In *Stuffing Birds, Pressing Plants, Shaping Knowledge: Natural History in North America, 1730–1860*, edited by Sue Ann Prince, 75–95. Philadelphia: American Philosophical Society, 2003.

"Chicago Academy of Sciences." *American Journal of Science* 41 (1866): 281.

Chicago Academy of Sciences. "Biography of Robert Kennicott." *Transactions of the Chicago Academy of Sciences* 1, part 2 (1869): 133–226.

William H. Dall and Henry M. Bannister. "List of the Birds of Alaska, with Biographical Notes." *Transactions of the Chicago Academy of Sciences* 1 (1869): 267–310.

Chickering, J. W. "The Potomac-Side Naturalists Club." *Science* 23 (1906): 264–65.

Chittenden, L. E. *Recollections of Abraham Lincoln and His Administration*. New York: Harper and Brothers, 1891.

[Circular on Chicago Academy of Sciences fire losses]. June 7, 1866.

Clark, Austin H., and Leila G. Forbes. "Science in Chicago." *The Scientific Monthly* 36, no. 6 (1933): 556–67.

Clark, Henry James. "Prodromus of the History, Structure, and Physiology of the order Lucernariae" *Boston Journal of Natural History* 7 (1859–1863): 531–67.

Cleevely, R. J. "Some Malacological Pioneers and their links with the transition of shell-collecting to conchology during the first half of the nineteenth century." *Archives of Natural History* 22, no. 3 (1995): 385–418.

Clemmer, Mary. *Memorial Sketch of Elizabeth Emerson Atwater, Written for Her Friends*. Buffalo: The Courier Company, 1879.

Coan, Eugene. *James Graham Cooper: Pioneer Western Naturalist*. Moscow: University Press of Idaho, 1981.

Colbert, Edwin H. *The Great Dinosaur Hunters and Their Discoveries*. New York: Dover, 1984.

Cole, Allan B. "The Ringgold-Rodgers-Brooke Expedition to Japan and the North Pacific, 1853–1859." *Pacific Historical Review* 16, no. 2 (1952): 152–62.

Cole, Allan B., ed. *Yankee Surveyors in the Shogun's Seas.* New York: Greenwood Press, 1947.

Collyer, Robert. "Talks about Life." *The Western Monthly* 3 (1870): 389–93.

Conant, Roger. *A Field Guide to Reptiles and Amphibians of Eastern/Central America.* 2nd edition. Boston: Houghton and Mifflin, 1975.

Conn, Steven. *Museums and American Intellectual Life, 1876–1926.* Chicago: University of Chicago Press, 1998.

Cooper, Lane. *Louis Agassiz as a Teacher.* Ithaca, NY: Comstock Publishing, 1945.

Croce, Paul Jerome. "Probabilistic Darwinism: Louis Agassiz vs. Asa Gray on Science, Religion, and Certainty." *Journal of Religious History* 22 (1998): 35–58.

Cromie, Robert. *The Great Chicago Fire, Illustrated Edition.* Nashville, TN: Rutledge Hill Press, 1994 (1958).

Cronon, William. *Nature's Metropolis: Chicago and the Great West.* New York: W. W. Norton, 1991.

Cropsey, Eugene H. *Crosby's Opera House: Symbol of Chicago's Cultural Awakening.* London: Fairleigh Dickinson University Press, Associated University Presses, 1999.

Crossette, George. *Founders of the Cosmos Club of Washington 1878.* Washington, DC: Cosmos Club, 1966.

Dall, Caroline H. "In Memoriam: Susan Wadden Turner, William Wadden Turner, Jane Wadden Turner," 1898, http://archive.org/stream/inmemoriamsusanw00dall#page/n5/mode/2up.

Dall, William H. "Baird, the Man." *Science* 57, no. 1468 (1923): 194–96.

———. "Some American Conchologists." *Proceedings of the Biological Society of Washington* 4 (1888): 129–33.

———. *Spencer Fullerton Baird.* Philadelphia: J. B. Lippincott, 1915.

———. "Theodore Nicholas Gill." *National Academy of Sciences Biographical Memoir* 8 (1916): 313–43.

Dance, S. Peter. *A History of Shell Collecting.* Leiden: E. J. Brill 1986.

Daniels, George H. *American Science in the Age of Jackson.* New York: Columbia University Press, 1968.

———. "The Process of Professionalization in American Science: The Emergent Period, 1820–1860." *Isis* 58 (1967): 63–78.

Darwin, Charles. *A Monograph of the Sub-class Cirripedia, with figures of all the species. The Balanidae (or Sessile Cirripedes); the Verrucidae, etc.* London: The Ray Society, 1854.

Davenport, F. Garvin. "Natural Scientists and the Farmers of Illinois, 1865–1900." *Journal of the Illinois State Historical Society* 51 (1958): 357–79.

Deiss, William A. "Spencer F. Baird and his Collectors." *Journal of the Society for the Bibliography of Natural History* 9, no. 4 (1980): 635–45.

Deiss, William A., and Raymond B. Manning, "The Fate of the Invertebrate Collections of the North Pacific Exploring Expedition, 1853–1856." In *History in the Service of Systematics,* edited by Alwyne Wheeler and James H. Price, 79–85. London: Society for the Bibliography of Natural History, Special Publication 1, 1981.

Desmond, Adrian. *Huxley: From Devil's Disciple to Evolution's High Priest.* Reading, MA: Perseus Books, 1999.

———. "Redefining the X Axis": 'Professionals,' 'Amateurs,' and the Making of Mid-Victorian Biology—A Progress Report." *Journal of the History of Biology* 34, no. 1 (2001): 3–50.

"Destruction of Scientific Museums by Fire." *American Journal of Science* 42 (1866): 135.

"Destruction of the Museum of the Chicago Academy of Sciences." *American Journal of Science* 2, 3rd series (1871): 387–88.

Detzer, David. *Dissonance: The Turbulent Days Between Fort Sumter and Bull Run.* Orlando, FL: Harcourt, 2006.

Dexter, Ralph W. "The Early American Naturalist as Revealed by Letters to the Founders." *The American Naturalist* 90, no. 853 (1956): 209–25.

———. "The 'Salem Secession' of Agassiz Zoologists." *Essex Institute Hist. Collections* 101, no. 1 (1965): 27–39.

Dolin, Eric Jay. *Leviathan: The History of Whaling in America.* New York: W. W. Norton, 2007.

"Dr. William Stimpson." *The Prairie Farmer* 43, no. 26 (June 29, 1872): 203.

Duis, Perry R. *Challenging Chicago: Coping with Everyday Life, 1837–1920.* Urbana: University of Illinois Press, 1998.

Dulles, Foster R. *Yankees and Samurai: America's Role in the Emergence of Modern Japan, 1791–1900.* New York: Harper & Row, 1965.

Dupree, A. Hunter. *Asa Gray: American Botanist, Friend of Darwin.* Cambridge, MA: Harvard University Press, 1959.

———. "The National Academy of Sciences and the American Definition of Science." In *The Organization of Knowledge in Modern America, 1860–1920,* edited by Alexandra Oleson and John Voss, 342–63. Baltimore: Johns Hopkins University Press, 1979.

———. *Science in the Federal Government: A History of Policies and Activities.* Baltimore: Johns Hopkins University Press, 1986 (1957).

Eastman, L. M. "The Portland Society of Natural History: The Rise and Fall of a Venerable Institution." *Northeastern Naturalist* 13, no. 1 (2006): 1–38.

Eiseley, Loren. *All the Strange Hours.* New York: Scribner's, 1975.

———. *The Man Who Saw Through Time.* Revised and enlarged edition of *Francis Bacon and the Modern Dilemma.* New York: Charles Scribner's Sons, 1973.

Eliot, Charles W. "Francis Humphreys Storer." *Proceedings of the American Academy of Arts and Sciences* 54 (1919): 415–18.

Engelmann, George. "Presidential Address 1872." *Transactions of the Academy of Sciences of St. Louis* 3 (1868–1877): xxxix.

———. "President's Annual Address, 1868." *Transactions of the Academy of Science of St. Louis* 2 (1861–1868): 581–82.

———. "President's Annual Address, 1867." *Transactions of the Academy of Science of St. Louis* 2 (1861–1868): 569–71.

Everts, William W. "The Church and Science." *The Chicago Pulpit* 1, no. 24 (1872): 221–28.

Fairchild, Herman L. *A History of the New York Academy of Sciences.* New York: Herman L. Fairchild, 1887.

Farber, Paul. *Finding Order in Nature: The Naturalist Tradition from Linnaeus to E. O. Wilson.* Baltimore: Johns Hopkins University Press, 2000.

Fortey, Richard. *Dry Storeroom No. 1: The Secret Life of the Natural History Museum.* New York: Knopf, 2008.

Foster, Mike. *Strange Genius: The Life of Ferdinand Vandeveer Hayden.* Niwot, CO: Roberts Rinehart Publishers, 1994.

Foster, Mike W. "Fielding Bradford Meek." *American National Biography* 14 (1999): 240–42.

Fox, Philip. "General Account of the Dearborn Observatory." In *Annals of the Dearborn Observatory of Northwestern University* 1 (1915): 1–20.

Ganong, W. F., ed. "John Robert Willis, the First Nova Scotian Conchologist: A Memorial." *Transactions of the Nova Scotian Institute of Natural Science* 7, pt. 4 (1889–90): 404–28, http://archive.org/stream/cihm_12507#page/n3/mode/2up.

Gardiner, Brian G. "Edward Forbes, Richard Owen and the Red Lions." *Archives of Natural History* 20, no. 3 (1993): 349–72.

Gibson, Terrica M. "Solomon G. Brown." In *The Smithsonian: 150 Years of Adventure, Discovery, and Wonder,* edited by James Conaway, 94–95. Washington, DC: Smithsonian Books, 1996.

Gill, Theodore. "The Doctrine of Darwin." *Proceedings of the Biological Society of Washington* 1 (1880–1881): 47–55.

———. "On a New Species of the Genus Macrorhinus." *Proceedings of the Chicago Academy of Sciences* 1 (1866): 33–34.

Gilman, Daniel C. *The Life of James Dwight Dana.* New York: Harper and Brothers, 1899.

Goetzmann, William H. *Exploration and Empire: The Explorer and the Scientist in the Winning of the American West.* New York: W. W. Norton, 1978 (1966).

———. *New Lands, New Men: America and the Second Great Age of Discovery*. New York: Viking, 1986.

———. "Paradigm Lost." In *The Sciences in the American Context: New Perspective*, edited by Nathan Reingold. Washington, DC: Smithsonian Institution Press, 1979.

Goldstein, Daniel. "Amos Henry Worthen." *American National Biography* 23 (1999): 885–86.

———. "Midwestern Naturalists: Academies of Science in the Mississippi Valley, 1850–1900." PhD diss., Yale University, 1989. University Microfilms International.

———. "Yours for Science: The Smithsonian Institution's Correspondents and the Shape of Scientific Community in Nineteenth-Century America." *Isis* 85, no. 4 (1994): 573–99.

Greenberg, Joel. *A Natural History of the Chicago Region*. Chicago: University of Chicago Press, 2002.

Gregory, Frederick. "The Impact of Darwinian Evolution on Protestant Theology in the Nineteenth Century." In *God and Nature: Historical Essays on the Encounter between Christianity and Science*, edited by David C. Lindberg and Ronald L. Numbers, 369–90. Berkeley: University of California Press, 1986.

Guarino, Jean. *Oak Park, A Pictorial History*. St. Louis: Bradley Publishing, 1988.

Habersham, Alexander W. *My Last Cruise*. Philadelphia: J. B. Lippincott, 1857.

Hafertepe, Kenneth. *America's Castle: The Evolution of the Smithsonian Building and its Institution, 1840–1878*. Washington, DC: Smithsonian Institution Press, 1984.

Haley, William D. *Philp's Washington Described. A Complete View of the American Capital, and the District of Columbia; with many notices Historical, Topographical, and Scientific, of the Seat of Government*. New York: Rudd and Carleton, 1861.

Hamlin, Christopher. "Robert Warington and the Moral Economy of the Aquarium." *Journal of the History of Biology* 19, no. 1 (1986): 131–53.

Harris, Harry. "Robert Ridgway with a Bibliography of his Published Writings." *The Condor* 30 (1928): 1–118.

Hawks, Graham P. "Increase A. Lapham, Wisconsin's First Scientist." PhD diss., University of Wisconsin, 1960. University Microfilms International.

Heine, Wilhelm. *Die Expedition in die Seen von China, Japan, Ochotsk unter Commando von Commodore Calw. Ringgold und Commodore John Rodgers, im Auftrage der Regierung der Vereinigten Staaten unternommen in den Jahren 1853 bis 1856*. Leipzig: Purfurst, 1858–1859.

Heitman, F. B. *Historical Register of the United States Army, from its Organization, September 29, 1789, to September 29, 1889*. Washington, DC: The National Tribune, 1890.

Hendrickson, Walter B. "Science and Culture in the American Middle West." *Isis* 64 (1973): 326–40.

Hendrickson, Walter B., and William J. Beecher. "In the Service of Science: The History of the Chicago Academy of Sciences." *Bulletin of the Chicago Academy of Sciences* 11, no. 7 (1972).

Henry, Joseph. "Report of the Operations of the Light-House Board Relative to Fog-Signals." In *Annual Report of the Light-House Board of the United States to the Secretary of the Treasury, for the fiscal year ending June 30, 1874*, 83–117. Washington, DC: Government Printing Office, 1874.

Henson, Pamela M. "Robert Edwards Carter Stearns." *American National Biography* 20 (1999): 596-97. New York: Oxford University Press, 1999.

———. "Spencer Baird's Dream: A U.S. National Museum." In *Cultures and Institutions of Natural History*, edited by Michael T. Ghiselin and Alan E. Leviton, 101–26. San Francisco: California Academy of Sciences, 2000.

Herber, Elmer C., ed. *Correspondence between Spencer Fullerton Baird and Louis Agassiz*. Washington, DC: Smithsonian Institution Press, 1963.

Hersey, John. *Blues*. New York: Vintage Books, 1988.

Higley, William K. "Historical Sketch of the Chicago Academy of Sciences." *Special Publication of the Chicago Academy of Sciences*, 1902.

Hine, Robert V. *Edward Kern and American Expansion*. New Haven, CT: Yale University Press, 1962.

Hinsley, Curtis M. *The Smithsonian and the American Indian: Making a Moral Anthropology in Victorian America*. Washington, DC: Smithsonian Institution Press, 1994 (1981).

Holton, Edgar. "A Chapter on the Coolie Trade." *Harper's New Monthly Magazine* 29 (1864): 1–11. Cornell University, Making of America website. http://ebooks.library.cornell.edu/m/moa/.

Hoogenboom, Olive. "Cadwalader Ringgold." *American National Biography* 18 (1999): 525–26.

Horowitz, Helen L. *Culture and the City: Cultural Philanthropy in Chicago from the 1880s to 1917.* Chicago: University of Chicago Press, 1976.

Hoy, Philo R. "Deep-Water Fauna of Lake Michigan." *Transactions of the Wisconsin Academy of Sciences, Arts, and Letters* 1 (1872): 98–101.

Hovenkamp, Herbert. *Science and Religion in America, 1800–1860.* Philadelphia: University of Pennsylvania Press, 1978.

Hughes, Robert. *The Fatal Shore: The Epic of Australia's Founding.* New York: Knopf, 1987.

Hung, Kuang-Chi. "'Plants that Remind Me of Home': Collecting, Plant Geography, and a Forgotten Expedition in the Darwinian Revolution." *Journal of the History of Biology* 50 (2017): 71–132.

Ilerbaig, Juan. "Pride in Place: Fieldwork, Geography, and American Field Zoology, 1850–1920." PhD diss., University of Minnesota, 2002.

Irmscher, Christoph. *Louis Agassiz: Creator of American Science.* Boston: Houghton Mifflin Harcourt, 2013.

Jackson, James R., and William C. Kimler. "Taxonomy and the Personal Equation: The Historical Fates of Charles Girard and Louis Agassiz." *Journal of the History of Biology* 32 (1999): 509–55.

Jeffreys, John Gwyn. *British Conchology; or An Account of the Mollusca which Now Inhabit the British Isles or Surrounding Seas*, vol. 4, *Marine Shells, in Continuation of the Gastropoda as Far as the Bulla Family.* London: John Van Voorst, 1867.

Johnson, Robert E. *Rear Admiral John Rodgers 1812–1882.* Annapolis, MD: United States Naval Institute, 1967.

Joiner, Gary D., ed. *Little to Eat and Thin Mud to Drink: Letters, Diaries, and Memoirs from the Red River Campaigns, 1863–1864.* Knoxville: University of Tennessee Press, 2007.

Jordan, David M. *"Happiness is Not My Companion": The Life of General G. K. Warren.* Bloomington: Indiana University Press, 2001.

Judd, Richard W. *The Untilled Garden: Natural History and the Spirit of Conservation in America, 1740–1840.* New York: Cambridge University Press, 2009.

Karamanski, Theodore J. *Rally 'Round the Flag: Chicago and the Civil War.* Chicago: Nelson-Hall Publishers, 1993.

Kay, E. Alison. "Darwin's Biogeography and the Oceanic Islands of the Central Pacific, 1859–1909." In *Darwin's Laboratory*, edited by Roy Macleod and Philip F. Rehbock, 49–69. Honolulu: University of Hawaii Press, 1994.

Kazar, John D. Jr. "The United States Navy and Scientific Exploration, 1837–1860." PhD diss., University of Massachusetts, 1973. University Microfilms edition.

Kellogg, G. M. "Self-Made Men." *Western Monthly* 2 (1869): 36–39.

Kisling Vernon N. Jr. "The Naturalists' Directory and the evolution of communication among American naturalists." *Archives of Natural History* 21, no. 3 (1994): 393–406.

Kohler, Robert. *All Creatures: Naturalists, Collectors, and Biodiversity, 1850–1950.* Princeton, NJ: Princeton University Press, 2006.

Kohler, Robert E. "Reflections on the History of Systematics." In *The Evolution of Phylogenetic Systematics*, edited by Andrew Hamilton, 17–46. Berkeley: University of California Press, 2014.

Kohlstedt, Sally G. "Creating a Forum for Science: AAAS in the Nineteenth Century." In *The Establishment of Science in America: 150 Years of the American Association for the Advancement of Science*, edited by Sally G. Kohlstedt, Michael M. Sokal, and Bruce V. Lowenstein, 7–49. New Brusnwick: Rutgers University Press, 1999.

———. *The Formation of the American Scientific Community: The American Association for the Advancement of Science 1848–1860.* Urbana: University of Illinois Press, 1976.

———. "From Learned Society to Public Museum: The Boston Society of Natural History." In *The Organization of Knowledge in Modern America, 1860–1920*, edited by Alexandra Oleson and John Voss, 386–406. Baltimore: Johns Hopkins University Press, 1979.

Kohlstedt, Sally Gregory, and Paul Brinkman. "Framing Nature: The Formative Years of Natural History Museum Development in the United States." *Proceedings of the California Academy of Sciences* 55, no. 2, supplement 1 (2004): 7–33.

Larsen, Anne. "Equipment for the Field." In *Cultures of Natural History*, edited by N. Jardine, J. A. Secord, and E. C. Spary, 358–377. Cambridge: Cambridge University Press, 1996.

Larsen, Anne L. "Not Since Noah: The English Scientific Zoologists and the Craft of Collecting, 1800–1840." PhD diss., Princeton University, 1993.

Laubacher, Matthew. "Cultures of Collection in late Nineteenth Century American Natural History." PhD diss., Arizona State University, 2011.

Lea, Isaac. *A Synopsis of the Family Unionidae*, 4th edition. Philadelphia: Henry C. Lea, 1870.

Leach, William. *Butterfly People: An American Encounter with the Beauty of the World*. New York: Pantheon Books, 2013.

Leech, Margaret. *Reveille in Washington*. New York: Carroll and Graf Publishers, 1991 (1941).

Lewis, Daniel. *The Feathery Tribe: Robert Ridgway and the Modern Study of Birds*. New Haven: Yale University Press, 2012.

Lewis, Lloyd, and Henry Justin Smith. *Chicago: A History of its Reputation*. New York: Harcourt Brace, 1929.

Lindsay, Debra. "Intimate Inmates: Wives, Households, and Science in Nineteenth-Century America." *Isis* 89, no. 4 (1998): 631–52.

———. *Science in the Subarctic: Trappers, Traders, and the Smithsonian Institution*. Washington, DC: Smithsonian Institution Press, 1993.

Long, E. B., ed. *Personal Memoirs of U. S. Grant*. New York: Da Capo Press, 1982.

Long, E. B., and Barbara Long. *The Civil War Day by Day: An Almanac 1861–1865*. New York: Da Capo Press, 1971.

Loomis, Chauncey C. *Weird and Tragic Shores: The Story of Charles Francis Hall, Explorer*. Lincoln: University of Nebraska Press, 1991.

Lowenthal, David. *George Perkins Marsh: Prophet of Conservation*. Seattle: University of Washington Press, 2000.

Lurie, Edward. *Louis Agassiz, A Life in Science*. Baltimore: Johns Hopkins University Press, 1988.

Lyell, Katherine, ed. *Life, letters, and journals of Sir Charles Lyell*. 2 vols. London: John Murray, 1881.

Lyman, Theodore. "Zoological Results of the Hassler Expedition, Ophiuridae and Astrophytidae." *Illustrated Catalogue of the Museum of Comparative Zoology*, no. 8. Cambridge University Press, 1874.

Mallis, Arnold. *American Entomologists*. New Brunswick, NJ: Rutgers University Press, 1971.

Manning, Raymond B. "The scientific contributions of William Stimpson, an early American naturalist and taxonomist." In *Crustacean Issues: History of Carcinology*, edited by Frank Truesdale, 8:109–18. Rotterdam: A. A. Balkema, 1993.

Mayer, Alfred G. "William Stimpson, 1832–1872." *National Academy of Sciences Biographical Memoirs* 8 (1918): 417–33.

Mayer, Harold M., and Richard C. Wade. *Chicago: Growth of a Metropolis*. Chicago: University of Chicago Press, 1969.

Mayr, Ernst. "The Naturalist in Leidy's Time and Today." *Proc. ANSP* 98 (1946): 271–76.

McCagg, Ezra B. "Address on Presentation of a Portrait." No date, typescript copy. McCagg Papers, Chicago History Museum.

———. "Beneath the Dust of a Generation." No date, typescript copy. McCagg Papers, Chicago History Museum.

McCarthy, Kathleen D. *Noblesse Oblige: Charity and Cultural Philanthropy in Chicago, 1849–1929*. Chicago: University of Chicago Press, 1982.

McCrady, John. "On the Lingula pyramidata described by Mr. W. Stimpson." *American Journal of Science* 30 (1860): 157–58.

McLure, John W. "A History of the Illinois State Geological Survey, 1851–1875." MA thesis, University of Illinois, 1962.

Meek, F. B. "Notes on the affinities of the Bellerophontidae." *Proceedings of the Chicago Academy of Sciences* 1 (1866): 9–11.

Meek, F. B., and A. H. Worthen. "Descriptions of Palaeozoic Fossils from the Silurian, Devonian, and Carboniferous rocks of Illinois, and other Western States." *Proceedings of the Chicago Academy of Sciences* 1 (1866): 11–23.

———. "Notice of some New Types of Organic Remains, from the Coal Measures of Illinois." *Proc. ANSP* 17 (1865): 41–53.

Merrill, Lynn L. *The Romance of Victorian Natural History*. Oxford: Oxford University Press, 1989.

Merton, Robert. *The Sociology of Science: Theoretical and Empirical Investigations*. Chicago: University of Chicago Press, 1973.

Mikkelsen, Paula M., and Rüdiger Bieler. *Seashells of Southern Florida: Living Marine Mollusks of the Florida Keys and Adjacent Regions: Bivalves,* vol. 1. Princeton, NJ: Princeton University Press, 2007.

Miller, Donald L. *City of the Century: The Epic of Chicago and the Making of America*. New York: Simon and Schuster, 1997.

Mills, Eric L. "Edward Forbes, John Gwyn Jeffreys, and British dredging before the *Challenger* expedition." *Journal of the Society for the Bibliography of Natural History* 8, no. 4 (1978): 507–36.

———. "Problems of Deep-Sea Biology: An Historical Perspective." In *The Sea,* vol. 8, *Deep Sea Biology,* edited by G. T. Rowe, 1–79. New York: Wiley and Sons, 1983.

Mooney, James L., ed. *Dictionary of American Naval Fighting Ships,* vol. 7. Washington, DC: Naval Historical Center, Dept. of the Navy, 1981.

Moore, Frank, ed. *The Rebellion Record, A Diary of American Events,* vol. 3. New York: G. P. Putnam, 1862.

Morison, Samuel E. *"Old Bruin": Commodore Matthew C. Perry, 1794–1858*. Boston: Atlantic Monthly Press, 1967.

———. *The Maritime History of Massachusetts, 1783–1860*. Boston: Northeastern University Press, 1979.

Musham, H. A. "The Great Chicago Fire, October 8–10, 1871." *Papers in Illinois History and Transactions for the Year 1940*. Illinois State Historical Society, 1940.

Numbers, Ronald L. *The Creationists: From Scientific Creationism to Intelligent Design*. Expanded edition. Cambridge, MA: Harvard University Press, 2006.

———. *Darwinism Comes to America*. Cambridge, MA: Harvard University Press, 1998.

Nyhart, Lynn K. *Biology Takes Form: Animal Morphology and the German Universities, 1800–1900*. Chicago: University of Chicago Press, 1995.

———. "Natural history and the 'new' biology." In *Cultures of Natural History,* edited by N. Jardine, J. A. Secord, and E. C. Spary, 426–443. Cambridge: Cambridge University Press, 1996.

Official Records of the Union and Confederate Navies in the War of Rebellion, Series 1, vol. 13, *South Atlantic Blockading Squadron, May 14, 1862–April 7, 1863*. Washington, DC: Government Printing Office, 1901. Cornell University, Making of America website.

Ordway, Albert. "Monograph of the Genus Callinectes." *Boston Journal of Natural History* 7, no. 4 (1863): 3–18.

Osborn, H. F. *Cope: Master Naturalist*. Princeton, NJ: Princeton University Press, 1931.

Orosz, Joel. *Curators and Culture: The Museum Movement in America, 1740–1870*. Tuscaloosa: University of Alabama Press, 1990.

Pachter, Marc, and Frances Wein, eds. *Abroad in America: Visitors to the New Nation, 1776–1914*. Reading, MA: National Portrait Gallery, Smithsonian Institution and Addison-Wesley, 1976.

Packard, Alpheus. "Notice of Hymenoptera & Nocturnal Lepidoptera, Collected in Alaska, By W. H. Dall, Director Sci. Corp, W. U. T. Exp, with a list of Neuroptera, by P. R. Uhler and Dr. H. Hagen." CAS Archives, original in Museum of Comparative Zoology, 1871.

Parshall, Karen H. "Truman Henry Safford." *American National Biography* 19 (1999): 190–91.

Parton, James. "Chicago." *Atlantic Monthly* 19 (1867): 325–45. Cornell University Making of America website.

The Past and Present of LaSalle County. Chicago: H. F. Kett & Co., 1877.

Pauly, Philip J. *Biologists and the Promise of American Life: From Meriwether Lewis to Alfred Kinsey.* Princeton, NJ: Princeton University Press, 2000.

Peattie, Donald C. *The Road of a Naturalist.* Boston: Houghton Mifflin, 1941.

Peck, Robert McCracken, and Patricia Tyson Stroud. *A Glorious Enterprise: The Academy of Natural Sciences of Philadelphia and the Making of American Science.* Philadelphia: University of Pennsylvania Press, 2012.

Pfeifer, Edward J. "United States." In *The Comparative Reception of Darwinism,* edited by Thomas F. Glick, 168–206. Chicago: University of Chicago Press, 1988.

Philbrick, Nathaniel. *Sea of Glory: America's Voyage of Discovery, the U.S. Exploring Expedition 1838–1842.* New York: Penguin Books, 2003.

Pierce, Bessie L. *A History of Chicago,* vol. 2, *From Town to City 1848–1871.* Chicago: University of Chicago Press, 1940.

Platt, Stephen R. *Autumn in the Heavenly Kingdom: China, the West, and the Epic Story of the Taiping Civil War.* New York: Knopf, 2012.

Ponko, Vincent Jr. *Ships, Seas, and Scientists: U.S. Naval Exploration and Discovery in the Nineteenth Century.* Annapolis, MD: Naval Institute Press, 1974.

Porter, Charlotte M. *The Eagle's Nest: Natural History and American Ideas, 1812–1842.* Tuscaloosa: University of Alabama Press, 1986.

Post, Robert C. "'Liberalizers' versus 'Scientific Men' in the Antebellum Patent Office." *Technology and Culture* 17 (1976): 24–54.

Prendergast, Michael L. "James Dwight Dana: The Life and Thought of an American Scientist." PhD diss., University of California, 1978.

Preston, Douglas J. *Dinosaurs in the Attic.* New York: Ballantine Books, 1986.

Pyne, Stephen J. *Fire in America: A Cultural History of Wildland and Rural Fire.* Princeton, NJ: Princeton University Press, 1982.

Rathbun, Richard "The American Initiative in Methods of Deep-Sea Dredging." *Science* 4, no. 76 (1884): 54–57.

Ravenel, Edmund. "Descriptions of New Recent Shells from the Coast of South Carolina." *Proc. ANSP* 13 (1861): 41–44.

Redfield, William C. "On the Cyclones or Typhoons of the North Pacific Ocean." *American Journal of Science* 24 (1857): 21–28.

Rehbock, Philip F. "The Early Dredgers: 'Naturalizing' in British Seas, 1830–1850." *Journal of the History of Biology* 12, no. 2 (1979): 293–368.

———. *The Philosophical Naturalists: Themes in Early Nineteenth-Century British Biology.* Madison: University of Wisconsin Press, 1983.

Reid, Anna M. "Charles Wright." *American National Biography* 24. New York: Oxford University Press, 1999.

Reingold, Nathan, ed. *Science in Nineteenth-Century America: A Documentary History.* Chicago: University of Chicago Press, 1985.

Rice, A. L., Harold L. Burstyn, A. G. E. Jones. "G. C. Wallich MD: Megalomaniac or Mis-used Oceanographic Genius?" *Journal of the Society for the Bibliography of Natural History* 7, no. 4 (1976): 423–50.

Rice, Tony. *Voyages of Discovery: Three Centuries of Natural History Exploration.* New York: Clarkson N. Potter, 1999.

Richards, I. A. *The Portable Coleridge.* New York: Penguin Books, 1986.

Riedman, Sarah R. *Trailblazer of American Science: The Life of Joseph Henry.* Chicago: Rand McNally, 1961.

Rivinus, E. F., and E. M. Youssef. *Spencer Baird of the Smithsonian.* Washington, DC: Smithsonian Institution Press, 1992.

Rodgers, John, and Anton Schönborn. "On the Avoidance of the Violent Portions of Cyclones; with Notices of a Typhoon at the Bonin islands." *American Journal of Science* 23 (1857): 205–11.

Rodriguez, Gilberto. "From Oviedo to Rathbun: The development of brachyuran crab taxonomy in the Neotropics (1535–1937)." In *Crustacean Issues: History of Carcinology*, vol. 8, edited by Frank Truesdale, 41–73. Rotterdam: A. A. Balkema, 1993.

Rosenberg, Charles E. *The Cholera Years: The United States in 1832, 1849, and 1866.* Chicago: University of Chicago Press, 1987.

Rothenberg, Marc. *The Papers of Joseph Henry*, vol. 10, *January 1858–December 1865, the Smithsonian Years.* Washington, DC: Smithsonian Institution Press, 2004.

Rothman, Sheila M. *Living in the Shadow of Death: Tuberculosis and the Social Experience of Illness in American History.* Baltimore: Johns Hopkins University Press, 1994.

Rozwadowski, Helen M. *Fathoming the Ocean: The Discovery and Exploration of the Deep Sea.* Cambridge, MA: Belknap Press of Harvard University Press, 2005.

———. "Small World: Forging a Scientific Maritime Culture for Oceanography." *Isis* 87, no. 3 (1996): 409–29.

Rudwick, Martin J. S. *Earth's Deep History: How it Was Discovered and Why it Matters.* Chicago: University of Chicago Press, 2014.

Ruschenberger, W. S. W. "A Sketch of the Life of Joseph Leidy." *Proceedings of the American Philosophical Society* 30, no. 138 (April 1892): 135–84.

Schorger, A. W. "Philo Romayne Hoy: Wisconsin's Greatest Pioneer Zoologist." *The Passenger Pigeon* 6, no. 3 (1944): 55–59.

Shabica, Charles W., and Andrew A. Hay, ed. *Richardson's Guide to the Fossil Fauna of Mazon Creek.* Chicago: Northeastern Illinois University, 1997.

Shaler, Nathaniel S. *The Autobiography of Nathaniel Southgate Shaler.* Boston: Houghton and Mifflin, 1909.

Scheltema, Rudolf S., and Amelie H. Scheltema. "Deep-sea Biological Studies in America, 1846 to 1872—Their Contribution to the *Challenger* Expedition." *Proceedings of the Royal Society of Edinburgh* 72, section B (1971–1972): 133–44.

Schlee, Susan. *The Edge of an Unfamiliar World: A History of Oceanography.* New York: E. P. Dutton, 1973.

Schouler, William. *A History of Massachusetts in the Civil War*, vol. 1. Boston: E. P. Dutton, 1868.

Schram, Frederick R. "Preface." In *Crustacean Issues: History of Carcinology* 8, edited by Frank Truesdale, i–viii. Rotterdam: A. A. Balkema, 1993.

Schultz, Rima L. "Kate Newell Doggett." In *Women Building Chicago 1790–1990, A Biographical Dictionary*, edited by Rima L. Schultz and Adele Hast, 224–229. Bloomington: Indiana University Press, 2001.

Sheets-Pyenson, Susan. *John William Dawson: Faith, Hope, and Science.* Montreal: McGill-Queen's University Press, 1996.

Sheppard, Carol A. "Benjamin Dann Walsh: Pioneer Entomologist and Proponent of Darwinian Theory." *Annual Review of Entomology* 49 (2004): 1–25.

Silliman, Robert H. "The Hamlet Affair: Charles Lyell and the North Americans." *Isis* 86 (1995): 541–61.

Silverman, Kenneth. *Edgar A. Poe: Mournful and Never-ending Remembrance.* New York: Harper Perennial, 1991.

Simmons, James R. Jr. "Robert Carter." *American National Biography* 4 (1999): 494–95.

Skrupskelis, Ignas K., and Elizabeth M. Berkeley, eds. *The Correspondence of William James*, vol. 1, *William and Henry 1861–1884.* Charlottesville: University Press of Virginia, 1992.

———. *The Correspondence of William James*, vol. 4, *1856–1877.* Charlottesville: University Press of Virginia, 1996.

Sloan, Thomas L. "William W. Boyington." In *Macmillian Encyclopedia of Architects*, vol. 1, edited by Adolf K. Placzek. London: The Free Press, 1982.

Slotten, Hugh R. *Patronage, Practice, and the Culture of American Science: Alexander Dallas Bache and the U.S. Coast Survey.* Cambridge: Cambridge University Press, 1994.

"The Smithsonian Aquarium at Washington." *Scientific American* 13, no. 15 (1857): 113. Cornell University Making of America online.

Sorensen, W. Conner. *Brethren of the Net: American Entomology, 1840–1880.* Tuscaloosa: University of Alabama Press, 1995.

Stanton, William. *The Great United States Exploring Expedition of 1838–1842.* Berkeley: University of California Press, 1975.

Stearns, Robert E. C. "Rambles in Florida." *American Naturalist* 3 (1869): 281–88; 349–356; 397–404; 455–70.

———. "Remarks of Robert E. C. Stearns on the Death of Colonel Ezekiel Jewett, before the California Academy of Sciences, June 18th, 1877." California Academy of Sciences, 1877.

———. "Some remarks on the death of William Stimpson." *Proceedings of the California Academy of Sciences* 4, series 1 (1872): 230–32.

Steinbeck, John. *The Log from the Sea of Cortez.* New York: Penguin Books, 1995 (1951).

Steinbeck, John, and Edward F. Ricketts. *Sea of Cortez.* Mamaroneck, NY: Paul A. Appel, 1971 (1941).

Stephens, Lester D. *Science, Race and Religion in the American South: John Bachman and the Charleston Circle of Naturalists.* Chapel Hill: University of North Carolina Press, 2000.

Stephens, Lester D., and Dale R. Calder. "John McCrady of South Carolina: Pioneer Student of North American Hydrozoa." *Archives of Natural History* 19 (1992): 39–54.

———. *Seafaring Scientist: Alfred Goldsborough Mayor, Pioneer in Marine Biology.* Columbia: University of South Carolina Press, 2006.

Sterling, Keir B. "Carl Robert Romanovich von Osten Sacken." *American National Biography* 16 (1999): 808–09.

———. "John Cassin." *American National Biography* 4 (1999): 556–57.

Stimpson, William H., and Richard W. Price. *A Stimpson Family in America.* North Salt Lake, UT: DMT Publishing, 2004.

Stresemann, Erwin. *Ornithology: From Aristotle to the Present.* Cambridge, MA: Harvard University Press, 1975.

Stroud, Patricia T. *Thomas Say: New World Naturalist.* Philadelphia: University of Pennsylvania Press, 1992.

Struik, Dirk J. *Yankee Science in the Making.* New York: Collier Books, 1962 (1948).

Sulloway, Frank J. *Born to Rebel: Birth Order, Family Dynamics, and Creative Lives.* New York: Pantheon Books, 1996.

Symonds, Craig L. "John Rodgers." *American National Biography* 18 (1999): 725-26.

Takarabe, Kae. "Samurai at the Smithsonian: First Japanese Visitors to Western Museum in the U.S." In *Cultures and Institutions of Natural History,* edited by Michael Ghiselin and Alan E. Leviton, 161–82. San Francisco: California Academy of Sciences, 2000.

Thomas, Phillip D. "Titian Ramsey Peale." *American National Biography* 17 (1999): 203–5.

"To the Friends and Correspondents of the Chicago Academy of Sciences." October 30, 1871.

Trautmann, Frederic, ed. *With Perry to Japan, A Memoir by William Heine.* Honolulu: University of Hawaii Press, 1990.

Troyer, James R. "John Charles Bowring (1821–1893): contributions of a merchant to natural history." *Archives of Natural History* 10, no. 3 (1982): 515–29.

Tryon, George W. "Monograph of the Terrestrial Mollusca of the United States." *American Journal of Conchology* 2, no. 3 (1866): 218–77.

———. "A Sketch of the History of Conchology in the United States." *American Journal of Science* 33 (1862): 161–80.

Tyler, Ron. "Illustrated Government Publications of the American West, 1848–1863." In *Surveying the Record: North American Scientific Exploration to 1930,* edited by Edward C. Carter II, 147–72. Philadelphia: American Philosophical Society, 1999.

Vasile, Ronald S. "The Early Career of Robert Kennicott, Illinois' Pioneering Naturalist." *Illinois Historical Journal* 87, no. 3 (1994): 150–70.

———. "The Megatherium Club." In *The Smithsonian: 150 Years of Adventure, Discovery, and Wonder,* by James Conaway, 86–87. Washington, DC: Smithsonian Books, 1995.

———. "Robert Kennicott." *American National Biography* 13 (1999): 583-85.

———. "William Stimpson." *American National Biography* 20 (1999): 786-87.

Vasile, Ronald S., Raymond B. Manning, and Rafael Lemaitre, eds. "William Stimpson's Journal from the North Pacific Exploring Expedition, 1853–1856." *Crustacean Research, Special Number 5*. Carcinological Society of Japan, 2005.

Verrill, Addison E. "Revision of the Polypi of the eastern coast of the United States." *Boston Journal of Natural History*, Memoirs Read Before the Boston Society of Natural History, vol. 1 (Boston: Society of Natural History, 1863): 1–45.

Viola, Herman, and Carolyn Margolis, eds. *Magnificent Voyagers*. Washington, DC: Smithsonian Institution Press, 1985.

Wallich, G. C. "Do Diatoms live on the Sea-Bottom at Great Depths?" *Annals and Magazine of Natural History* 12, 3rd series (1863): 166.

Walsh, Jane MacLaren. "Collections as Currency." In *Anthropology, History, and American Indians: Essays in Honor of William Curtis Sturtevant*, edited by William L. Merrill and Ives Goddard. Smithsonian Contributions to Anthropology, no. 44. Washington, DC: Smithsonian Institution Press, 2002.

Warren, Leonard. "George Washington Tryon Jr." *American National Biography* 21 (1999):.883-84.

———. *Joseph Leidy: The Last Man Who Knew Everything*. New Haven, CT: Yale University Press, 1998.

Warren, Robert E., and Russell W. Graham. "*Cervalces*: An Ice Age Discovery." *Living Museum* 50, no. 3 (1988): 38–41.

Washburn, Wilcomb E. "The Influence of the Smithsonian Institution on Intellectual Life in Mid-Nineteenth-Century Washington." *Records of the Columbia Historical Society of Washington, D.C., 1963-1965*. Washington, DC: Columbia Historical Society, 1966.

Wayman, Dorothy G. *Edward Sylvester Morse: A Biography*. Cambridge, MA: Harvard University Press, 1942.

Welch, Margaret. *The Book of Nature: Natural History in the United States 1825–1875*. Boston: Northeastern University Press, 1998.

White, Charles A. "Biographical Memoir of Fielding Bradford Meek." *National Academy of Sciences Biographical Memoirs* 4. Washington, DC: National Academy of Sciences, 1902.

Williams, Austin B. "Reflections on crab research in North America since 1758." In *Crustacean Issues: History of Carcinology*, edited by Frank Truesdale, 259–73. Rotterdam: A. A. Balkema, 1993.

Winsor, Mary P. *Reading the Shape of Nature*. Chicago: University of Chicago Press, 1991.

———. *Starfish, Jellyfish, and the Order of Life: Issues in Nineteenth-Century Science*. New Haven, CT: Yale University Press, 1976.

Worster, Donald. *A River Running West: The Life of John Wesley Powell*. New York: Oxford University Press, 2001.

Wyman, Jeffries. "Notice of the Life and Writings of the late Dr. Waldo Irving Burnett." *Proc. BSNH* 5 (1854–1856): 64–74.

Yoon, Carol Kaesuk. *Naming Nature: The Clash Between Instinct and Science*. New York: W. W. Norton, 2009.

Zwinger, Ann H. "John Xantus." *American National Biography* 24 (1999): 97–98.

———, ed. *Xantus: The Letters of John Xantus to Spencer Fullerton Baird from San Francisco and Cabo San Lucas 1859-1861*. Los Angeles: Dawson's Book Shop, 1986.

OTHER PRIMARY SOURCES

1850 Washington, DC, census
1860 Massachusetts census
1870 Illinois census
American Journal of Conchology
American Naturalist
Annual Report of the National Academy of Sciences, 1879
Appleton's Journal
Congressional Globe
Harper's Weekly
Museum of Comparative Zoology, Annual Reports for 1863–1864
The Nation
Proceedings of Association for the Advancement of Science, Sixteenth Meeting, held at Burlington, Vermont, August 1867, Cambridge: Joseph Lovering, 1868.
Proceedings of the Essex Institute
Proceedings of the National Academy of Sciences 1870
Report of the Superintendent of the United States Coast Survey
Smithsonian Annual Reports 1852, 1859, 1860, 1862, 1863, 1864, 1865, 1867, 1870
The War of the Rebellion: A Compilation of the Official Records of the Union and Confederate Armies, Series 1, vol. 25, part 1, 1889, pp. 539–41; Series 1, vol. 27, part 1, 1889, pp. 599–600, Cornell University, Making of America series
Who Was Who in America, vol. 1, 1897–1942
Who Was Who in America, vol. 5, 1969–1973

INDEX OF SUBJECTS

Page numbers in *italics* denote illustrations.
Also see index of names beginning on page 295.

A

Index of names

Page numbers in *italics* denote illustrations.
Also see index of subjects beginning on page 285.

A